Technological Advances and Trends in Cheese Making

Syed Mansha Rafiq
Editor

Technological Advances and Trends in Cheese Making

Springer

Editor
Syed Mansha Rafiq
Department of Food Science and Technology
National Institute of Food Technology Entrepreneurship and Management
Kundli, Haryana, India

ISBN 978-3-031-94475-8 ISBN 978-3-031-94476-5 (eBook)
https://doi.org/10.1007/978-3-031-94476-5

This Springer imprint is published by the registered company Springer Nature Switzerland AG
The registered company address is: Gewerbestrasse 11, 6330 Cham, Switzerland

Preface

Cheese has long held a unique place in the culinary and cultural fabric of societies around the world. From ancient traditions passed down through generations to state-of-the-art innovations in food science, the journey of cheese making is as rich and diverse as the varieties of cheese themselves. This book was born out of a desire to bridge the timeless art of cheese making with the rapidly evolving landscape of modern dairy technology.

As both a craft and a science, cheese making demands a deep understanding of microbiology, chemistry, and process engineering. Through this comprehensive volume, we explore everything from the historical roots and cultural significance of cheese, to the detailed mechanisms of milk coagulation, starter cultures, and ripening processes. Special attention is given to the functional and rheological properties of different cheese types—including low-fat, low-sodium, and vegan varieties—along with the advanced techniques now transforming the field, such as precision fermentation, membrane filtration, and real-time quality evaluation tools.

The chapters are designed to serve as both a foundational resource for students and a practical guide for professionals working in dairy science, food processing, and product innovation. Readers will also find valuable insights into sustainability-focused practices, novel packaging, and the incorporation of emerging technologies into traditional workflows.

Ultimately, this book aims to honor the heritage of cheese making while embracing its future—a future that balances flavor, nutrition, efficiency, and innovation. Whether you are a researcher, a dairy technologist, a cheese enthusiast, or someone exploring new frontiers in food science, I hope this work inspires curiosity and deepens your appreciation for one of the world's most beloved foods.

I wish to thank sincerely the authors who contributed to the text and whose cooperation made my task as editor a pleasure.

Kundli, Haryana, India | Syed Mansha Rafiq

Contents

1 **History and Culture of Cheese Making** 1
Syed Mansha Rafiq, Syed Insha Rafiq, and Bazilla Gayas

2 **Basic Cheese Making Process** 11
Ridhi Pandey and Yogesh Khetra

3 **Quality of Milk for Cheese Making** 33
Reshab Majumder and Syed Mansha Rafiq

4 **Starter Culture** 47
Reshab Majumder, Chandrakanta Sen, Subhadip Manik, and Akshay Ramani

5 **Coagulation of Milk** 63
Subhadip Manik, Akshay Ramani, Rakendhu Saji, Chandrakanta Sen, and Reshab Majumder

6 **Accelerated Cheese Ripening** 81
Ronit Mandal, Partha Pratim Debnath, Payal Karmakar, Rashmi Kumari, Anindita Debnath, and Jui Lodh

7 **Overview of the Manufacturing Processes** 107
Shivam Kayasth and Gauri Thakur

8 **Functional Properties and Rheology** 135
Reshab Majumder and Saptadip Miatur

9 **Quality Evaluation of Cheese** 151
Tridip Boruah, Morjina Parveen, David Daimary, and Gulzar Ahmad Nayik

10 **Advanced Testing Techniques** 169
Kuntal Roy, V. P. Aparnna, and Harsh Jindal

11 Advances in Packaging and Storage 191
Syed Insha Rafiq, Bazilla Gayas, Deepak Joshi, Ayush Pandey, and Syed Mansha Rafiq

12 Low-Fat Cheese 207
Pavel Rout and Shreya Saha

13 Low Sodium Cheese 229
Rajat Singh and Bazilla Gayas

14 Vegan Cheeses 249
Ridhi Pandey, Sonam Kumari, and Yogesh Khetra

15 Novel Cheese Making Technologies 265
Deepak Joshi and Syed Insha Rafiq

16 Dairy Ingredients in Cheesemaking 293
Syed Mansha Rafiq, Shamim Hossain, and Syed Irshaan

Index 305

Chapter 1
History and Culture of Cheese Making

Syed Mansha Rafiq, Syed Insha Rafiq, and Bazilla Gayas

1.1 Introduction

Cheese is a broad term used for fermented milk-based food products. The word "cheese" is a Latin derived word "caseus" meaning correct or perfect thing. Cheese may be defined as a solid mass which is separated from the whey.

FAO/WHO defines cheese as the fresh product or mature product either solid or semi-solid obtained by skim milk, coagulating milk, or partly skimmed milk, cream, butter milk or whey, or any combination of these materials, through the action of coagulating agents or rennet and by partially drained whey resulting from coagulation; or the product which is obtained by processing technologies involving milk coagulation and/or materials obtained from milk (provided that the whey protein casein ratio does not exceed that of milk) and which gives a final product having similar chemical, physical, or organoleptic properties as the product defined above. Cheese or varieties of cheeses shall have flavours free from rancidity and off-flavours as well as have pleasant taste. It may contain permitted food additives and shall conform to the microbiological requirements prescribed in the regulation.

These regulations also define cheese on the basis of ripening as follows:

Ripened cheese: Cheese which is not ready for consumption shortly after manufacture but which must be held for some time at such temperature and under such

S. Mansha Rafiq (✉)
Department of Food Science and Technology, National Institute of Food Technology Entrepreneurship and Management, Sonipat, Haryana, India

S. I. Rafiq
Department of Food Engineering, Bhaigurdas College of Engineering and Technology, Sangrur, Punjab, India

B. Gayas
Department of Food Technology, Chandigarh University, Chandigarh, Punjab, India

S. Mansha Rafiq (ed.), *Technological Advances and Trends in Cheese Making*,
https://doi.org/10.1007/978-3-031-94476-5_1

other conditions as will result in necessary biochemical and physical changes characterizing the cheese in question.

Mould ripened cheese: Cheese in which the ripening has been accomplished primarily by the development of characteristic mould growth through the interior and/or on the surface of the cheese.

Unripened cheese: Cheese including fresh cheese which is ready for consumption shortly after manufacture.

1.2 History of Cheese Making

Cheese is the broad term used for a group of fermented milk-based products and is believed to be evolved in Iraq over 8000 years ago. It was an important food item of the Roman and Greek diet. Cheese likely originated by accident through the preservation techniques of nomadic tribes aiming to maintain the nutritional value of milk. In hot climates, the natural souring of milk due to lactic acid bacteria would occur, causing the pH to drop and eventually reach the isoelectric point of caseins. This process led to the creation of fermented milks. When this acidified milk was cut or broken, it separated into curds and whey. People quickly discovered that whey could be consumed as a refreshing drink, while curds could be eaten fresh or stored for later use. Salting or dehydrating the curds helped preserve them longer. This likely led to the development of acid-curd cheeses.

In contrast, the majority of other cheeses are made by coagulating milk using rennet. This may have stemmed from the observation of curds in the stomachs of young animals slaughtered for food or from seeing curds in the vomit of infants. Such discoveries, combined with the use of stomach bags for holding milk, likely led to the use of gastric enzymes (rennet) in cheesemaking. Over time, cheese production spread globally, especially with the migration of settlers from Europe and the Middle East to regions like North and South America, Oceania, and Africa.

Cheese making remained a local farm industry until the nineteenth century and the first cheese factory was built in 1851 by Jesse Williams in New York. Cheese industry showed growth in later half of 1800s (Scott, 1998). Earlier cheese making was all art, but towards the end of eighteenth century the inclusion of science and technology and replacement of traditional cheese making process with mechanization has made tremendous progress in cheese making technology. The primary objective of cheese making was to extend the shelf life of milk and conserve its nutritive value. However, with the gradual acquisition of knowledge on the chemistry and microbiology of milk and cheese, it became possible to direct the changes involved in cheese making in a more controlled fashion (Fox & McSweeney, 2004).

During the course of time the cheese making has undergone many developments from the use of refrigeration and pasteurized milk to the use of coagulants other than enzyme rennin. The more recent development is the accelerated ripening process which gives desired flavour and texture in very less time. With the increased milk production all over the world, the efficiency of cheese plants increased.

Implementation of new technologies like membrane concentration and separation processes came into existence and greatly enhanced plant capacity however, full implementation of membrane processing is yet to be achieved (Johnson, 2017). The increasing demand for nutritious and functional foods has driven the researchers and scientist towards the production of healthy and functional cheeses. Nowadays low fat, low sodium, low calorie, omega 3 fortified, fibre enriched, PUFA enriched, cheeses are being manufactured.

1.3 Cheese Classification

The classification of cheese is extremely complicated because of the wide range of variants. There are 18 cheese family groups namely Cheddar, Edam, Swiss, Gouda, Camembert, Limburger, Parmesan, Provolone, Romano, Roquefort, Brick, Herve, Sapsago, Cottage, Neufchatel, Trappist, Cream and Whey cheeses that contain around thousands of cheese varieties. Cheese is generally classified on the basis of age, type of milk, composition, country of origin and ripening process or agents.

Type of milk: On the basis of type of milk used cheeses are classified as cow milk cheese, goat milk cheese, buffalo milk cheese.

Composition: On the basis of composition cheeses are classified as:

(a) Very hard: The cheeses having moisture less than 35% are classified as very hard. These include grating cheeses like Parmesan and Romano cheese.
(b) Hard: These have moisture less than 40% and include Cheddar and Swiss cheeses.
(c) Semi-hard: These cheeses have moisture in the range of 40–47% and include Brick, Limburger, Camembert, Gorgonzola, Blue, Roquefort cheese.
(d) Soft: Soft cheeses have moisture more than 47% e.g. Cottage and Neufchatel cheese.

Country of origin: On the basis of country of origin cheeses are classified as

(a) Italian cheeses: The cheeses that originated in Ialy include Pecorino Romano, Gorgonzola, Parmigiano-Reggiano, Grana Padano, Roquefort, Stilton and *Pasta filata* cheeses like Provolone and Mozzarella. These use milk from all species cow, buffalo, goat and sheep or a mixture of all. Goat milk cheeses are called "caprini", Sheep milk cheese "pecorini", and buffalo milk cheese "bufalini".
(b) Swiss-type cheeses: These originated from Switzerland and include Emmentaler, Gruyere and Appenzeller. These cheeses are characterized by round regular eyes and the body & texture corresponds to hard and semi hard cheeses.
(c) Danish cheese: Danish cheeses originated from Denmark and include Danbo, Havarti and Danablu.

(d) Dutch-type cheeses: The cheeses produced in Netherlands constitute the important types of cheese produced in the world. Examples are Gouda and Edam.
(e) Cheese from Balkan counties: Cheeses that originated from Balkan countries like Bulgaria and Greece include Kashkaval and Feta. Kashkaval is manufactured from cow's, sheep's, goat's milk or mixed milk whereas, Feta cheese is manufactured from sheep's milk or from a mixture of sheep's and goat's milk.

Ripening process: On the basis of ripening process cheeses are classified as:

(a) Bacterial ripened: These are characterized by the development of bacteria on their surfaces during ripening e.g. Parmesan and Romano cheese.
(b) Mould ripened cheeses: These are further classified as:

Externally ripened. Where the mould (*Penicillium camemberti*) grows on the surface of the cheese e.g. Camembert cheese.
Internally ripened. Where the mold (*Penicillium roqueforti*) grows within the cheese e.g. Gorgonzola, Blue, Roquefort and Stilton cheese.

1.4 Starter Cultures for Cheese Making

Starter cultures contain selected LAB (lactic acid bacteria) which converts lactose in milk to lactic acid, along with minimum amounts of by-products such as Carbon di oxide, acetic acid, and diacetyl, acetaldehyde, in few cases. In milk almost 1% of lactose is entirely converted to lactic acid. This acid production at a controlled rate is an quintessential step in the good quality cheese manufacturing. A better starter culture must be capable of producing a clean lactic acid flavor and must not produce any off flavour. In addition to regular growth, it must produce a consistent and controlled rate of acid. It must contain only desired lactic acid bacteria. In the usual starter culture, the lactic acid bacteria include *Lc. lactis* subsp. *diacetylactis*. *Lc. lactis* subsp. *cremoris and Lactococcus lactis* subsp. *lactis*.

Starter cultures have been classified in many ways. One of the classifications includes primary and secondary starters depends on the activity of starter microorganisms. *Primary starters are those* whose primary function is the conversion of lactose into lactic acid with or without production of flavour and textural changes. *Secondary starters:* Microorganisms that do not aid acid production but only produce flavour and textural changes are called secondary starters. On the basis of required temperature for growth. They have been classified into *Mesophilic starters* include *Lactococcus lactis* subsp. *cremoris, Lactococcus lactis* subsp. *lactis, Pediococcus pentosaceus, Leuconostoc mesenteroides* subsp. *cremoris, Lactobacillus casei* subsp. *Casei, Lactococcus lactis* subsp. *lactis* biovar. *diacetylactis* and *Thermophilic starters* include *Lactobacillus helveticus, Streptococcus thermophilus, Lactobacillus delbrueckii* subsp. *bulgaricus Enterococcus faecium,*

Lactobacillus acidophilus and Lactobacillus lactis. On the basis of physical form *Liquid starter cultures, Frozen starter cultures, Concentrated freeze-dried cultures. Depending on the absence or presence of flavour bacteria and their type. O cultures: B cultures:* These do not ferment citrate e.g. *leuconostoc* and *lactococci* spps. These are also called L cultures. *D cultures:* These cultures contain both citrate fermenting and non-fermenting *Lactococci* but no *Leuconostoc* species. *BD cultures:* These contain *Leuconostoc* sps. as well as *Lactococci* found in D culture. *Artisanal cultures Natural milk cultures:* These cultures which are used as starter are made under controlled conditions by incubating milk that favour the growth of naturally occurring thermophilic lactic acid bacteria. And also cheese specific cultures.

1.5 Enzymes for Cheese Making

A complex of enzymes used in the preparation of most cheeses is the rennet. Rennet contains enzymes like chymosin, pepsin and lipase. Chymosin also termed rennin is the active enzyme involved in milk coagulation. All rennets are proteolytic in nature and belonging to the group acid proteinases. These cleave the proteins so that they become insoluble and form a coagulated mass enclosing the other milk ingredients causing it to separate into whey and curd. Rennet from lambs, kids, or calves, have been used traditionally but the most widely used enzyme preparation is calf rennet. There are non-animal sources of rennet also suitable for vegetarian consumption called rennet substitutes. These include plant proteinases, coagulants from bacteria and fungi and recombinant proteases metabolized by genetically modified micro-organisms.

1.6 History of Cheese Manufacture in India

In India, demand for cheese was very scarce before independence and it was met either by few cottage scale cheese plants around hill stations or by imports. After independence, some cooperative sector and private agencies started manufacturing cheese to meet the rising demand. The technology for cheese manufacture was traditional and no mechanization or automation was involved. The quality of the cheese was compromised and due to the lack of cold chain the marketing of natural cheese was also put on a stop. Most of the natural cheese was processed for the ease of marketing. The promising players in the cheese market were Amul, Vijaya and Verka. These players continued to expand their cheese manufacturing capacities to meet the growing cheese demand. During this period many small scale industries starting manufacturing cheese manufacturers in various parts of the country to meet the local demand from restaurants, hotels, and consumers around their operation area. In the recent years many players like Dynamix, Dabon, Vadilal, Nandini,

Vintage, Parag, Prabhat, Sudharsan, Mother Dairy and Britannia have entered the cheese market and have come up with wide varieties of cheeses including the foreign ones. This has increased the domestic cheese sales and decreased the import to about 5% of total cheese market in India.

1.7 Cheese Production and Market in India

In India, the market is dominated by paneer-consuming consumers however, in the recent years there has been a surge in the demand for cheese which can be attributed to changing food habits and rise in food service outlets like Domino's and Pizza Hut throughout the country. The Indian fast-food market is growing annually at 25–30%. Cheese is the prominent ingredient in a number of fast foods such as pizzas, burgers, pasta, fondues, sandwiches, and garlic breads. In addition to this, cheese is used as a flavour enhancer in many traditional Indian recipes such as parathas, dosa, and uttapam. Rising disposable income levels and government initiatives are also major factors contributing to the growth of cheese in the country (Dairy Times Report, 2015). Cheese production in India is estimated to be growing at the rate of 16–17% per year in terms of value and 10–12% in terms of volume per year (Jayadevan, 2013).

Over the years the production and export of cheese has increased immensely. Eighty percent of the total cheese market in India is processed cheese which is mostly used by restaurants and fast food chains. Imported cheeses form a fraction of total Indian market and over the years the domestic cheese sales have increased abruptly. The retail sales of cheese in India have increased from 99 million dollars in 2010 to 244 million dollars in 2014 with maximum purchase by restaurants and hotels. Nowadays, many Indian dairy industries manufacture special foreign cheese varieties to meet up the demands of cheese loving consumers. Major players include Amul, Britania, Gowardhan, Milky mist and Mother dairy. Amul dominates the processed cheese market with a 75% share which contributes about 2000 tons a year. Processed cheese is sold as slices, spreads, cubes, wedges, blocks and sauces to diversify the cheese market in India. The demand for natural cheese is also rising in the Indian cheese market. The total cheese production has been growing by about 15% per year. Currently the market for cheese in India is worth Rs 1250 crore. The urban per capita consumption of cheese is 700 g per year which is much lower than the world average of 7 kg/year. In India over 60% of the cheese production is consumed in metropolitan cities viz., Mumbai, Delhi, Kolkata and Chennai. Mumbai accounts for 30% followed by Delhi (20%), Kolkata (7%) and Chennai (6%) of cheese sold in the country (Lagos & Intodia, 2015; Jana, 2015).

1.8 Global Cheese Market

The global cheese market has witnessed considerable growth in recent years, driven by a variety of factors, including changing consumer preferences towards protein-rich, convenient food options. These shifting dietary trends are key drivers of the increasing demand for cheese. Additionally, the surge in culinary shows, impactful marketing strategies, and the growing influence of social media platforms are all contributing to the global popularity of cheese. Europe remains the dominant region in the cheese market, owing to the wide availability of a diverse range of cheese varieties. Furthermore, the cultural importance of cheese in European cuisines enhances its demand both within the region and internationally.

In 2023, the global cheese market was valued at US$ 88.5 billion, with projections to reach US$ 143.5 billion by 2032, reflecting a compound annual growth rate (CAGR) of 5.3% between 2024 and 2032. Technological advancements in cheese production, along with the expansion of the food service industry, are primarily driving this market growth. Within Europe, the cheese market is further bolstered by the easy access to various cheese options and the growing presence of prominent online retailers such as Love Cheese, Italia Regina, La Gourmeta, and Frank and Sal. Major cheese-producing countries like Italy, France, Germany, and the Netherlands play a critical role, contributing to 39.53% of the total cheese consumption in the region in 2022. Cheese holds particular value in France, where its diverse flavors and cultural significance lead to high daily consumption, with almost 40% of the population eating cheese daily, and over 95% including it in their regular diet. Furthermore, the increasing popularity of regional Italian dishes made with cheeses like Parmesan, Mozzarella, and Pecorino is expected to drive continued growth in the European cheese market. The growing global adoption of conventional cheese made from animal-based milk sources such as cow, buffalo, and goat milk is also a significant contributor to market expansion. Cow milk, with its mild flavor and versatility, remains the most widely used source for cheese production. Buffalo milk is valued for its unique flavor profile, attracting those seeking distinct cheese varieties, while goat milk-based cheeses appeal to consumers looking for nuttier and tangier flavor options.

Natural cheese continues to hold the largest market share, particularly in Western Europe, where it constitutes 87% of retail sales. Consumer preferences for clean-label, minimally processed products are expected to sustain growth in this segment. Among cheese types, Mozzarella holds the largest market share, driven by its widespread use in pizzas and as a topping in a variety of dishes. Sliced cheese is a popular choice for sandwiches and burgers, while shredded cheese is commonly used in baking and cooking to enhance the flavor of various dishes. Blocks of cheese are versatile, allowing for slicing or grating as needed, and are often used for both culinary and entertaining purposes. Cheese spreads, commonly enjoyed on crackers or in sandwiches, are appreciated for their creamy texture. Liquid cheese, or cheese sauce, is increasingly popular as a dip or topping for snacks such as chips, nachos, and fries, further propelling the global market in this segment (IMRAC Report, 2024).

1.9 Nutritional Value of Cheese

Cheese has a high food and nutritive value. The composition of some of the cheese varieties in given in Table 1.1. It is a rich source of high-quality protein containing all the essential amino acids. During proteolysis bioactive peptides are released from proteins which provide specific health benefits like antihypertensive and immune-modulatory effects. During cheese manufacture, fat and casein is retained in cheese curd, while water soluble constituents (whey protein, lactose, water soluble vitamins) partition into whey. Cheese from whole milk is known to contribute fatty acid but there has been an effort to introduce more unsaturated fatty acids (linolenic) into the diet of the animal in order to augment those in normal milk. Cheese is a rich source of calcium, phosphorus and iron. Calcium and phosphorus are required for bone and teeth formation and may prevent osteoporosis and iron is important for red blood cell formation. B_{12} and to a lesser extent, folate. Cheese has been found to be anticariogenic and antiosteoporotic.

1.10 Mode of Utilization

Natural cheese can be enjoyed as is or used as a topping on bread, biscuits, and similar items. Over the past century, advancements in melting techniques have led to the creation of processed cheese products, which are made from natural cheese at varying stages of maturity. These products are developed with controlled flavors, textures, and improved shelf stability. Additionally, various forms such as slices, shapes, and configurations have been introduced, offering diverse applications. Consumers can incorporate these products into a variety of dishes as cooking ingredients or enjoy them directly as snacks. They can also be used as spreads, sandwich slices, dips, or snacks toppings. Cheese can also be dried to have extended shelf life and dried cheese has many applications in bakery items or as cheese alternatives. Cheese-based snacks like Cheese Pakoura are gaining popularity in our region. The demand for cheese is growing rapidly in our country, with processed cheese spreads and slices emerging as excellent companions for bread. The rising consumption of bread has further contributed to this trend. Traditional spreads like butter are now less favored due to their high fat content, which has been linked to obesity and cardiovascular issues. Similarly, conventional spreads like jams and jellies are increasingly regarded as unhealthy options. The introduction of pizza to the Indian market has significantly boosted the demand for Mozzarella cheese.

Table 1.1 Composition of some of the cheese varieties (g/100 g)

Cheese type	Moisture	Protein	Fat	Lactose	Ash	Salt	Calcium	Phosphorus	Vitamin A (I.U.)	Thiamin (mg/100 g)	Riboflavin (mg/100 g)
Cheddar	37.0	25.0	32.2	2.1	3.7	1.5	0.75	0.48	1310	0.03	0.46
Gouda	38.5	25.5	28.5	–	2.5	1.5	0.74	0.49	–	0.03	0.3
Parmesan	30.0	36.0	26.0	2.9	–	–	1.20	0.81	1060	0.02	0.73
Swiss	39.0	27.0	28.0	–	2.0	1.5	0.90	0.75	1140	0.01	0.40
Feta	54.34	17.39	20.2	0.53	4.12	4.39	0.36	0.28	–	0.04	0.21
Ricotta	76.4	13.2	8.4	3.6		–	0.24	0.17	–	0.02	0.19
Mozzarella	54.0	22.1	18.0	–	2.3	0.7	0.59	0.42	–	0.03	0.31
Cottage (creamed)	78.3	13.6	4.2	2.9	1.0	0.8	0.07	0.16	170	0.03	0.24
Camembert	52.2	17.5	24.7	1.8	–	–	0.10	0.18	1010	0.04	0.75
Roquefort	39.5	22.0	32.0	–	2.0	4.0	0.65	0.45	–	0.04	0.65
Gorgonzola	44.0	22.0	29.0	2.0	–	3.0	0.50	0.37	1240	0.03	0.41
Cream cheese	60.0	8.0	30.0	2.1	2.0	0.75	0.09	0.1	1540	0.02	0.24
Processed cheese	40.0	24.0	28.0	1.8	4.0	3.0	0.60	0.8	1220	0.02	0.41

Source: Kosikowski (1982)

References

Dairy Times Report. (2015). *India – The next big cheese market*. download.ameft.com/Cheese.pdf

Fox, P. F., & McSweeney, P. L. H. (2004). Cheese: An overview. In P. F. Fox, P. L. H. McSweeney, T. M. Cogan, & T. P. Guinee (Eds.), *Cheese: Chemistry, physics and microbiology. Vol. 1, general aspects* (3rd ed., pp. 1–18). Elsevier.

IMRAC Report. (2024). Global Cheese Market. https://www.imarcgroup.com/global-cheese-market.

Jana, A. (2015). Indian cheese industry—Status and future scope. *Food and Beverage News*. FnBnews.com

Jayadevan, G. R. (2013). A strategic analysis of cheese and cheese products market in India. *Indian Journal of Research, 2*(3), 247–250.

Johnson, M. E. (2017). Cheese production and quality. *Journal of Dairy Science, 100*(12), 9952–9965.

Kosikowski, F. V. (1982) in Cheese and Fermented Milk Foods. 2nd edition by F. V. Kosikowski LLC, F. V. Kosikowski & Associates.

Lagos, J. E., & Intodia, V. (2015). *Cheese demand rising – New market opportunities*. Global Agricultural Information Network Report.

Scott, R. (1998). *Cheesemaking practice*. Aspen Publishing.

Chapter 2
Basic Cheese Making Process

Ridhi Pandey and Yogesh Khetra

2.1 Introduction

Cheese stands out as the most diverse category of dairy products and is arguably among the most intriguing and complex from an academic perspective. Unlike many dairy products that remain biologically, biochemically, chemically, and physically stable when properly produced and stored, cheeses are biologically and biochemically dynamic and inherently unstable. The cheese-making process involves a meticulously coordinated sequence of biochemical reactions. When these processes are well-timed and balanced, they produce cheeses with highly desirable aromas and flavours. Conversely, an imbalance can lead to undesirable off-flavours and odours. It is widely believed that cheese originated in the "Fertile Crescent," an area extending from the Tigris and Euphrates rivers through present-day southern Turkey to the Mediterranean coast, around 8000 years ago. This region was pivotal during the "Agricultural Revolution," marked by the domestication of plants and animals. Early humans recognized the nutritional value of milk from domesticated animals and began to develop methods to process and preserve it. Initially, goats and sheep, known for their gregarious and docile nature, were the first dairy animals to be domesticated. Over time, cattle became the dominant dairy species in many parts of the world. It has an extensive variety of flavours and forms across the globe, shaped by regional cultures and available resources. It can be considered a bio-complex ecosystem inhabited by diverse microorganisms, collectively referred to as cheese microflora. These microorganisms, introduced through raw milk, starter cultures, and adjunct cultures, are crucial in defining the sensory attributes of different cheeses. They interact with milk proteins, carbohydrates, and fats during the critical technological process of ripening (Forde & Fitzgerald, 2000). The transformation of milk into cheese begins with the coagulation of milk proteins, primarily casein,

R. Pandey (✉) · Y. Khetra
Dairy Technology Division, ICAR-National Dairy Research Institute, Karnal, Haryana, India

S. Mansha Rafiq (ed.), *Technological Advances and Trends in Cheese Making*,
https://doi.org/10.1007/978-3-031-94476-5_2

which is triggered by enzymatic action or acidification. This step is essential as it sets the stage for the textural and flavour properties of the cheese. Subsequent processes, including curd cutting, cooking, and draining, influence the moisture content and texture of the final product. The fermentation of lactose by added microorganisms whether starter cultures or adjunct bacteria plays a key role in developing the flavour profile and texture of the cheese during ripening. The chemical composition of the milk—particularly its fat, proteins, and minerals significantly affect the properties of the curd and the final cheese. Each step in the cheese-making process, from coagulation to ripening, reinforces these interactions and determines the cheese's quality in terms of flavour and texture. This chapter aims to provide a foundational understanding of cheese production, paving the way for further exploration of various cheese types and their distinctive characteristics.

2.2 Aspects of Cheese Curd Production

Cheese curd production involves a complex process influenced by various factors. It is basically a dehydration process where liquid milk is converted into curd with different steps involved in expulsion of moisture. The general steps involved in most cheeses, includes selection of milk, milk coagulation, curd formation, syneresis of moisture and ripening (Huppertz, 2019). Milk coagulation can be achieved by acid, rennet or heat along with acid addition. Moisture expulsion from curd is done by combination of different techniques like cutting, cooking and drainage of whey and pressing. Not all cheese are subjected to ripening. Majority of acid coagulated, acid-heat coagulated and few rennet coagulated are consumed fresh and so their final flavour and texture are achieved at the end of curd formation. Factors affecting milk quality like composition and microbial quality are to be taken in consideration during cheese manufacturing (Law & Tamime, 2010). A more detailed description of all the steps are given as follow:

2.2.1 Selection and Pre-treatment of Milk

An essential requirement for producing high grade cheese is to begin with high quality milk.

2.2.1.1 Milk Quality

Milk quality and its type has direct effect on quality of cheese produced. Milk of different species like cow, sheep, goats, ewe and buffalo, are used for making cheese. It is crucial to take care of quality milk starting from taking care of nutrition and medications given to animal. Animal nutrition system differs according to animals and environmental factors and its aim is to maintain optimum health of animal and optimum milk production. Feeds which give taints to milk will make it unfit for cheese production as the taint from milk will probably be transferred to cheese (Scott et al., 1998). Milk from diseased animals should not be mixed with healthy animal milk. Also, early and late lactation milk should be excluded given that there composition differs from normal milk. Late lactation milk has lower casein and higher level of FFA and a high somatic cell count. These have negative effect on cheese such as high moisture and pH, reduced yield and have effect on degradation of αs1-casein in cheese during ageing (Guinee & O'Brien, 2010). Proteolytic and lipolytic activity of cheese increases in presence of somatic cells which can lead to development of biogenic amines and off-flavours (Moradi et al., 2021). Late lactation milk and mastitis milk have high activity of proteinases, like plasmin and cathepsin D and may contribute to proteolysis in cheese during ripening (O'Mahony et al., 2013). Milk must be free from residues of antibiotics. These antibiotics are given to animal to treat them against various bacterial disease. Since residue of these drugs passes to milk, it is recommended to use milk only after withdrawal period of the drug is over. If used, they can cause starter culture failure or delayed action of starter culture which will have effect on flavour and texture of cheese or even development of resistant pathogens. Apart from residues of antibiotic, milk should also be free from aflatoxins as it is toxic to human. Aflatoxins are secondary metabolites produced by mold such as *Aspergillus flavus* and *Aspergillus parasiticus*. These metabolites are heptacarcinogens, mutagens, and immunosuppressive agents. (Badea et al., 2004; Singh et al., 2013). The aflatoxin found in milk is M1. In milk, aflatoxin M1 is formed when animal ingest feed infected with alflatoxin B1. The dairy industry conducts regular and rapid testing of milk for presence of aflatoxins at reception and any milk that exceeds the legal limit, is discarded. After milking, milk is cooled to 4 °C and at this temperature it is transported to cheese plant for further processing of milk.

2.2.1.2 Thermization

This temperature treatment of 65 °C for 15 s is given to raw milk when it is stored for long time at refrigeration temperature. This treatment destroys psychotropic organism which may have proliferated due to long storage at refrigeration temperature. It is an essential treatment because it also prevents the production of lipases and proteinases by psychotropic organism which otherwise would result in undesirable proteolysis during ripening of cheese.

2.2.1.3 Microfiltration and Bactofugation

To improve the bacteriological quality of cheese milk, it may be passed through microfiltration membrane. Microfiltration membrane have pore size of 0.01 to 10 mm and so retains fat particles, endospores and vegetative bacterial cells which decrease the bacterial load of milk to a very low level. The retentate can be combined with cheese milk after UHT treatment (Scott et al., 1998). Examples of cheeses that are made with micro filtrated milk are the Swedish Greve, Herrgård and the Swiss Raclette.

Alternative process to reduce bacterial load is bactofugation. It is widely used to reduce spore forming bacteria in milk and it operates on centrifugal separation based on difference of densities between spore and milk. It reduces by centrifugation of milk at 8000G–1000G. The sludge (fat-rich) that contains the endospores amounts to 2–3% of initial feed and is treated at 130 °C–140 °C for a few seconds, then added back to the milk. A defect in cheese called "late blowing" can be prevented by use of bactofugation as it is caused by bacterial spore of *Clostridium tyrobutyricium*. This act by anaerobic metabolism of lactate to butyrate, carbon dioxide and hydrogen. Excessive gas formation can lead to the development of numerous large holes, splits, and cracks, ultimately compromising the integrity of the cheese.

2.2.1.4 Standardization

Standardization milk is done to adjust the compositional parameters of milk to desired level, mainly fat and protein. It helps in ensuring that legal specification of cheese is met by the manufacturer and also produces uniform characteristic of cheese. It is often done by use of centrifugal separator which give skim milk and cream, which can be blended in different proportion to get desired ratio of constituents mostly within an automated in-line process (Kelly, 2007). Protein content can be adjusted by using retentate of ultra-filtered milk.

2.2.1.5 Homogenization

Homogenization is done to reduce the fat globule size to prevent creaming. In rennet curd cheese like Cheddar and Tilsit, homogenization decrease the quality of finished cheese while when done for blue cheese, homogenization produces a superior product. It can done for producing fresh, acid curd cheese, as it helps create a smooth, thick, and creamy texture in the final product. Homogenization of milk has certain advantages like higher yield, lower fat losses in whey, reduced shrinkage during ripening and storage and reduced fat leakages at high temperature (Peters, 1964).

2.2.1.6 Pasteurization

Pasteurization of milk is a common process during the manufacturing of cheese which has a purpose to kill pathogenic microbes, like *Listeria monocytogenes*, *E. coli*, and *Salmonella spp*. The temperature used is either 72 °C–75 °C for 15–20 s or 63 °C–65 °C for 30 min. While pasteurization is important from the standpoint of public health, it alters the milk's indigenous microflora, which impact the flavour and texture of the cheese produced. Cheese made from pasteurized milk is typically safer and more consistent in quality than cheese made from raw milk. On the other hand, pasteurization can negatively impact the coagulant ability of rennet and characteristic features of curd formation in milk. Further, the killing of desirable indigenous LAB and denaturation of indigenous milk enzymes by heat treatment contribute to cheese ripening at a slower rate with a milder flavour in pasteurized cheese. Adjunct culture can be added to ensure proper flavour production during ripening. Pasteurization causes partial denaturation of whey protein and its aggregation with casein which causes increase in yield by 0.1%–0.4%. This heat denaturation of whey is done for production of fresh acid curd cheese like, Quark and Fromage Frais (Bintsis & Papademas, 2017). High heat temperature treatment (HTT) of milk has also been used to produce rennet curd cheese as it can increase the yield and improve texture in low fat cheese by increased recovery of whey protein through denaturation and interaction of whey proteins with k casein. This complex of k casein and whey proteins remain with the para k casein micelle after rennet action. Though this interaction impede rennet gelation and curd syneresis as a result of precipitation of calcium phosphate, reduction of $CaCl_2$ resulting in poor texture and flavour (Guinee, 2021).

2.2.1.7 Ultrafiltration

Ultrafiltration is used for standardization of cheese milk and provides a way for concentration protein and fat in milk and this reduces the need of whey syneresis and whey proteins are retained in the curd. These proteins have good water absorbing capacities and so some of the water soluble components are also retained with whey proteins. In a study by Gursoy et al. (2013), concentrated retentate of UF had been applied to different type of cheese, in which cheddar cheese showed an increase in yield by 6–8% and Feta cheese showed 14% without disturbing the quality of cheese. The ultrafiltration of milk before the cheese making process leads to a concentration of protein and to a curd that is heavier and more prone to tearing. This problem can be alleviated by reducing the coagulation temperature from 31 °C to 27 °C, which slows down the second stage of coagulation and gives a softer, more pliable curd. Cheeses made from high protein levels also exhibit lower levels of proteolysis, a defect that can be offset by increasing the rennet-to-casein ratio in the cheese making process (Guinee et al., 1994).

2.2.2 Acidification of Milk

Acidification of cheese milk is done either by addition of starter culture of or by use of acid directly. The development of acidity inhibits the growth of unwanted bacteria and once the desired acidity level is reached, rennet is added for most varieties of cheese to form curd with the desired properties. Acidification starts before all operation and continues throughout all stages of production. The production of acid at the correct rate and timing is a crucial factor in producing high-quality cheese. It has directly or indirectly impact on coagulant activity, curd strength, syneresis, growth of non-starter microbes and final curd property. pH significantly influences the solubilisation rate of colloidal calcium phosphate, which in turn affects the proteolytic susceptibility of casein and impacts the cheese rheological properties. When the pH is too acidic, the cheese becomes crumbly. Conversely, if the pH is too basic, the cheese develops a pasty and sticky texture, resulting in an undesirable consistency (Holsinger et al., 1995). Acidification inhibits the growth of spoilage microbes and has role in providing optimum pH for enzymatic activity. It also promotes syneresis and has an impact on cheese composition and ultimately affecting microbial activity during ripening and yield. It determines retention of coagulant activity during ripening (Fox & Guinee, 2013). Acidification by starter culture is done by in-situ production of lactic acid. The incorporation of microbial cultures, including bacteria, yeasts, and molds, is a fundamental aspect of cheese production. These cultures have two roles, first is the decreasing the pH during manufacturing of cheese, second is aid production of desired flavour and texture through their residual activity during ripening. Starter culture for cheese varieties heated not more than 40 °C are *L. lactis subsp lactis*, *L. lactis subsp cremoris* and for varieties cooked to higher temperature such Swiss and hard variety cheeses, *S. thermophilus* and Lactobacillus spp. like *L. delbrueckii subsp. bulgaricus*, *L. delbrueckii subsp. lactis* or *L. helveticus* are used. Typically, the ultimate pH of cheese curd ranges between 4.6 and 5.1, though the time required to achieve this varies, ranging from 5 h in Cheddar cheese to 6–12 h in Blue, Dutch, and Swiss varieties. These differences are influenced by factors such as the amount of starter culture added to the cheese milk (ranging from 0.2% to 5%), the cooking temperature and time, which can inhibit starter microorganism growth, and the rate at which the curd is subsequently cooled (Fox & McSweeney, 2017). Direct acidification make use of citric acid and other acidogen for few varieties of cheese like Mozzarella, UF Feta-type and Cottage cheese. Starter cultures of lactic acid producing bacteria are used in most cheese. Direct acidification has advantage of greater control when compared to starter cultures and, unlike starter culture, it is not bacteriophage attacked. However, its limited use is primarily attributed to the absence of the additional biochemical contributions from starter cultures, which are essential for proper ripening in cheese production (Fox & Guinee, 2013).

2.2.3 Coagulation of Milk

Coagulation of milk is an essential step in cheese manufacturing which results in formation of casein gel which entraps fat and other components, having significant impact on yield and quality of cheese. Milk coagulation occurs by destabilizing casein micelles, either through acidification, which neutralizes the negative charge on the micelle surface, or through enzymatic action, where rennet removes the hydrophilic glycomacropeptide. These two methods form the basis for classifying cheeses: (1) acid-coagulated and (2) rennet-coagulated, with rennet-coagulated cheeses making up over 75% of global cheese production (Fox & Guinee, 2013). Traditionally, rennet has been used in most varieties of cheese. Rennet is extracted from inner mucosa in the fourth stomach of milk-fed calves, in which the principal active ingredient is chymosin. However, rennet is not able to meet high market demand caused by rising in cheese consumption around the world (Murlidhar et al., 2017). Along with unavailability, high cost, ethical and religion concerns have let to use of rennet substitutes. A brief of different type of rennet substitute is given in Fig. 2.1. Plant proteases for milk coagulation can be sourced from various plant parts, including fruits, leaves, flowers, seeds, latex, and rhizomes. (Shah et al., 2014). However, challenges include bitter taste from excessive proteolytic activity and dealing with variability in enzyme concentration and activity across different plant sources (Mohsin et al., 2024). Recombinant rennet are rennet in which calf chymosin gene has been cloned. Recently it has been cloned in *Kluyveromyces lactis*, Escherichia coli, and *Aspergillus niger* and chymosin extracted from these organisms are used in cheese production (Fox & McSweeney, 2017). Microbial rennet substitutes are proteases extracted from microorganisms. They have advantages of having similar clotting activity as rennet and also are accepted by

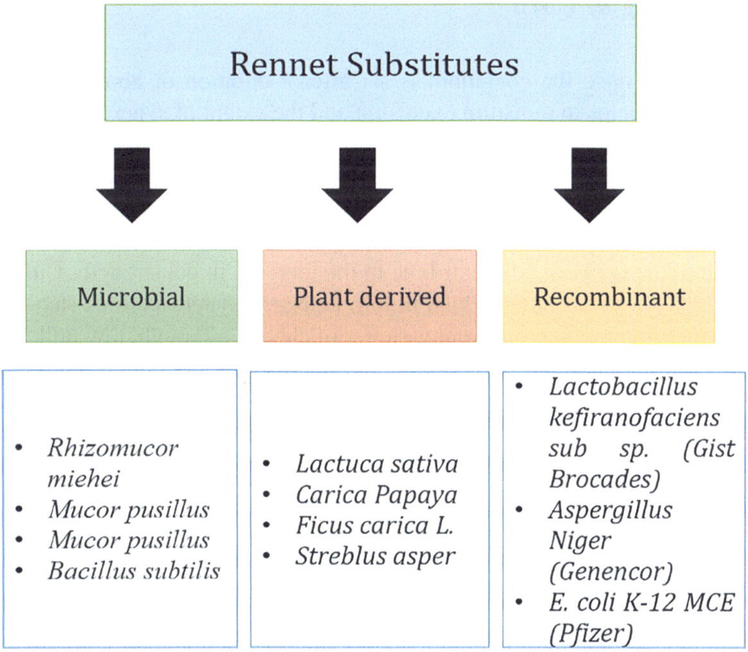

Fig. 2.1 Different types of rennet substitutes used for cheese production. (Liu et al., 2021)

vegetarians. Since the enzymes also play a role in the ripening process, the amount used and their specific characteristics are crucial to the quality of the cheese being made.

Rennet coagulation is a process involving two successive steps: one enzymatic and one non-enzymatic. In the first step, enzymes catalyse hydrolysis of the casein, mainly at the surface of the micelle. The enzymatic step is followed by a second process in which the originally spherical micelles progressively lose their ability to interact with water. This reduction in the interaction of water is due to the fact that κ-casein's macropeptide has been removed, which reduces the steric and charge repulsion allowing the micelles to interact with each other, forming aggregates and chains (Lucey et al., 2003). The micellar chains continued to grow in length and thickness and interlock to form a three-dimensional, net-like structure as coagulation proceeds further. This matrix entraps water and other important milk constituents such as lactose, fat, whey proteins, and minerals (Kindstedt, 1995). Rennet-curds are quite different from acid-precipitated curds and, in particular, their syneresis properties. Rennet curds possess optimum syneresis for the purposes of manufacturing low-moisture curd without its becoming hard (Holsinger et al., 1995). The production of acid-coagulated cheeses (such as Cottage, Quark) relies on the fact that caseins become insoluble at their isoelectric point (pH 4.6), forming a gel that traps fat (Guinee & O'Brien, 2010).

2.2.4 Cutting of Curd

Cutting is done once the coagulum is set after a duration of about 25 min-2 h. Cutting of curd helps in moisture expulsion and the extent of syneresis of moisture determines the composition of final cheese. Apparently, this change in moisture content likely reflects the degree of bonding between and within casein particles, which strengthens over time (Lucey et al., 2003). Additional processes, including particle fusion, structural rearrangements, and the continued incorporation of micelles into the network, all contribute to the increase in gel strength. Cutting the curd expands its surface area, which in turn enhances syneresis. This step has significance in differentiating different varieties of cheese, although, raw milk composition, acidification, type of starter and rennet are also equally important. Again, the rate and extent of syneresis is influenced by the size of cubes, pH of whey, composition of raw milk, cooking temperature, rate of stirring of curd (Fox & McSweeney, 2017). Determining the precise time to cut the curd is crucial for ensuring cheese quality. Cheese makers typically test this by inserting a hand, rod, or thermometer below the surface of the coagulum and lifting it, creating a break along a cleavage line. If the break is clean with green whey at the base, the curd is ready for cutting. However, if the cleavage is soft and irregular with white whey, it indicates that the curd is still too soft to cut. Several factors affect the rate and extent of syneresis such as milk composition, calcium concentration, cooking temperature and time, and stirring rate. These parameters are critical for the development of the texture and moisture level of the curd and represent the first step in the differentiation process between cheese varieties. While these factors are very important, the type and quantity of starter culture, coagulant, and the makeup of the cheese milk also play a crucial role that determines the final characteristics of the cheese (Holsinger et al., 1995).

2.2.5 *Cooking of Curd*

Cooking is essential for controlling syneresis by influencing shrinkage of the curd and development of acid. At the beginning, curds are soft with open coats on particles. Gentle stirring of the curd during the expulsion of the first whey prevents excessive curd crashing, loss of fat, and curd dust. As the curd coats become more membrane-like, stirring can be intensified. Cooking or scalding of the curd causes shrinkage of the protein matrix and whey to be expressed. Higher temperatures also increase bacterial metabolism in the curd causing increased lactic acid production and lower pH, which in turn promotes further whey expulsion. Cooking temperatures vary from 30 °C for high-moisture cheeses (e.g Camembert) to 55 °C for low-moisture cheeses (e.g. Emmental and Parmigiano-Reggiano). After heat treatment, curds are normally transferred to molds for further draining and acidification.

2.2.6 *Whey Drainage and Curd Treatment*

When appropriate degree of syneresis has occurred, the curds are separated from the whey, typically by allowing the material to drain through a perforated metal screen. Remaining manufacturing steps may cause additional syneresis to occur, depending on the variety of cheese. Small, high-moisture curd cheeses are scooped from the vat into various-sized perforated molds, where additional syneresis occurs. During molding, and after whey drainage, the curd particles become joined or "knit" together in a single mass. This knitting action continues the dehydration and demineralization initiated during cooking. Usually knitting is followed by pressing, which includes the application of force to the curd to force out more whey, furthering syneresis. Pressing further contributes to a finer fusing of the curd particles, making the texture denser and the surface with fewer openings.

2.2.7 *Salting*

Salt is an essential ingredient of cheese and it can be added to cheese curd as in cheddar and related varieties or used as brine, in which cheese is immersed for short time, as for Gouda, feta and related varieties. Salting helps in shelf life extension by preventing growth of undesirable bacteria by its ability to reduce water activity (Albarracín et al., 2011) and promoting growth of desirable bacteria like lactic acid bacteria. Apart from this preservation effect, salt has role in whey separation and control of the final moisture content, influence on the secondary microorganisms that may grow and produce volatile compounds during the aging process, enzyme activity control as well as body of the final cheese product when NaCl replaces Ca in the cheese matrix, flavour, and taste improvement of the cheese. Since salt can influence these factors, the amount and the method of incorporation of salt has major influence on final characteristic of cheese (El-Bakry et al., 2011). The concentration of NaCl in most varieties of cheese is high enough to prevent the growth of starter bacteria. In many varieties, dry salt is mixed with the curd near the end of manufacture and hence the pH of the curd must be very nearly at its final value when the curd is salted. In other varieties, salt is applied after molding, either by immersion in brine or by application of dry salt to the surface (Fox & McSweeney, 2017).

2.2.8 Pressing

Pressing serves to eliminate the last portions of whey from the curd. This operation also gives some varieties of cheese their conventional shape. The pressure applied during pressing varies based on the cheese type, and it should be gradual at first, as high pressure too soon can compress the surface, trapping moisture inside the cheese. The temperature of the curd before pressing should be below the liquid fat temperature that is 23.9 °C in summer and 26 °C in winter. Otherwise, fat will leak from the curd and be lost in the whey, or will fill spaces in between the curds and give a greasy cheese. After pressing, paraffin coating is applied in some varieties of cheese, and the cheese is stored at controlled temperature and humidity for ripening.

2.2.9 Ripening of Curd

While the processing conditions and chemical composition of milk determine the extent of cheese dehydration, the moisture content, amount of salt, pH, and cheese microbiota will in turn determine the extent of various biochemical reactions during ripening. Due to these reactions, flavour, aroma, texture, and functionality are attained in the final cheese product. In light of these, it would be appropriate to state that quality and characteristics in the finished cheese are imparted by the manufacturing process. Although some cheeses are consumed fresh and form a major part of cheese consumption in several countries, notably acid-coagulated varieties; in the case of most varieties, the flavour and texture that distinguish ripened cheeses from their fresh uncured states develop during the course of ripening. The phase of ripening itself is thus important in the maturation and identity of diversities (Fox & McSweeney, 2017).

2.3 Effect of Compositional Parameters on Various Aspects of Curd Formation

Every aspect of cheese production, including rennet coagulation, firmness and syneresis, are influenced by the chemical composition of cheese milk, especially levels of casein, calcium, and pH. This section will explore the specific impacts these compositional factors have on the different aspects of curd formation.

2.3.1 Protein

Casein makes up about 80% of the total protein content in milk, with the remaining 20% being whey protein. The size of casein micelles in milk from individual cows is not affected by lactation stage or protein content (De Kruif & Huppertz, 2012). Casein micelles are characterized by κ-casein on their surface, which imparts a net negative charge and stabilizes the micelles, preventing flocculation. A study found that milk with smaller casein micelles had longer coagulation time (Ekstrand, 1980). However, more recent studies have shown that smaller micelles improve coagulation, resulting in shorter coagulation time and firmer gels. In fact, smaller native casein micelles were observed to form stronger gels, suggesting that selecting for smaller micelles and lower milk pH could enhance gelation properties (Glantz et al., 2010). Milk with low content of κ-casein and a low proportion of κ-casein in relation to total casein showed poor coagulating properties (Wedholm et al., 2006). The effect of compositional parameters are interconnected. The initial fat and protein content of milk has influence on the final protein and fat content of cheese which consecutively influence different properties. Casein to fat ratio of 0.7 is recommended for manufacturing cheese with 50% fat in dry matter. High protein content in milk results in higher protein in cheese but a lower fat content. A high fat in milk gave high fat and low protein in cheese but it is also associated with high fat losses in whey and lower retention when compared to higher protein content of milk (Lou & Ng-Kwai-Hang, 1992).

2.3.2 Fat

The texture and structure of cheese are significantly influenced by its fat content and the distribution of fat globules (Lopez et al., 2007). Fat can be integrated into the cheese matrix as either an inert filler or as a copolymer within the casein structure. Fat globules can exist as individual globules, coalesced clusters, or free fat (Everett & Auty, 2017). Fat retention is crucial for texture, cheese yield, and mouthfeel, as higher fat retention results in a less dense protein matrix, more available water, and a softer texture with better meltability (Everett & Auty, 2008). The quantity of fat encapsulated within the cheese matrix is affected by several variables, including the fat-to-protein ratio of the milk utilized and the dimensions of curd particles, both of which influence moisture retention and, in turn, fat concentration (Johnston et al., 1998). Studies have shown that cheese produced from milk with differing sizes of milk fat globules (MFG) can exhibit variations in both composition and texture. For example, Emmental cheese manufactured using milk with smaller milk fat globules (3.3 μm) had a higher moisture content, which led to increased yield and stronger flexibility, yet less fracture stress, compared to cheese made from milk with bigger milk fat globules (5.6 μm). This cheese was also softer, and exhibited a lower fracture stress compared with the cheese from the large MFG milk (Michalski et al., 2007). Additionally, high-pressure processing of milk results in smaller, more evenly dispersed fat globules, producing firmer, more elastic cheese that melts and flows more easily (Buffa et al., 2001).

2.3.3 *Lactose*

Lactose are the main carbohydrate of milk and its content varies from species to species. Also, its concentration is affected by udder infections like mastitis and lactation. In cheese, the fermentation of lactose by lactic acid bacteria into lactic acid, is an important feature of acidification by starter culture (Fox et al., 1996). The quantity and rate of formation of lactic acid differ among varieties. Thus, for Cheddar-type cheeses, most of the lactic acid is produced before moulding; while for several other types, acidification occurs primarily after moulding. Bovine milk contains about 4.6% lactose of which about 98% is lost in the whey either as lactose or as lactic acid. The remaining 2% is fermented by starter bacteria over the course of the curd formation and early stages of ripening. With the formation of lactic acid, the pH of the curd reduces-which in turn effects factors such as the retention and activity of coagulants during ripening, firmness of curd, syneresis, mineral content, and growth of non-starter lactic acid bacteria (Fox et al., 1990).

2.3.4 *Salts*

2.3.4.1 Calcium (Ca) and Phosphorus ($P0_4$)

Calcium plays a significant role in milk coagulation. In milk, calcium exists in a dynamic equilibrium between soluble and colloidal phases. Colloidal calcium phosphate (CCP) is bound to phosphorylated serine residues in casein, contributing to the stabilization of casein micelles (Gaucheron, 2005). The addition of calcium results in a slight reduction in average micelle size, an increase in smaller micelles, and the formation of larger micellar aggregates in the coagulum (Müller-Buschbaum et al., 2007). Calcium also reduces rennet coagulation time by lowering pH and increasing free Ca^{2+} ion concentration (Hooydonk et al., 1986), which neutralizes negative charges on casein, enhancing micelle aggregation and gel formation (Gastaldi et al., 1994). Low levels of calcium increase gel firmness (Lucey & Fox, 1993). In a study by Ong et al. (2013), $CaCl_2$ added during cheese making showed an increase in the final cheese hardness and a decreases in whey fat losses due to the cross-linking with casein, which entraps fat globules in the matrix.

Milk also supplies organic phosphate (PO_4), which is covalently bound to phosphoserine residues in casein micelles. CCP likely acts as a binding agent, stabilizing micelle structure. As pH decreases, CCP solubilizes, with the rate increasing dramatically below pH 5.6 and completing at around pH 5.0. At pH 5.3, typical for many cheese varieties at the end of production, nearly all PO_4 in milk is solubilized (Schmidt, 1982). Cheese's acid-base buffering capacity is influenced by insoluble calcium phosphate, which exists in colloidal forms similar to those in milk and plays a significant role in cheese's buffering properties (Lucey & Fox, 1993).

2.3.4.2 Citrate

In milk, citrate functions as a chelating agent (Gaucheron, 2005), binding some of the free calcium ions present in the milk serum. Variations in citrate levels in raw milk have been suggested to affect milk processing characteristics by altering citrate's interactions with other milk components (Faulkner & Peaker, 1982), and are therefore likely to influence coagulation.

2.4 Complex Chemistry of Cheese Making

2.4.1 Chemistry of Rennet Coagulation

Coagulation by rennet is a critical step in the development of protein matrix during the manufacture of cheese. Casein micelles are complex structures composed of casein molecules that are interconnected by salt bridges, hydrophobic interactions, electrostatic forces, and hydrogen bonding (Euston et al., 2002). The casein micelles are draped with a layer of casein macropeptides, commonly termed the "hairy layer," which consists of the hydrophilic ends of κ-casein molecules (Horne & Lucey, 2017). These κ-caseins comprise 15% of total casein in bovine milk and contain the most hydrophilic regions; for that reason, interactions with water are enabled by the abundance of hydrophilic amino acids. κ-casein is not strongly bound to cations like other caseins, so it can stabilize caseins sensitive to calcium and allow the process of micelle formation (Fox et al., 2000). The destabilization of casein micelles due to aspartic proteinases in rennet causes coagulation in milk and the formation of curd. This is particularly facilitated by the addition of calcium, either intrinsically present in milk or externally added as $CaCl_2$, which would have reduced the surface potential of para-casein micelles. Calcium may bind to micelles through electrostatic cross-linking or interact directly with the carboxylic acid groups of α- and β-caseins, neutralizing its charges and, consequently, facilitating aggregation among the renneted micelles. Increased levels of calcium lower the rennet coagulation time and improve the curd firmness (Ong et al., 2013). One common strategy for increasing the moisture and protein contents of cheese is to pre-acidify the cheese milk to pH 5.8–6.2 before renneting. This process decreases charge repulsion between caseins and increases solubilisation of colloidal calcium, thereby shortening rennet coagulation time because more κ-casein is available to interact with the coagulant (Kelly et al., 2008). Such a technology also counteracts the effects of high heat treatments that damage coagulation properties of milk, and increases the strength of rennet-induced gels and gel-forming ability. The coagulation time is prolonged by heating above the pasteurization temperatures (72–74 °C) due to the formation at the micelle surfaces of β-lactoglobulin/κ-casein complexes, which hinder κ-casein hydrolysis by chymosin contained in rennet (Guinee et al., 1995). Effect of pH at renneting on yield and micro-structure of cheddar cheese has been studied by Ong et al. (2012), and it was suggested that renneting at pH in between 6.3 and 6.5 will result in increase in yield by 13% on dry matter basis without significant changes in cheese texture while at a high pH of 6.7 resulted in lower yield and a low pH of 6.1 showed a coarser, irregular and more porous structure with uneven fat distribution. In terms of retention of fat, low pH cheese has better retention ability when compared to higher pH, although the fat retained is lost during pressing. Pressing has mechanical stress which increase loss of fat globules smaller in size and free fat. Another study by Ong et al. (2011), showed the effect of temperature on coagulation of milk and on fat retention and cheese structure. Significantly high levels of fat and fat in dry matter is observed for milk coagulated at low temperature (27–30 °C) than those coagulated at higher temperature (33–36 °C). Also, lower temperature (27 °C) leads to formation of fine protein network while high temperature gave an irregular and less continuous cheese matrix. The reason may be increased size of casein aggregates at high temperature which may be as a result of hydrophobic interaction between casein particles. Also due to reduced capacity to retain fat would have an effect on aggregation of casein.

2.5 Chemistry of Cheese Ripening

Ripening is a fundamental technological process in cheese production, involving a series of intricate biochemical and microbiological transformations. These changes are driven by the metabolic activities of both primary and adjunct microbial cultures. The organoleptic properties of cheese, such as aroma and flavour, are significantly influenced by these processes during ripening. The extent of ripening is particularly critical, as it directly impacts the development of characteristic sensory attributes by altering the cheese's chemical composition.

2.5.1 Metabolism of Carbohydrates During Ripening of Cheese

Milk is obtained from cows, goats, or sheep and is subjected to pasteurization for the elimination of the pathogenic organisms. The milk is fermented by starter bacteria. Chymosin acts on casein proteins by aggregating them to form a solid-gel mass of milk. Once curdling is done, it undergoes pressure applied to get the extra moisture out from the curd. Flavour development involves the utilization of starter bacteria-consisting mainly of lactic acid bacteria. Fermentation of lactose to lactate occurs through glycolysis. Lactic acid bacteria are non-starter bacteria, further fermenting the lactate into by-products like acetate and CO_2. Approximately 1–2% of lactose is not metabolized by the starter cultures, but rather processed during ripening. Metabolism of lactate may take several paths: racemization to DL-lactate by NSLAB, conversion to butyrate and H_2 by Clostridium species, which may result in the formation of undesirable cracks and off-flavours, or its conversion to the short-chain flavour compounds acetate, acetoin, diacetyl, ethanol, and acetaldehyde via pyruvate (Melchiorsen et al., 2002). Species of Propionibacterium can convert lactate to propionate, acetate, and CO_2, and this contributes to the development of eyes in Swiss-type cheeses. Species of Penicillium metabolize lactate to CO_2 and H_2O (Haasnoot et al., 1989). This fermented lactose forms lactic acid, which in turn aids in the cheese becoming more acidic, giving it a tangy taste that is a characteristic of most cheeses (Iwasawa et al., 2014). Starter bacteria include Lactobacillus species, *Lactococcus lactis*, and *Streptococcus thermophilus*, for production of cheeses such as Feta, Cheddar, and Swiss, to name just a few. The starter bacteria, like *Lactococcus*, are also capable of taking in glutathione that contains sulphur, adding to the flavour of the cheese. Adjunct cultures and NSLAB play an important role in the ripening process, where the typical cheese flavour develops. Individual strains of probiotics also contribute to the various flavours and aromas of the diverse types of cheeses. Citrate-positive LAB includes *S. diacetylactis*, *L. lactis*, and *L. cremoris* converts residual citrate present in the curd to flavour compounds, such as acetoin, diacetyl, and butanediol responsible for eye formation in Dutch and Cottage cheeses.

2.5.2 Metabolism of Protein During Ripening of Cheese

Proteolysis is a key biochemical process in cheese maturation, involving the breakdown of various casein types such as αs_1, αs_2, β-casein, and κ-casein—into smaller peptides and free amino acids. This breakdown is facilitated by both endogenous milk proteases and proteolytic enzymes produced by lactic acid bacteria (LAB) used as starter cultures (Gan et al., 2016). Chymosin, residual from the coagulation process, initiates milk clotting by cleaving casein into short hydrophobic peptides, which can impart bitterness. This proteolytic activity transforms the cheese texture from a rubbery curd to a smoother, creamier consistency (Clark et al., 2009).

Proteinases and peptidases further degrade peptide chains into free amino acids, which then participate in various biochemical reactions to form flavour compounds. Aminotransferases convert these free amino acids into corresponding α-keto acids, which are subsequently broken down into volatile organic compounds (VOCs) through different metabolic pathways. Branched-chain amino acids like leucine, isoleucine, and valine can be decarboxylated into amines such as ketoisocaproate, α-keto-L-methyl valerate, and α-keto isovalerate—that have strong, unpleasant odors (McSweeney et al., 2004). Microorganisms capable of decarboxylation, including *Enterobacteria*, *Pseudomonas spp.*, *Enterococci*, and LAB, are involved in this process (Stratton et al., 1991). These amino acids can also be deaminated, resulting in carboxylic acids and ammonia. For instance, butyric acid contributes a rancid flavour, while pentanoic acid imparts a fatty taste (Gan et al., 2016). Aromatic amino acids are similarly converted into α-keto acids and further degraded. Casein contains small amounts of sulfur-containing amino acids, such as methionine, which are metabolized during milk and cheese ripening (Burbank & Qian, 2008).

2.5.3 Metabolism of Lipid During Ripening of Cheese

Milk triglycerides undergo hydrolysis by bacterial and endogenous milk enzymes, resulting in the formation of short- and intermediate-chain fatty acids as well as free fatty acids through the process of lipolysis (Collins et al., 2003). Short-chain fatty acids are critical for flavour development; however, excessive lipolysis can be problematic in certain cheese varieties due to rancidity, as observed in Cheddar, Gouda, and Swiss cheeses (Forde & Fitzgerald, 2000). In contrast, extensive lipolysis is advantageous for cheeses such as Blue, Emmental, Parmesan, and Italian varieties like Romano and Provolone (Clark et al., 2009). The microflora responsible for lipolytic activity includes *Propionibacterium freudenreichii*, *Geotrichum candidum*, and *Penicillium spp.*, which are prevalent in mold-ripened cheeses where extensive lipolysis is desired (McSweeney, 2011). During lipolysis, enzymes break the ester bonds between fatty acids and glycerol in triacylglycerides. Esterases target shorter acyl ester chains (C_2 –C_8), while lipases act on longer chains exceeding 10 carbon atoms (Collins et al., 2003). Short-chain fatty acids such as hexanoic, octanoic, and decanoic acids are known to influence cheese flavour and aroma significantly. Hexanoic acid imparts sweaty, pungent, and rancid notes (Clark et al., 2009), octanoic acid provides a goaty, waxy flavour, and decanoic acid contributes fatty, citrus aromas (Gan et al., 2016). Dodecanoic acid is associated with fatty odor (Clark et al., 2009). Free fatty acids, the products of the secondary phase of lipolysis, serve as precursors to a range of volatile organic compounds (VOCs) that directly affect cheese flavour due to their low flavour threshold. Variations in milk fat content can impact the profile of volatile compounds produced; for example, low-fat Feta cheese has been reported to contain higher aldehyde levels compared to full-fat Feta, which has increased concentrations of alcohols, ketones, and esters (Andiç et al., 2015). Additionally, full-fat Feta cheese tends to have a more intense flavour. Lipolysis is crucial in cheese ripening, particularly in blue cheese varieties, as it leads to the formation of free fatty acids that act as precursors for flavour compounds such as methyl ketones, alcohols, and lactones (Smit et al., 2005). Furthermore, packaging methods can influence the texture and fatty acid content of cheese. A comparative study of aluminium foil and vacuum packaging revealed that vacuum packaging, by preventing oxidation and the release of CO_2 and H_2O, promotes the formation of more fatty esters due to the availability of alcohols, fatty acid substrates, and a moist environment (Duval et al., 2018).

2.6 Conclusion

This chapter has outlined the basic principles of cheese manufacture, describing in some detail the sequence of complex processes through which milk is transformed into a cheese. The key steps in cheese production, namely coagulation, curd cutting, cooking, draining, and pressing, all of which are critical to the development of texture and quality in the finished cheese. Of more importance are the compositional factors, hence the concentration of milk proteins, fat, and minerals, as these elements come into direct contact with the texture and yield of cheese. Another important subject discussed by this chapter was the role of rennet and biochemical changes during ripening, which showed how microbial flora and enzymatic activities contribute to cheese flavour and aroma. Looking ahead, technology and scientific research continue to advance the art of cheese making. Future developments may be expected to include novel starter cultures and adjunct bacteria for new flavour profiles and textural qualities, along with an increasing interest in the optimization of the use of milk composition by genetic and dietary manipulations of dairy animals to further improve the quality and consistency in cheeses. New analytical techniques continue to provide novel insights into the biochemical and microbial interactions that control the process of cheese ripening; these may lead to more targeted control of the manufacturing process and a higher consistency of product.

References

Albarracín, W., Sánchez, I. C., Grau, R., & Barat, J. M. (2011). Salt in food processing; usage and reduction: A review. *International Journal of Food Science & Technology, 46*(7), 1329–1336. https://doi.org/10.1111/j.1365-2621.2010.02492.x

Andiç, S., Tunçtürk, Y., & Boran, G. (2015). Changes in volatile compounds of cheese. In *Processing and impact on active components in food* (pp. 231–239). Elsevier. https://doi.org/10.1016/B978-0-12-404699-3.00028-7

Badea, M., Micheli, L., Messia, M. C., Candigliota, T., Marconi, E., Mottram, T., Velasco-Garcia, M., Moscone, D. M., & Palleschi, G. (2004). Aflatoxin M1 determination in raw milk using a flow-injection immunoassay system. *Analytica Chimica Acta, 520*, 141–148.

Bintsis, T., & Papademas, P. (2017). An overview of the cheesemaking process. In P. Papademas & T. Bintsis (Eds.), *Global cheesemaking technology* (1st ed., pp. 120–156). Wiley. https://doi.org/10.1002/9781119046165.ch0f

Buffa, M. N., Trujillo, A. J., Pavia, M., & Guamis, B. (2001). Changes in textural, microstructural, and colour characteristics during ripening of cheeses made from raw, pasteurized or high-pressure-treated goats' milk. *International Dairy Journal, 11*(11–12), 927–934. https://doi.org/10.1016/S0958-6946(01)00141-8

Burbank, H., & Qian, M. C. (2008). Development of volatile sulfur compounds in heat-shocked and pasteurized milk cheese. *International Dairy Journal, 18*(8), 811–818. https://doi.org/10.1016/j.idairyj.2007.12.006

Clark, S., Costello, M., Drake, M., & Bodyfelt, F. (Eds.). (2009). *The sensory evaluation of dairy products*. Springer. https://doi.org/10.1007/978-0-387-77408-4

Collins, Y. F., McSweeney, P. L. H., & Wilkinson, M. G. (2003). Lipolysis and free fatty acid catabolism in cheese: A review of current knowledge. *International Dairy Journal, 13*(11), 841–866. https://doi.org/10.1016/S0958-6946(03)00109-2

De Kruif, C. G. K., & Huppertz, T. (2012). Casein micelles: Size distribution in milks from individual cows. *Journal of Agricultural and Food Chemistry, 60*(18), 4649–4655. https://doi.org/10.1021/jf301397w

Duval, P., Chatelard-Chauvin, C., Gayard, C., Rifa, E., Bouchard, P., Hulin, S., Delile, A., Pollet, B., Montel, M. C., & Picque, D. (2018). Changes in biochemical and sensory parameters in industrial blue-veined cheeses in different packaging. *International Dairy Journal, 77*, 89–99. https://doi.org/10.1016/j.idairyj.2016.10.004

Ekstrand, B. (1980). Casein micelle size and composition related to the enzymatic coagulation process. *Biochimica et Biophysica Acta (BBA) - General Subjects, 630*(3), 361–366. https://doi.org/10.1016/0304-4165(80)90284-6

El-Bakry, M., Duggan, E., O'Riordan, E. D., & O'Sullivan, M. (2011). Effect of chelating salt type on casein hydration and fat emulsification during manufacture and post-manufacture functionality of imitation cheese. *Journal of Food Engineering, 102*(2), 145–153. https://doi.org/10.1016/j.jfoodeng.2010.08.012

Euston, S. R., Piska, I., Wium, H., & Qvist, K. B. (2002). Controlling the structure and rheological properties of model cheese systems. *Australian Journal of Dairy Technology, 57*(2), 145.

Everett, D. W., & Auty, M. A. E. (2008). Cheese structure and current methods of analysis. *International Dairy Journal, 18*(7), 759–773. https://doi.org/10.1016/j.idairyj.2008.03.012

Everett, D. W., & Auty, M. A. E. (2017). Cheese microstructure 1. In *Cheese* (pp. 547–569). Elsevier. https://doi.org/10.1016/B978-0-12-417012-4.00021-1

Faulkner, A., & Peaker, M. (1982). Reviews of the progress of dairy science: Secretion of citrate into milk. *Journal of Dairy Research, 49*(1), 159–169. https://doi.org/10.1017/S002202990002224X

Forde, A., & Fitzgerald, G. F. (2000). Biotechnological approaches to the understanding and improvement of mature cheese flavour. *Current Opinion in Biotechnology, 11*(5), 484–489. https://doi.org/10.1016/S0958-1669(00)00130-0

Fox, P. F., & Guinee, T. P. (2013). Cheese science and technology. In Y. W. Park & G. F. W. Haenlein (Eds.), *Milk and dairy products in human nutrition* (1st ed., pp. 357–389). Wiley. https://doi.org/10.1002/9781118534168.ch17

Fox, P. F., & McSweeney, P. L. H. (2017). Cheese: An overview. In *Cheese* (pp. 5–21). Elsevier. https://doi.org/10.1016/B978-0-12-417012-4.00001-6

Fox, P. F., Lucey, J. A., & Cogan, T. M. (1990). Glycolysis and related reactions during cheese manufacture and ripening. *Critical Reviews in Food Science and Nutrition, 29*(4), 237–253. https://doi.org/10.1080/10408399009527526

Fox, P. F., O'Connor, T. P., Mcsweeney, P. L. H., Guinee, T. P., & O'Brien, N. M. (1996). Cheese: Physical, biochemical, and nutritional aspects. In *Advances in food and nutrition research* (Vol. 39, pp. 163–328). Elsevier. https://doi.org/10.1016/S1043-4526(08)60075-3

Fox, P. F., McSweeney, P. L. H., Cogan, T. M., & Guinee, T. P. (2000). *Fundamentals of cheese science*. Springer.

Gan, H. H., Yan, B., Linforth, R. S. T., & Fisk, I. D. (2016). Development and validation of an APCI-MS/GC–MS approach for the classification and prediction of Cheddar cheese maturity. *Food Chemistry, 190*, 442–447. https://doi.org/10.1016/j.foodchem.2015.05.096

Gastaldi, E., Pellegrini, O., Lagaude, A., & De La Fuente, B. T. (1994). Functions of added calcium in acid milk coagulation. *Journal of food science, 59*(2), 310–312.

Gaucheron, F. (2005). The minerals of milk. *Reproduction Nutrition Development, 45*(4), 473–483. https://doi.org/10.1051/rnd:2005030

Glantz, M., Devold, T. G., Vegarud, G. E., Lindmark Månsson, H., Stålhammar, H., & Paulsson, M. (2010). Importance of casein micelle size and milk composition for milk gelation. *Journal of Dairy Science, 93*(4), 1444–1451. https://doi.org/10.3168/jds.2009-2856

Guinee, T. P. (2021). Effect of high-temperature treatment of milk and whey protein denaturation on the properties of rennet–curd cheese: A review. *International Dairy Journal, 121*, 105095. https://doi.org/10.1016/j.idairyj.2021.105095

Guinee, T. P., & O'Brien, B. (2010). The quality of milk for cheese manufacture. In B. A. Law & A. Y. Tamime (Eds.), *Technology of cheesemaking* (1st ed., pp. 1–67). Wiley. https://doi.org/10.1002/9781444323740.ch1

Guinee, T. P., Pudja, P. D., & Mulholland, E. O. (1994). Effect of milk protein standardization, by ultrafiltration, on the manufacture, composition and maturation of Cheddar cheese. *Journal of Dairy Research, 61*(1), 117–131. https://doi.org/10.1017/S0022029900028119

Guinee, T. P., Pudja, P. D., Reville, W. J., Harrington, D., Mulholland, E. O., Cotter, M., & Cogan, T. M. (1995). Composition, microstructure and maturation of semi-hard cheeses from high protein ultrafiltered milk retentates with different levels of denatured whey protein. *International Dairy Journal, 5*(6), 543–568. https://doi.org/10.1016/0958-6946(94)00021-G

Gursoy, O., Kesenkas, H., & Yilmaz, Y. (2013). 12. White cheese. In V. R. Preedy, R. R. Watson, & V. B. Patel (Eds.), *Human health handbooks* (Vol. 6, pp. 181–192). Wageningen Academic Publishers. https://doi.org/10.3920/978-90-8686-766-0_12

Haasnoot, W., Stouten, P., & Venema, D. P. (1989). High-performance liquid chromatography determination of the extent of proteolysis in Gouda cheese. *Journal of Chromatography A, 483*, 319–329. https://doi.org/10.1016/S0021-9673(01)93131-0

Holsinger, V. H., Smith, P. W., & Tunick, M. H. (1995). Overview: Cheese chemistry and rheology. In E. L. Malin & M. H. Tunick (Eds.), *Chemistry of structure-function relationships in cheese* (Vol. 367, pp. 1–6). Springer. https://doi.org/10.1007/978-1-4615-1913-3_1

Hooydonk, A. C. M. van., Hagedoorn, H. G. & Boerrigter, I. J. (1986). The effect of various cations on the renneting of milk. *Netherlands Milk and Dairy Journal, 40*(4) 369–390.

Horne, D. S., & Lucey, J. A. (2017). Rennet-induced coagulation of milk. In *Cheese* (pp. 115–143). Elsevier. https://doi.org/10.1016/B978-0-12-417012-4.00005-3

Huppertz, T. (2019). Curd formation, syneresis and drainage: Technological aspects. In *Reference module in food science* (p. B9780081005965213796). Elsevier. https://doi.org/10.1016/B978-0-08-100596-5.21379-6

Iwasawa, A., Suzuki-Iwashima, A., Iida, F., & Shiota, M. (2014). Effects of flavor and texture on the desirability of Cheddar cheese during ripening. *Food Science and Technology Research, 20*(1), 23–29. https://doi.org/10.3136/fstr.20.23

Johnston, K. A., Luckman, M. S., Lilley, H. G., & Smale, B. M. (1998). Effect of various cutting and stirring conditions on curd particle size and losses of fat to the whey during Cheddar cheese manufacture in Ost Vats. *International Dairy Journal, 8*(4), 281–288. https://doi.org/10.1016/S0958-6946(98)00050-8

Kelly, A. L. (2007). Preparation of cheesemilk. In *Cheese problems solved* (pp. 11–29). Elsevier. https://doi.org/10.1533/9781845693534.11

Kelly, A. L., Huppertz, T., & Sheehan, J. J. (2008). Pre-treatment of cheese milk: Principles and developments. *Dairy Science and Technology, 88*(4–5), 549–572. https://doi.org/10.1051/dst:2008017

Kindstedt, P. S. (1995). Factors affecting the functional characteristics of unmelted and melted mozzarella cheese. In E. L. Malin & M. H. Tunick (Eds.), *Chemistry of structure-function relationships in cheese* (Vol. 367, pp. 27–41). Springer. https://doi.org/10.1007/978-1-4615-1913-3_4

Law, B. A., & Tamime, A. Y. (Eds.). (2010). *Technology of cheesemaking* (1st ed.). Wiley. https://doi.org/10.1002/9781444323740

Liu, X., Wu, Y., Guan, R., Jia, G., Ma, Y., & Zhang, Y. (2021). Advances in research on calf rennet substitutes and their effects on cheese quality. *Food Research International, 149*, 110704. https://doi.org/10.1016/j.foodres.2021.110704

Lopez, C., Camier, B., & Gassi, J.-Y. (2007). Development of the milk fat microstructure during the manufacture and ripening of Emmental cheese observed by confocal laser scanning microscopy. *International Dairy Journal, 17*(3), 235–247. https://doi.org/10.1016/j.idairyj.2005.12.015

Lou, Y., & Ng-Kwai-Hang, K. F. (1992). Effects of protein and fat levels in milk on cheese and whey compositions. *Food Research International, 25*(6), 445–451. https://doi.org/10.1016/0963-9969(92)90169-6

Lucey, J. A., & Fox, P. F. (1993). Importance of calcium and phosphate in cheese manufacture: A review. *Journal of Dairy Science, 76*(6), 1714–1724. https://doi.org/10.3168/jds.S0022-0302(93)77504-9

Lucey, J. A., Johnson, M. E., & Horne, D. S. (2003). Invited review: Perspectives on the basis of the rheology and texture properties of cheese. *Journal of Dairy Science, 86*(9), 2725–2743. https://doi.org/10.3168/jds.S0022-0302(03)73869-7

McSweeney, P. L. H. (2011). Cheese | biochemistry of cheese ripening. In *Encyclopedia of dairy sciences* (pp. 667–674). Elsevier. https://doi.org/10.1016/B978-0-12-374407-4.00080-7

McSweeney, P. L. H., Ottogalli, G., & Fox, P. F. (2004). Diversity of cheese varieties: An overview. In *Cheese: Chemistry, physics and microbiology* (Vol. 2, pp. 1–23). Elsevier. https://doi.org/10.1016/S1874-558X(04)80037-X

Melchiorsen, C., Væver, K., & Jo, R. (2002). The level of pyruvate-formate lyase controls the shift from homolactic to mixed-acid product formation in Lactococcus lactis. *Applied Microbiology and Biotechnology, 58*(3), 338–344. https://doi.org/10.1007/s00253-001-0892-5

Michalski, M.-C., Camier, B., Gassi, J.-Y., Briard-Bion, V., Leconte, N., Famelart, M.-H., & Lopez, C. (2007). Functionality of smaller vs control native milk fat globules in Emmental cheeses manufactured with adapted technologies. *Food Research International, 40*(1), 191–202. https://doi.org/10.1016/j.foodres.2006.09.011

Mohsin, A. Z., Norsah, E., Marzlan, A. A., Abd Rahim, M. H., & Meor Hussin, A. S. (2024). Exploring the applications of plant-based coagulants in cheese production: A review. *International Dairy Journal, 148*, 105792. https://doi.org/10.1016/j.idairyj.2023.105792

Moradi, M., Omer, A. K., Razavi, R., Valipour, S., & Guimarães, J. T. (2021). The relationship between milk somatic cell count and cheese production, quality and safety: A review. *International Dairy Journal, 113*, 104884. https://doi.org/10.1016/j.idairyj.2020.104884

Müller-Buschbaum, P., Gebhardt, R., Roth, S. V., Metwalli, E., & Doster, W. (2007). Effect of calcium concentration on the structure of casein micelles in thin films. *Biophysical journal, 93*(3), 960–968.

Murlidhar, M., Anusha, R., & Bindhu, O. S. (2017). *Plant-based coagulants in cheese making: Review*. Apple Academic Press.

O'Mahony, J. A., Fox, P. F., & Kelly, A. L. (2013). Indigenous enzymes of Milk. In P. L. H. McSweeney & P. F. Fox (Eds.), *Advanced dairy chemistry* (pp. 337–385). Springer. https://doi.org/10.1007/978-1-4614-4714-6_12

Ong, L., Dagastine, R. R., Auty, M. A. E., Kentish, S. E., & Gras, S. L. (2011). Coagulation temperature affects the microstructure and composition of full fat Cheddar cheese. *Dairy Science & Technology, 91*(6), 739–758. https://doi.org/10.1007/s13594-011-0033-6

Ong, L., Dagastine, R. R., Kentish, S. E., & Gras, S. L. (2012). The effect of pH at renneting on the microstructure, composition and texture of Cheddar cheese. *Food Research International, 48*(1), 119–130. https://doi.org/10.1016/j.foodres.2012.02.020

Ong, L., Dagastine, R. R., Kentish, S. E., & Gras, S. L. (2013). The effect of calcium chloride addition on the microstructure and composition of Cheddar cheese. *International Dairy Journal, 33*(2), 135–141. https://doi.org/10.1016/j.idairyj.2013.03.002

Peters, I. I. (1964). *Homogenized milk in cheesemaking.* https://www.cabidigitallibrary.org/doi/pdf/10.5555/20063034160

Schmidt, D. G. (1982). *Association of caseins and casein micelle structure.* https://www.cabidigitallibrary.org/doi/full/10.5555/19820480355

Scott, R., Robinson, R. K., & Wilbey, R. A. (1998). *Cheesemaking practice*. Springer. https://doi.org/10.1007/978-1-4615-5819-4

Shah, M. A., Mir, S. A., & Paray, M. A. (2014). Plant proteases as milk-clotting enzymes in cheesemaking: A review. *Dairy Science & Technology, 94*(1), 5–16. https://doi.org/10.1007/s13594-013-0144-3

Singh, N. A., Kumar, N., Raghu, H. V., Sharma, P. K., Singh, V. K., Khan, A., & Raghav, N. (2013). Spore inhibition-based enzyme substrate assay for monitoring of aflatoxin M1 in milk. *Toxicological & Environmental Chemistry, 95*, 765–777. https://doi.org/10.1080/02772248.2013.807540

Smit, G., Smit, B. A., & Engels, W. J. M. (2005). Flavour formation by lactic acid bacteria and biochemical flavour profiling of cheese products. *FEMS Microbiology Reviews, 29*(3), 591–610. https://doi.org/10.1016/j.fmrre.2005.04.002

Stratton, J. E., Hutkins, R. W., & Taylor, S. L. (1991). Biogenic amines in cheese and other fermented foods: A review. *Journal of Food Protection, 54*(6), 460–470. https://doi.org/10.4315/0362-028X-54.6.460

Wedholm, A., Larsen, L. B., Lindmark-Månsson, H., Karlsson, A. H., & Andrén, A. (2006). Effect of protein composition on the cheese-making properties of milk from individual dairy cows. *Journal of Dairy Science, 89*(9), 3296–3305. https://doi.org/10.3168/jds.S0022-0302(06)72366-9

Chapter 3
Quality of Milk for Cheese Making

Reshab Majumder and Syed Mansha Rafiq

3.1 Introduction

In the science and the art of cheesemaking, the quality of the milk plays a fundamental role in determining the finished product's sensory and nutritional qualities as well as its production efficiency (Campbell & Marshall, 2016). The microbiological profile, physical attributes, and nutritional value of milk are all important factors in influencing the quality of cheese. The chemical, biological, and physical characteristics of milk interact intricately to influence several stages of cheese manufacturing, from coagulation to aging, and this interaction is at the core of the process. In order to consistently produce high-quality cheese, it is crucial to comprehend the particulars of milk quality and how they affect the process of making cheese (Bansal & Veena, 2024).

The interaction of milk's chemical, biological, and physical properties dictate the course of several cheese-making stages (Enab et al., 2012). During coagulation, the balance of proteins and fats in the milk determines how well it forms curds when rennet and acid are added. The subsequent cutting, cooking, and draining of curds are influenced by the milk's initial composition and texture. The aging process, where microbial and enzymatic activities continue to transform the cheese, is also affected by the milk's quality (Fusco et al., 2020). For instance, the development of flavor and texture during aging depends on the activity of specific microorganisms and the milk's fat and protein content. Understanding these interactions allows cheese makers to tailor their techniques and conditions to achieve consistent and high-quality results. To consistently produce high-quality cheese, it is imperative to

R. Majumder (✉)
Jharkhand Milk Federation, Ranchi, Jharkhand, India

S. Mansha Rafiq
Department of Food Science and Technology, National Institute of Food Technology Entrepreneurship and Management, Sonipat, Haryana, India

S. Mansha Rafiq (ed.), *Technological Advances and Trends in Cheese Making*,
https://doi.org/10.1007/978-3-031-94476-5_3

have a thorough understanding of the particulars of milk quality and its impact on the cheesemaking process (Fox et al., 2017). This involves not only assessing the milk's microbiological, physical, and nutritional attributes but also recognizing how these factors interact during the various stages of cheese production. Advances in dairy science and technology provide deeper insights into these interactions, enabling cheese makers to refine their practices and improve product consistency. By applying scientific knowledge and technical expertise to manage milk quality, cheesemakers can optimize production processes, enhance flavor and texture profiles, and ensure the production of superior cheese (Khattab et al., 2019).

The quality of milk is a fundamental determinant of cheese quality, influencing every aspect of the cheesemaking process from coagulation to aging. The complex interplay of milk's microbiological, physical, and nutritional characteristics underscores the importance of understanding and controlling these factors to produce high-quality cheese (Murphy et al., 2016). As science and technology continue to advance, the ability to analyse and manage milk quality will remain central to achieving excellence in cheesemaking.

3.2 Chemical Composition of Milk

Milk is a complex mixture of minerals, water, fat, proteins, and lactose. The main constituent of milk is water, which accounts for around 87% of its volume, with fat making up the remaining 4%. About 3.3% of the total is made up of proteins, and the main sugar, lactose, makes up about 4.8%. These ingredients' concentration and balance are vital since they influence how milk behaves when producing cheese (Kailasapathy, 2015). For example, the texture and flavor profile of cheese are greatly influenced by its fat level. Cheese made from high-fat milk has a creamier, richer texture, which improves the mouthfeel and flavor. Lower fat milk, on the other hand, makes cheeses that are stiffer and have a less creamy texture, which affects the end product's sensory qualities as well as its yield (Truong et al., 2016).

Casein and whey proteins, which are found in milk, are essential for the development of curds to produce cheese. Approximately 80% of milk proteins are casein, which is essential to the coagulation process. Casein proteins in milk aggregate to form curds when exposed to rennet or acid; the curds are then separated from the whey. The amount and kind of casein present have a direct impact on the procedure' effectiveness as well as the curds' quality (Stocco et al., 2018). Cheese consistency and texture can vary depending on the type and amount of casein present. For example, soft cheeses like Brie require lower amounts of casein to be creamy and spreadable, whereas hard cheeses like Parmesan require higher levels to be crumbly and rigid (da Silva et al., 2020). The sugar found in milk, called lactose, is another important factor affecting the quality of cheese. In the process of producing cheese, bacteria break down lactose into lactic acid. As a result of this acidification, the milk's pH is lowered, which encourages curd formation and enhances flavor development. Inadequate lactose can result in poor curd growth and inadequate fermentation, while much lactose can cause excessively acidic curds that may severely affect texture and flavor. Furthermore, the remaining lactose in cheese influences the final flavor profile and should be taken into account by those who are lactose intolerant (Smit et al., 2005).

Making cheese also heavily depends on the minerals in milk, such as calcium and phosphorus. Since calcium helps to stabilize the casein network and makes the curds firmer, it is crucial for the coagulation process. Low calcium levels can cause weak or poorly formed curds, which will impact the cheese's yield and texture (Santos et al., 2013). Phosphorus interacts with casein and influences the final qualities of the cheese; however, it has a smaller effect than calcium. Achieving the right cheese qualities requires balancing these minerals, and varying amounts might produce uneven results (Upreti et al., 2006).

3.3 Microbiological Quality

The safety, flavor, and texture of cheese are all significantly influenced by the microbiological quality of the milk. Because microbial load can have a substantial impact on these aspects, this component of milk quality is essential to the production process as well as the final attributes of the cheese. It is imperative to guarantee the absence of harmful bacteria in milk to prevent spoiling, health hazards, and adverse impacts on the quality of cheese. *Salmonella, Listeria monocytogenes*, and *Escherichia coli* are the three pathogens in milk that should worry people the most (Sonnier et al., 2018). All of these infections have different hazards associated with them and can affect the process of manufacturing cheese in different ways, which emphasizes the need for strict hygiene procedures during the milking and handling phases. *Salmonella, Listeria monocytogenes*, and *Escherichia coli* are well-known pathogens that can be present in milk and lead to severe health risks if not properly controlled (Sonnier et al., 2018). *Salmonella* is a type of bacteria that can cause salmonellosis, a foodborne illness characterized by symptoms such as diarrhea, fever, and abdominal cramps (Gourama, 2020). In dairy production, *Salmonella* contamination can occur from infected animals or through improper handling and sanitation practices. The bacteria can persist in the environment and cross-contaminate milk if strict hygiene measures are not followed. Once in milk, *Salmonella* can survive and proliferate, potentially leading to the contamination of cheese and posing significant health risks to consumers (Mkangara, 2023). *Listeria monocytogenes* is another pathogenic bacterium that can be particularly problematic in dairy products. It is known for causing listeriosis, a serious infection that can lead to symptoms such as fever, muscle aches, and gastrointestinal issues. *Listeria* is unique in its ability to grow at refrigeration temperatures, which makes it a concern in both raw and pasteurized milk. Contamination often arises from poor sanitation practices, environmental contamination, or the presence of infected animals. In cheese making, *Listeria monocytogenes* can lead to spoilage and health risks, especially in soft cheeses where the bacteria can thrive if not properly controlled

(Falardeau et al., 2021). *Escherichia coli*, particularly pathogenic strains like *E. coli* O157, is known for causing severe foodborne illnesses, including hemorrhagic colitis and kidney failure (Mesele & Abunna, 2019). This bacterium can be introduced into milk through fecal contamination or from infected animals. The presence of pathogenic *E. coli* in milk can lead to significant health risks, and controlling this pathogen is critical to ensuring the safety of dairy products.

Our knowledge of pathogen control and milk quality is constantly being improved by developments in microbiological research and dairy technology. New technologies help manage milk microbiology more effectively. Examples include fast pathogen detection techniques, better milking equipment, and sophisticated cleaning solutions. Reducing the danger of pathogen contamination also involves advances in feed composition and genetic selection of dairy breeds for disease resistance. Improved methods for managing infections and enhancing cheese quality may result from a better understanding of the intricate relationships between these microbes. The development of focused therapies, like antibacterial treatments and probiotic cultures, presents viable ways to control microbiological quality and improve cheese's safety and sensory qualities.

Lactic acid bacteria are a group of microorganisms that are naturally present in milk and are essential for the cheese-making process (Coelho et al., 2022). A type of bacterium called lactic acid bacteria is found naturally in milk and is necessary for the production of cheese. Their ability to convert lactose, the main sugar in milk, into lactic acid is widely recognized. The pH of the milk is lowered during this fermentation process, which is important for a number of reasons. First, the pH drops, which makes it easier for the casein proteins to coagulate and form curds. Because casein proteins are sensitive to pH variations, when the pH falls, they clump together to form a network that resembles gel, which traps fat and other ingredients to produce curd. Lactic acid bacteria contribute to the development of cheese's flavor and texture in addition to aiding curd form. The sensory qualities of the cheese are influenced by the metabolic by-products that LABs make during fermentation, including acids, peptides, and flavor compounds (Ianni et al., 2020). For example, many cheeses have an acidic taste due to the formation of lactic acid, while other by-products like acetaldehyde and diacetyl can provide buttery or fruity implications. These flavors can be greatly influenced by the particular strains of LAB employed in the production of cheese.

Additionally, LABs contribute to the development of the cheese's texture. The firmness and consistency of the cheese are impacted by the changes in milk proteins brought about by the acids produced during fermentation. Certain LAB strains, for instance, generate enzymes that aid in the digestion of proteins and lipids, giving cheeses like Brie and Camembert a smoother texture. On the other hand, by creating a dense protein network, LABs aid in the development of a harder texture in hard cheeses like Parmesan and Cheddar (Ong et al., 2022).

The source of the milk and the methods used during processing can have an impact on the diversity and concentration of LABs in milk. Variations in farm situations, animal diets, and dairy environments can result in differences in the microbial populations found in milk. For example, due to variations in local microbiomes and environmental conditions, milk from different geographical regions could contain unique populations of LAB (Guo et al., 2021). The microbial profile of milk is significantly shaped by handling habits as well. The kinds and amounts of LAB present can be affected by procedures such processing techniques, storage conditions, and hygiene during milking. For instance, inadequate sanitation procedures may encourage the growth of undesirable microbes, which may eventually outcompete advantageous LABs and lower the cheese's overall quality. Cheese manufacturers can achieve the desired cheese characteristics and standardize the fermentation process by introducing certain LAB strains in specified quantities into the milk. The precise control over the fermentation process made possible by this regulated inoculation ensures that the cheese acquires the desired texture, and flavor profiles (Chandan & Kapoor, 2011).

3.3.1 Microbiological Quality Assessment

A number of tests are used to evaluate the microbiological quality of milk, and the results offer important insights into the total microbial load and possible quality problems. The somatic cell count (SCC) and the total bacterial count (TBC) are two important tests used to assess the quality of milk (Kuchtík et al., 2021). The number of live bacteria in milk is determined by the total bacterial count (TBC). A high TBC may be a sign of low-quality milk as well as problems with handling or hygienic standards. Inadequate cleaning of the milking equipment, improperly maintained temperatures, or environmental contamination can all lead to elevated bacterial counts.

Another important measure of milk quality is the somatic cell count. When dairy cows have mastitis, an infection of the mammary gland, their milk contains more somatic cells, mainly white blood cells. High SCC levels can have a negative impact on milk quality and cheese supply since they are frequently an indication of mastitis (Pegolo et al., 2022). Mastitis-affected cows' milk may have different texture and flavor, a decreased production of cheese, and a higher microbial burden. As a result, keeping an eye on SCC levels is essential to preserving milk quality and guaranteeing the production of cheese of the finest quality.

3.4 Physical and Sensory Attributes

The physical characteristics of milk, such as its color, odor, and viscosity, are essential to the process of making cheese and have an impact on the final result. It's crucial to examine every characteristic and how it affects cheese quality in order to completely understand its significance. The primary characteristic of milk is its color, which informs a lot about its quality and the diet of the cow that produced it. Fresh milk typically has a clean, white to slightly yellow color, depending on what the animal was fed. For example, the presence of carotenoids such as beta-carotene in rich, green pastures might cause the milk produced by the cows to have a yellower color. The natural coloring of cheese is typically considered acceptable and has no negative impact on its quality. But if milk exhibits any odd colors, such pink or grayish tones, it may be contaminated or spoiled, which could affect the quality of the cheese made from it. To avoid having a detrimental effect on the quality of the cheese, such discolored milk should be closely inspected (Ritota & Manzi, 2020).

Another important component in evaluating the quality of milk is odor. To indicate its freshness and purity, fresh milk should have a mild, pleasant smell. Any unpleasant or acidic smells point to bacterial infection or spoiling, which can seep into the cheese and impart unwanted flavors in the finished product. Furthermore, the nutrition that the animal receives can affect the scent of the milk (Kumar & Patyal, 2024).

When producing cheese, viscosity, or the thickness of the milk, is largely influenced by the amount of fat and protein in the milk. Viscosity has an impact on the milk's potential to coagulate and mix with other substances like rennet and cultures. The proper viscosity of milk should result in curds that are evenly stiff and soft, which will add to the cheese's intended texture. Too-thick milk can produce difficult-to-manage, excessively stiff curds that can turn into crumbly, dry cheese. On the other hand, overly thin milk might result in soft curds that aren't solid in structure, which can give cheese a weak texture (Caldwell, 2016).

Cheese producers frequently have to modify their procedures to account for changes in milk viscosity in order to guarantee that the finished product lives up to quality standards. For example, to get the right curd consistency, one might need to adjust the amount of rennet or change the way you chop and heat the milk if it's too thick. However, if the milk is too thin, it may be necessary to add more milk solids or make other changes to enhance the texture of the cheese overall and the development of curds (Caldwell, 2016).

3.5 Handling and Storage

Proper handling and storage of milk are essential processes that directly impact its quality, safety, and shelf life. These practices encompass a range of scientific and technical considerations aimed at preserving the milk's freshness and preventing spoilage. Adhering to rigorous protocols for collecting, cooling, storing, and protecting milk ensures that it remains free from contaminants and retains its nutritional value. Collection of milk is the initial step where quality control begins. It is crucial that milk is collected using clean, sanitized containers to prevent the introduction of microorganisms. The containers should be made of materials that are non-reactive, such as stainless steel or certain food-grade plastics, to avoid any chemical interactions with the milk that could compromise its quality. Sanitation protocols should be strictly followed to eliminate any residues from previous uses and ensure that the containers are free from bacteria, molds, or other contaminants that could affect the milk. Additionally, the milking equipment, including the udder liners and milking machines, must be meticulously cleaned and sanitized to prevent cross-contamination (Rashid, 2020). Rapid cooling of milk post-collection is critical in controlling bacterial growth. The temperature at which milk is cooled should be managed to inhibit the proliferation of microorganisms, particularly those that can cause spoilage or health risks. Ideally, milk should be cooled to a temperature below 4 °C (39 °F) within a short period after milking. This rapid cooling is achieved through the use of plate coolers, rapid chillers, or bulk tanks equipped with refrigeration systems. By maintaining milk at this temperature, the growth of psychotropic bacteria, which can thrive at lower temperatures, is minimized. Failure to cool milk promptly can result in the exponential growth of bacteria, leading to souring and spoilage, which deteriorates the milk's sensory and nutritional qualities (Lu & Wang, 2017).

Storage temperature is another critical factor in maintaining milk quality. Raw milk must be kept at temperatures consistently below 4 °C (39 °F) to preserve its freshness and prevent microbial contamination (Perin et al., 2019). Storage at higher temperatures accelerates bacterial growth and enzymatic activity, which can lead to off-flavors, changes in texture, and nutritional degradation. For longer-term storage, ultra-low temperatures are employed in freezing, but this is less common for raw milk intended for immediate consumption. Ensuring that storage facilities, such as milk tanks and refrigerated transport units, are equipped with accurate temperature control systems and regular monitoring is vital to maintaining the required storage conditions. Protection from light and air is equally important in preserving milk quality. Exposure to light, particularly ultraviolet (UV) light, can cause photochemical reactions in milk that lead to the breakdown of vitamins, particularly riboflavin (vitamin B2), and result in the formation of off-flavors (Potts, 2016). This phenomenon is known as light-induced oxidation. To mitigate these effects, milk storage containers should be opaque or coated with materials that block light. Additionally, air exposure can lead to oxidative reactions and the deterioration of milk's lipids, which can cause rancidity and off-flavors. Thus, milk containers should be sealed effectively to minimize contact with air and prevent the ingress of contaminants (Gaceu, 2020).

Container materials play a significant role in maintaining milk quality. The materials used for milk storage must be non-reactive and able to withstand the cold temperatures required for effective storage. Stainless steel is commonly used due to its non-reactive properties, durability, and ease of cleaning. Food-grade plastics, such as high-density polyethylene (HDPE) or polypropylene, are also used but must be carefully selected to avoid any interactions with milk components that could impart undesirable flavors or affect its quality. Containers must be smooth and easy to clean to avoid any bacterial growth or residue build up that could compromise the milk (Cadwallader, 2020). Use of preservatives and additives is a practice employed to extend the shelf life of milk and maintain its quality. Preservatives must be used judiciously and in accordance with regulatory guidelines to ensure they are safe for consumption. Common additives include stabilizers and antimicrobials that help inhibit bacterial growth and prevent spoilage. The use of these additives must be carefully regulated to avoid any potential adverse effects on the milk's nutritional profile or consumer safety. Regulatory agencies set strict limits on the types and quantities of additives allowed, ensuring that their use does not compromise the overall quality and safety of the milk. To maintain the quality and safety of milk, it is essential to handle and store it properly. Milk needs to be carefully handled at every stage, starting at the moment of collection and continuing through rapid cooling, suitable storage, and shielding from light and air exposure. The freshness of the milk is further preserved and its shelf life is increased by the storage containers chosen and the controlled use of additives and preservatives. Delivering premium milk that satisfies customer expectations and safety regulations requires adherence to certain scientific and technical principles.

3.6 Influence of Diet and Environment

The diet and environment of dairy animals profoundly influence milk quality, impacting both its nutritional profile and sensory characteristics. Understanding these effects requires a comprehensive examination of how feed composition and environmental conditions contribute to the milk's overall quality and its suitability for dairy products, including cheese. Diet is a primary factor affecting milk quality, as the nutritional content of an animal's feed directly influences the composition of the milk it produces. Dairy animals, particularly cows, can have their milk's fatty acid profile and nutrient content altered based on their diet. For instance, cows that are fed a forage-based diet, which includes high-quality forages such as alfalfa or clover, typically produce milk with higher concentrations of beneficial fatty acids. These forages are rich in essential fatty acids, such as omega-3 and omega-6, which can significantly enhance the nutritional profile of the milk. Additionally, the presence of carotenoids in these forages can result in milk with a more desirable flavor and color, which can be particularly beneficial when producing cheese, as these attributes can improve the sensory qualities of the cheese (Kilcawley et al., 2018).

Conversely, a diet high in grains, such as corn or soybean meal, leads to different changes in milk composition. Grain-rich diets are generally associated with a higher concentration of saturated fatty acids and a different balance of proteins compared to forage-based diets. This shift in fatty acid composition can affect the flavor profile of the milk, sometimes leading to a less complex taste. The protein content and type in milk can also be influenced by the diet, as grains can alter the casein-to-whey protein ratio, which affects the milk's coagulation properties and, consequently, the cheese-making process. Therefore, the choice of feed can have significant implications for the texture, yield, and flavor of cheese (Kilcawley et al., 2018). Environmental factors further impact milk quality by influencing the health and well-being of dairy animals. The housing conditions of dairy cows, including ventilation, cleanliness, and space, play a crucial role in their overall health and productivity. Cows housed in well-ventilated, clean environments are less likely to suffer from respiratory infections or other health issues that could affect milk quality. Proper ventilation helps control humidity and temperature, reducing the risk of heat stress and related health problems that can negatively impact milk production and composition. Heat stress, in particular, can lead to reduced milk yield and alterations in milk composition, including changes in fat and protein levels (Aggarwal et al., 2013). On the other hand, stressful conditions, such as extreme temperatures, poor ventilation, and overcrowding, can have detrimental effects on milk quality. Cows exposed to high temperatures may experience heat stress, which can lead to decreased milk production and alterations in milk composition, such as reduced fat and protein content. Extreme cold can also impact milk production by affecting the cows' health and feeding behaviour. Overcrowding can lead to increased competition for feed and reduced access to water, both of which can adversely affect milk yield and quality. Stressful conditions can also lead to behavioural changes in cows that may impact their feeding patterns and overall well-being, further influencing the quality of milk produced.

Comfort and care are integral to maintaining high milk quality. Ensuring that dairy animals are comfortable involves providing adequate shelter, a balanced diet, and proper healthcare. Regular veterinary care is essential for preventing and managing diseases that can affect milk quality. Additionally, the management practices related to milking routines, such as maintaining clean equipment and adhering to hygiene protocols, contribute to the overall quality of the milk. Stress reduction strategies, such as providing ample space, minimizing handling stress, and ensuring a comfortable living environment, are vital for sustaining the health and productivity of dairy animals (Vyas et al., 2020).

Dairy animals' environments and diets are closely related to the quality of their milk. The fatty acid profile, protein level, and overall flavor of milk are influenced by the feed's composition, whether it is grain-rich or forage-based. These factors ultimately impact the quality of dairy products like cheese. The health and welfare of dairy animals are contingent upon environmental factors, such as housing quality and management techniques, which in turn guarantee consistent milk output and quality. Dairy producers can improve the qualities of their products and maximize milk quality by carefully controlling environmental and dietary parameters. This will ultimately result in an improved finished product that satisfies high requirements for nutrition and flavor.

3.7 Innovations and Future Directions

Advances in dairy science and technology are transforming the landscape of milk production and cheese making, leading to improved quality and efficiency across the industry. The continuous evolution of scientific knowledge and technological innovations is significantly enhancing our understanding of milk quality and its effects on cheese production. These advancements encompass genetic selection, feed formulation, milking and processing technologies, and insights into microbial communities, all of which contribute to producing superior milk and cheese products. Additionally, sustainable practices are emerging as crucial components of modern dairy farming, aligning environmental stewardship with high-quality milk production.

Genetic selection of dairy breeds represents one of the most significant advancements in enhancing milk quality. Through selective breeding programs, scientists can now identify and propagate genetic traits that optimize milk composition. Genetic markers associated with higher milk yield, improved protein and fat content, and better overall health are used to guide breeding decisions. For instance, dairy breeds such as the Holstein, Jersey, and Guernsey have been selectively bred for specific milk qualities, including increased butterfat content and more favorable protein ratios. Advances in genomics and molecular biology have enabled the development of more precise breeding strategies, allowing for the enhancement of traits that directly impact cheese-making processes. By selecting for genetic traits that produce milk with optimal coagulation properties and desirable flavor profiles, cheese makers can achieve consistent and high-quality cheese production (Fox & Cogan, 2004).

Improved feed formulations are another area where technological advancements are making a significant impact. Modern feed formulations are designed to balance the nutritional needs of dairy animals while optimizing milk composition. Research into the effects of different feed components, such as forages, grains, and supplements, has led to the development of more effective diets that enhance the nutritional profile of milk. For example, incorporating high-quality forages and precision feeds can increase the concentration of beneficial fatty acids in milk, positively affecting cheese flavor and texture. Additionally, advances in feed technology, such as the use of additives like probiotics and prebiotics, can improve gut health in dairy animals, leading to better nutrient absorption and overall milk quality. The precise control of feed ingredients and their impact on milk composition allows for tailored diets that meet the specific needs of different dairy breeds and production systems (Roche et al., 2018).

Milking and processing technologies have also seen considerable advancements, contributing to higher milk quality and more efficient cheese production. Modern milking systems, such as automated milking robots, ensure that milk is collected under optimal conditions, reducing the risk of contamination and improving overall milk quality. These systems are equipped with sensors and technology that monitor milk yield, quality parameters, and animal health, providing real-time data that can be used to manage herd performance and ensure high-quality milk production. In processing, innovations such as microfiltration and ultrafiltration techniques are used to improve milk quality by removing undesirable components and concentrating beneficial ones. These technologies enhance the efficiency of cheese production by ensuring that the milk used is of consistent quality and composition, which is critical for achieving the desired characteristics in the final cheese product (Johnson, 2017).

Research into milk microbiomes is offering new insights into the role of microbial communities in cheese production. The milk microbiome consists of a diverse array of microorganisms, including bacteria, yeasts, and molds, that can influence milk quality and cheese characteristics. Advances in microbiology and metagenomics have allowed researchers to identify specific microbial strains that contribute to desirable cheese attributes, such as flavor, texture, and aroma. By understanding the interactions between these microorganisms and milk components, scientists can develop strategies to control and optimize cheese fermentation processes. For instance, selecting starter cultures that contain beneficial bacteria can enhance the fermentation process, leading to consistent and high-quality cheese production. Additionally, research into the microbial ecosystems of different dairy environments helps in developing methods to manage microbial populations, reducing the risk of contamination and ensuring the desired quality of cheese (Tilocca et al., 2020).

Sustainable practices in dairy farming are becoming increasingly important as the industry seeks to balance productivity with environmental responsibility. Reducing the carbon footprint of dairy operations involves implementing strategies such as improving feed efficiency, optimizing manure management, and adopting renewable energy sources. For example, feeding strategies that reduce methane emissions from enteric fermentation can contribute to lower greenhouse gas emissions. Additionally, practices such as precision agriculture and conservation tillage help in managing resources more efficiently and minimizing environmental impact. Improving animal welfare is another critical aspect of sustainability, as healthy and well-cared-for animals produce higher-quality milk. Ensuring that dairy animals have access to comfortable housing, proper nutrition, and veterinary care contributes to better milk quality and overall farm productivity.

As the dairy industry continues to evolve, ongoing research and technological advancements will remain crucial in ensuring the quality of milk and the excellence of cheese. Innovations in genetics, feed formulation, milking technologies, and microbiome research are driving improvements in milk quality and cheese production. Furthermore, the integration of sustainable practices addresses environmental and ethical considerations, aligning the goals of high-quality milk production with broader sustainability objectives. The future of dairy science will likely bring continued progress in these areas, leading to even more refined techniques and practices that enhance both the quality and sustainability of dairy products. By embracing these advancements, the dairy industry can continue to meet the growing demands for high-quality, nutritious cheese while contributing to a more sustainable and responsible agricultural system.

3.8 Conclusion

In summary, the quality of milk is a pivotal factor in determining the quality of cheese, influencing each stage of cheese production, from coagulation to aging. The intricate relationship between milk composition, microbiological quality, physical attributes, and handling practices underscores the importance of ensuring that the milk used in cheese making adheres to the highest standards. These factors collectively shape the characteristics of the final cheese product, impacting its flavor, texture, and overall quality. Handling practices are integral to maintaining milk quality throughout the production process. As technology and scientific knowledge continue to advance, the future of cheese making is poised to benefit from deeper insights into milk quality and production processes. Innovations in genetic selection, feed formulation, and milking technologies enhance milk composition and processing efficiency. Advances in microbiology provide better understanding and control of microbial cultures, leading to more consistent and desirable cheese qualities. Sustainable practices further support high-quality milk production while addressing environmental and ethical concerns. These ongoing advancements ensure that cheese makers can produce a diverse range of cheeses with exceptional quality and consistency, meeting the evolving demands of consumers and the industry.

References

Aggarwal, A., Upadhyay, R., Aggarwal, A., & Upadhyay, R. (2013). Heat stress and milk production. In *Heat stress and animal productivity* (pp. 53–77). Springer.

Bansal, V., & Veena, N. (2024). Understanding the role of pH in cheese manufacturing: General aspects of cheese quality and safety. *Journal of Food Science and Technology, 61*(1), 16–26.

Cadwallader, D. C. (2020). *Sensory and instrumental analysis of flavors due to scalping and migration from packaging in fluid milk.* Master's thesis. North Carolina State University.

Caldwell, G. (2016). *Mastering basic cheesemaking: The fun and fundamentals of making cheese at home*. New Society Publishers.

Campbell, J. R., & Marshall, R. T. (2016). *Dairy production and processing: The science of milk and milk products*. Waveland Press.

Chandan, R. C., & Kapoor, R. (2011). Principles of cheese technology. In *Dairy ingredients for food processing* (pp. 225–265). Wiley.

Coelho, M. C., Malcata, F. X., & Silva, C. C. (2022). Lactic acid bacteria in raw-milk cheeses: From starter cultures to probiotic functions. *Foods, 11*(15), 2276.

da Silva, V. B., da Bahia, U. F., Barreto Bezerra, L. S. M., de Oliveira Alves, B. P., & dos Santos, B. C. (2020). Rheology and texture applied to goat and sheep milk cheeses: Principles, applications and future trends. In *Avanços em Ciência e Tecnologia de Alimentos* (Vol. 2, pp. 409–436). Editora Científica Digital, Editora Científica Digital.

Enab, A. K., Hassan, F. A., & Abd, M. A. M. G. E. (2012). Effect of manufacture steps on cheese structure. *International Journal of Academic Research, 4*(6), 79–89.

Falardeau, J., Trmčić, A., & Wang, S. (2021). The occurrence, growth, and biocontrol of Listeria monocytogenes in fresh and surface-ripened soft and semisoft cheeses. *Comprehensive Reviews in Food Science and Food Safety, 20*(4), 4019–4048.

Fox, P. F., & Cogan, T. M. (2004). Factors that affect the quality of cheese. In *Cheese: Chemistry, physics and microbiology* (Vol. 1, pp. 583–608). Academic Press.

Fox, P. F., Guinee, T. P., Cogan, T. M., McSweeney, P. L., Fox, P. F., Guinee, T. P., et al. (2017). Factors that affect cheese quality. In *Fundamentals of cheese science* (pp. 533–542). Springer.

Fusco, V., Chieffi, D., Fanelli, F., Logrieco, A. F., Cho, G. S., Kabisch, J., et al. (2020). Microbial quality and safety of milk and milk products in the 21st century. *Comprehensive Reviews in Food Science and Food Safety, 19*(4), 2013–2049.

Gaceu, L. (2020). *Design principles for assuring food safety in Romanian mountain dairy farms*. CABI.

Gourama, H. (2020). Foodborne pathogens. In *Food safety engineering* (pp. 25–49). Springer.

Guo, X., Yu, Z., Zhao, F., Sun, Z., Kwok, L. Y., & Li, S. (2021). Both sampling seasonality and geographic origin contribute significantly to variations in raw milk microbiota, but sampling seasonality is the more determining factor. *Journal of Dairy Science, 104*(10), 10609–10627.

Ianni, A., Bennato, F., Martino, C., Grotta, L., & Martino, G. (2020). Volatile flavor compounds in cheese as affected by ruminant diet. *Molecules, 25*(3), 461.

Johnson, M. E. (2017). A 100-year review: Cheese production and quality. *Journal of Dairy Science, 100*(12), 9952–9965.

Kailasapathy, K. (2015). Chemical composition, physical, and functional properties of milk and milk ingredients. In *Dairy processing and quality assurance* (pp. 77–105). Wiley.

Khattab, A. R., Guirguis, H. A., Tawfik, S. M., & Farag, M. A. (2019). Cheese ripening: A review on modern technologies towards flavor enhancement, process acceleration and improved quality assessment. *Trends in Food Science and Technology, 88*, 343–360.

Kilcawley, K. N., Faulkner, H., Clarke, H. J., O'Sullivan, M. G., & Kerry, J. P. (2018). Factors influencing the flavour of bovine milk and cheese from grass based versus non-grass-based milk production systems. *Foods, 7*(3), 37.

Kuchtík, J., Šustová, K., Sýkora, V., Kalhotka, L., Pavlata, L., & Konečná, L. (2021). Changes in the somatic cells counts and total bacterial counts in raw goat milk during lactation and their relationships to selected milk traits. *Italian Journal of Animal Science, 20*(1), 911–917.

Kumar, A., & Patyal, A. (2024). An insight on microbial flora of milk and milk products. In *The microbiology, pathogenesis and zoonosis of milk borne diseases* (pp. 69–94). Academic Press.

Lu, M., & Wang, N. S. (2017). Spoilage of milk and dairy products. In *The microbiological quality of food* (pp. 151–178). Woodhead Publishing.

Mesele, F., & Abunna, F. (2019). Escherichia coli O157: H7 in foods of animal origin and its food safety implications. *Advances in Biology Research, 13*(4), 134–145.

Mkangara, M. (2023). Prevention and control of human salmonella enterica infections: An implication in food safety. *International Journal of Food Science, 2023*(1), 8899596.

Murphy, S. C., Martin, N. H., Barbano, D. M., & Wiedmann, M. (2016). Influence of raw milk quality on processed dairy products: How do raw milk quality test results relate to product quality and yield? *Journal of Dairy Science, 99*(12), 10128–10149.

Ong, L., Li, X., Ong, A., & Gras, S. L. (2022). New insights into cheese microstructure. *Annual Review of Food Science and Technology, 13*(1), 89–115.

Pegolo, S., Tessari, R., Bisutti, V., Vanzin, A., Giannuzzi, D., Gianesella, M., et al. (2022). Quarter-level analyses of the associations among subclinical intramammary infection and milk quality, udder health, and cheesemaking traits in Holstein cows. *Journal of Dairy Science, 105*(4), 3490–3507.

Perin, L. M., Pereira, J. G., Bersot, L. S., & Nero, L. A. (2019). The microbiology of raw milk. In *Raw milk* (pp. 45–64). Academic Press.

Potts, D. M. (2016). Factors Affecting Flavor Quality of Bovine Milk and Dairy Products (Doctoral dissertation, University of Minnesota).

Rashid, R. (2020). Transmission of Staphylococcus aureus in milk and milk products Chattogram Veterinary and Animal Science University. [a mini review] A, 1–38.

Ritota, M., & Manzi, P. (2020). Natural preservatives from plant in cheese making. *Animals, 10*(4), 749.

Roche, J. R., Berry, D. P., Delaby, L., Dillon, P. G., Horan, B., Macdonald, K. A., & Neal, M. (2018). New considerations to refine breeding objectives of dairy cows for increasing robustness and sustainability of grass-based milk production systems. *Animal, 12*(s2), s350–s362.

Santos, B. N. C., Silva, C. C. C., Domingues, J. R., Cortez, M. A. S., Freitas, D. D. G. C., Chiappini, C. C. J., & Araújo, K. G. L. (2013). Effect of calcium addition and pH on yield and texture of Minas cured cheese. *Arquivo Brasileiro de Medicina Veterinária e Zootecnia, 65*, 601–609.

Smit, G., Smit, B. A., & Engels, W. J. (2005). Flavour formation by lactic acid bacteria and biochemical flavour profiling of cheese products. *FEMS Microbiology Reviews, 29*(3), 591–610.

Sonnier, J. L., Karns, J. S., Lombard, J. E., Kopral, C. A., Haley, B. J., Kim, S. W., & Van Kessel, J. A. S. (2018). Prevalence of Salmonella enterica, Listeria monocytogenes, and pathogenic

Escherichia coli in bulk tank milk and milk filters from US dairy operations in the National Animal Health Monitoring System Dairy 2014 study. *Journal of Dairy Science, 101*(3), 1943–1956.

Stocco, G., Cipolat-Gotet, C., Gasparotto, V., Cecchinato, A., & Bittante, G. (2018). Breed of cow and herd productivity affect milk nutrient recovery in curd, and cheese yield, efficiency and daily production. *Animal, 12*(2), 434–444.

Tilocca, B., Costanzo, N., Morittu, V. M., Spina, A. A., Soggiu, A., Britti, D., et al. (2020). Milk microbiota: Characterization methods and role in cheese production. *Journal of Proteomics, 210*, 103534.

Truong, T., Palmer, M., Bansal, N., & Bhandari, B. (2016). *Effect of milk fat globule size on the physical functionality of dairy products* (pp. 31–34). Springer.

Upreti, P., Metzger, L. E., & Hayes, K. D. (2006). Influence of calcium and phosphorus, lactose, and salt-to-moisture ratio on Cheddar cheese quality: Proteolysis during ripening. *Journal of Dairy Science, 89*(2), 444–453.

Vyas, D., Nelson, C. D., Bromfield, J. J., Liyanamana, P., Krause, M., & Dahl, G. E. (2020). MILK symposium review: Identifying constraints, opportunities, and best practices for improving milk production in market-oriented dairy farms in Sri Lanka. *Journal of Dairy Science, 103*(11), 9774–9790.

Chapter 4
Starter Culture

Reshab Majumder, Chandrakanta Sen, Subhadip Manik, and Akshay Ramani

4.1 Introduction

Starters are a class of beneficial and active microorganisms that can use for fermentation to alter milk and milk products in a way as desired. These are specifically chosen microorganisms that are added to milk in order to "start" and continue the intended fermentation process, which results in fermented milk products. Generally, starters mainly alter the ultimate product's appearance, texture, body, and flavor. The heat treatments applied to the milk cause the natural microflora to become ineffective, unpredictable, and out of control, or they completely eradicate it. When introduced to milk, starter culture can give specific results in a more regulated and predictable fermentation (Parente et al., 2017).

The most popular application for starter cultures is the synthesis of lactic acid from lactose, or milk sugar. This process lowers the pH of milk, which typically causes or aids in the coagulation of milk protein. Lactic acid bacteria (LAB) are the common term for the cultures that generate lactic acid. They also produce flavor compounds especially diacetyl during fermentation process. In addition to their effect on pH, starter organisms can also affect the texture of aged and/or cultured goods by breaking down proteins, lipids, and other components of milk. While inhibition is linked to other growth-related by products of starters, including production of toxins, organic acids, pigments etc., the lower pH of cultivated items can suppress some spoilage microbes (Bintsis, 2018).

R. Majumder
Jharkhand Milk Federation, Ranchi, Jharkhand, India

C. Sen · A. Ramani
Dairy Chemistry Division, ICAR-National Dairy Research Institute, Karnal, Haryana, India

S. Manik (✉)
Dairy Technology Division, ICAR-National Dairy Research Institute, Karnal, Haryana, India

S. Mansha Rafiq (ed.), *Technological Advances and Trends in Cheese Making*,
https://doi.org/10.1007/978-3-031-94476-5_4

The relevance of fermented foods is in their ability to maintain large amounts of nutrient-dense foods with a diverse range of flavors, scents, and textures, hence adding value to human diets. There are more than 3500 traditional fermented foods in the world and from the ancient times, humans have consumed fermented foods, and these fermented milks have long been a staple of diets and nutrition (Tamang et al., 2016). The original purpose of fermented milks was to preserve nutrients of milk and other food products. Left to stand at room temperature, milk naturally fermented and may be consumed straight away or diluted with water (Gemechu, 2015).

The industrialization of cheese and butter production began about 200 years ago. To scale up the traditional production process of dairy products into industrial level, manufacturing process of cheese was standardized and altered. Similarly, the standardization of production process for starter culture is essential as the size of the operations increased. During late nineteenth century, companies were focused on the production of different starter culture. Initially the system was restricted may be due to presence of undefined and unidentified strains of appropriate species. Afterwards, starters underwent analysis and definition, which made it possible for them to be assembled from particular strains with particular qualities chosen to produce a standard product. Compared to undefined beginnings, defined starters often consist of a smaller number of specialized strains with a lower species diversity (Carafa et al., 2019).

Hence, this chapter provide a discussion related to different starter culture, the factors affecting its quality and activity, utilisation and selection of starter culture, probiotic starter culture, and adjunct starter culture as well as starter culture stimulants.

4.2 Classification of Starter Culture

One can categorize starter cultures according to their purpose, growing temperature, or bacterial composition. The primary starters play a major role in the early stages of cheese manufacturing when lactose is converted to lactic acid. Consequently, a large quantity of active cells is added to the cheese milk. Many of them produce small volatile compounds, such as CO_2 from lactose (heterofermentative species) and citrate (homofermentative and heterofermentative species), which helps in some cheeses during typical development in open texture, and diacetyl from citrate, which is a flavorful component of fresh cheese. The development of flavor and aroma in ripening cheeses is also influenced by their proteolytic systems and amino acid metabolism. Additionally, they support the microbiological safety of cheese by reducing the pH, competing with pathogenic and rotting microbes, and generating antimicrobial chemicals. From a taxonomical and functional perspective, the secondary microbiota of cheese such as *propionibacteria*, *staphylococci*, *coryneforms*, nonstarter LAB (NSLAB), yeasts, and molds may impact on the organoleptic qualities of cheeses. Since, these microbes are essential to the ripening process, large initial concentrations are not required; instead, many cheese varieties still rely on natural contamination from milk and the surrounding cheese environment (Beresford et al., 2001).

The two main starters types are typically categorized as thermophilic or mesophilic. The latter are traits of cheese varieties from Italy (grana, pecorino, mozzarella) and Switzerland (emmentaler, sbrinz, gruyère), where early in the cheese-making process, a high temperature (>37 °C, reaching 48–52 °C for hard-cooked cheese variations) is typical. All cheese variants (such as cheddar, gouda, edam, camembert, etc.) in which the curd temperature during the early stages of acid generation does not exceed around 40 °C use mesophilic starters. However, mesophilic and thermophilic species are frequently used together in both defined and undefined starters for the production of cheeses, like mozzarella (Limsowtin et al., 1996; Parente, 2006) and cheddar (Beresford et al., 2001), this distinction is becoming less meaningful.

Every starter culture on the market today is generated in one way or another from natural (or artisanal) starters that are continuously recreated in cheese factories through back slopping. These starters have an unknown composition, meaning they comprise a mixture of diverse strains and/or species. The use of natural whey cultures i.e., adding whey from a previous batch of cheese to cheese for the production of Grana was first introduced in Italy in 1888, and reports on adding sour cream or buttermilk to cream to improve the quality of butter date back to the 1860s (Bottazzi, 1993).

Commercial undefined starters, also known as mixed-strain starters (MSS), have taken the place of the "best" natural starters for many cheeses. These starters are produced under controlled conditions by specialized institutions, such as dairy research centres or commercial starter companies, and are then supplied to cheese plants, where they are used to either directly inoculate the milk in the vat or build up bulk starter. In actuality, four businesses were selling vague strain starters to Danish creameries in 1891 (Leisner, 2005). The composition of MSS is unknown, although the inherent unpredictability connected to the use of artisanal starters is lessened when they are reproduced in more controlled settings with fewer transfers.

Compared to defined-strain starters (DSS), natural starter cultures (NSC) and commercial MSS are referred to as traditional starters due to their lengthy history (Limsowtin et al., 1996) as shown in Fig. 4.1. DSS were initially employed in Australia and New Zealand for the production of cheddar cheese in the 1930s. They are made up of one or more strains (cultures including up to 13 strains have been utilized in Switzerland) (O'Toole, 2004). Similar to MSS, DSS are chosen, developed, maintained, and disseminated by institutions with specific expertise. Many cheese kinds, including certain protected designation of origin (PDO) European varieties, are now produced using DSS instead of traditional starters due to their high phage resistance, highly reproducibility and optimization.

The usage of these strains is unrestricted and they are not considered genetically modified organisms under the current criteria in the US and Europe. The commercial usage of genetically engineered strains of starter which are developed to show improved autolysis, increased aroma, over-expression of peptidases, novel phage resistance mechanisms, etc. It has been restricted due to concerns about consumer

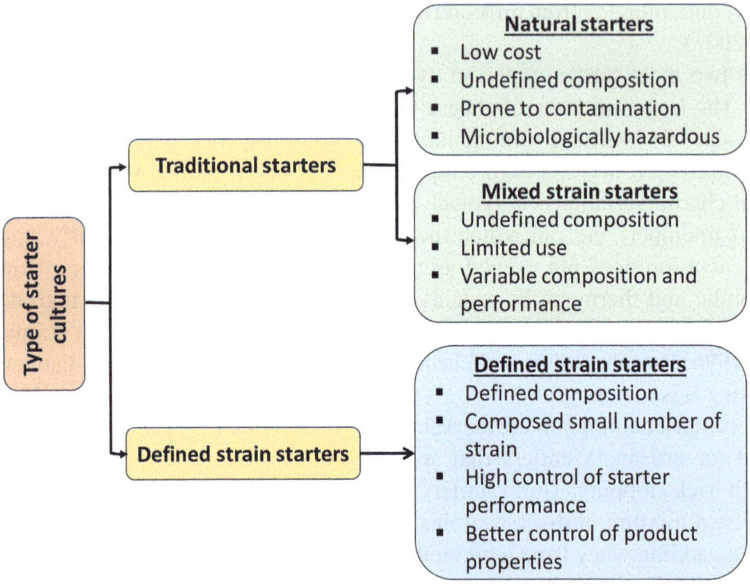

Fig. 4.1 Types of starter culture

acceptability (Pedersen et al., 2005). Many startups still rely on selection programs, which have benefited tremendously from recent developments in the "-omics" which indicates the complete biomolecular study of LAB (Derkx et al., 2014).

In recent times, the preferred method for studying the ecology of NSC has been culture-independent approaches. They have verified that the community and population structure are shaped by random contamination that occurs during incubation (and consequently, cheese technology). Contrary to *S. thermophilus* strains isolated from Asiago d'Allevo, Montasio, and Monte Veronese (which have similar technologies), which cannot be separated by RAPD-PCR, strains isolated from Grana, Parmigiano Reggiano, and Provolone can be easily separated based on both phenotypic and genotypic characteristics (Gatti et al., 2004). By using cultivation-dependent methodologies, only a small portion of the community may be identified (Bottari et al., 2013; Rossetti et al., 2008; Santarelli et al., 2008). According to a recent study using "next generation sequencing" (NGS) techniques, there is a core microbiome for every type of cheese, the diversity of the NSC is quite limited, and some species discovered in the NSC may not necessarily be important for cheese (De Filippis et al., 2014). Nonetheless, the modest quantity of NSLAB in NSC may aid in the ripening of cheese, particularly matured kinds (Bottari et al., 2013). The stability of the composition and performance of undefined cultures comprising many bacterial strains is significantly enhanced when they are grown under controlled circumstances with a low number of subcultures, while maintaining the benefit of resistance to phage infection (Stadhouders & Leenders, 1984). Initially, these undefined MSS were obtained through a meticulous selection process of natural starters. They are commonly utilized for the manufacture of cheese and are usually maintained, propagated, and distributed by start-up enterprises and research organizations. Citrate fermentation and composition can be used to further classify mesophilic MSS, which are typically grown at or below 30 °C.

Like artisanal starters, MSS are consisting of a mixture unknown strain that vary in terms of their technological characteristics, such as phage resistance. Temperate phages are found in the majority of strains, however virulent phages can also be found. The divergence of strains in MSS has been estimated using a range of techniques. In order to assess the diversity of *Lact. lactis* subsp. *cremoris* in seven MSS used in the production of Cheddar cheese in Canada. Bissonnette et al. (2000) isolated and typed a sizable number of isolates (30) from each culture using plasmid profiles and phenotypic characteristics. Two MSS showed a high diversity, with 18–24 different strains, whereas three were dominated by 7–9 strains.

Two instances of the successful development and long-term application of MSS are the development of MSS for the production of Dutch cheeses by NIZO (Stadhouders & Leenders, 1984) and thermophilic MSS for the production of Swiss cheese varieties by the Swiss Federal Dairy Research Station (Glättli, 1990; Limsowtin et al., 1996).

Variable starter activity and the presence of open texture defects (caused by Citrate+ strains) in Cheddar cheese made with MSS led to the development of mesophilic specified strain starters. The technological performance of DSS is very reproducible because the strain and/or species ratio is determined. This is undoubtedly a very desirable attribute in contemporary cheese operations with huge milk throughputs and strict production schedules. Thermophilic DSS can also be purchased commercially and used to make a range of cheeses that are Swiss and Italian in style. In Italy, associations of *S. thermophilus* and *L. delbrueckii* subsp. *bulgaricus* (rod:coccus starter cultures) are utilized to produce low-moisture mozzarella cheese, but starters made of one or more strains of *S. thermophilus* are still preferred for producing high-moisture mozzarella cheese.

Considerable research interest has been drawn to the finding of CRISPR/Cas systems as phage resistance mechanisms in BIMs (Deveau et al., 2010). Traditionally, research facilities have been the primary source of DSS in the form of bulk or seed starter inoculum. However, commercial providers of direct-to-vat culture concentrates are gradually embracing the defined-culture method as a means of achieving consistent and characterized culture performance.

4.3 Factors Affecting Starter Activity

When it comes to dairy fermentation, starter cultures are the magicians who turn milk into a variety of delicious products that are loved all over the world. The key to all dairy fermentation processes, from the complex aromas of cheese to the creamy tang of yogurt, is the action of starter cultures. The effectiveness and uniformity of starting culture activity in milk and dairy products depend on a wide range of variables, including substrate composition, processing methods, ambient circumstances, and microbial dynamics. This portion sheds light on the critical elements that affect the results of ferm entation by navigating the complicated prevalence of starter culture activity in dairy products. The different factors affecting the performance of starter culture is briefly presented in Table 4.1.

Table 4.1 Factors affecting performance of starter culture

Factors	Description	References
Selection of microbial strains	Microbial strain selection profoundly impacts starter culture performance, dictating flavor profiles, texture, and shelf life of fermented foods. It's a critical factor in achieving desired product characteristics and maintaining consistency in quality	Vinicius De Melo Pereira et al. (2020)
Milk composition	Milk composition influences starter culture activity, affecting fermentation rate, flavor development, and texture in dairy products. Variations in fat, protein, and lactose content can impact microbial growth and metabolic pathways, ultimately shaping product quality	Mohammadi et al. (2012)
Processing parameters	Milk processing parameters such as pasteurization temperature and duration, homogenization, and storage conditions profoundly influence starter culture efficacy. These factors can impact microbial viability, activity, and subsequent fermentation dynamics, affecting product attributes like flavor, texture, and shelf life	Sodini et al. (2002)
Environmental impact	Environmental factors like temperature, pH, oxygen levels, and available nutrients significantly impact starter culture behavior. Optimal conditions promote microbial growth, metabolism, and desired fermentation outcomes, while deviations can lead to inconsistencies in flavor, texture, and product quality	Li et al. (2016)
Interaction with milk microbiota	Interactions with indigenous milk microbiota can influence starter culture performance, affecting fermentation kinetics, metabolite production, and product characteristics. Understanding and managing these interactions are crucial for achieving consistent quality and desired outcomes in fermented dairy products	Blaya et al. (2018)
Development of technology	Technological advancements in biotechnology and microbiology have enabled the development of more robust and specialized starter cultures. These innovations enhance fermentation efficiency, product consistency, and the ability to tailor flavor profiles to meet consumer preferences effectively	de Vuyst (2000)
Quality control and assurance	Quality control and assurance procedures ensure the purity, viability, and consistency of starter cultures. Rigorous testing and monitoring protocols safeguard against contamination, guaranteeing optimal performance in fermentation processes and maintaining product integrity and safety standards	Hutkins (2006)

4.3.1 Microbial Strain Selection

Choosing the right microbial strains is crucial to the process of dairy fermentation and has a significant impact on the quality characteristics of final product. The symbiotic interaction between *Lactobacillus bulgaricus* and *Streptococcus thermophilus* is critical to yogurt manufacturing as it drives taste development and acidity. Texture, aroma, and maturation dynamics are all determined by the ripening cultures and LAB used in cheesemaking. Furthermore, the significance of strain selection in dairy fermentation is highlighted by the genetic variety and metabolic versatility of particular strains, which impact their ability to respond to varying milk compositions and processing circumstances (Vinicius De Melo Pereira et al., 2020).

4.3.2 Environmental Conditions

Microbial growth kinetics, enzymatic activity, and product properties are all significantly influenced by the environmental effects, which also has an impact on starter culture activity. Temperature is one of the important environmental factors. Mesophilic cultures function best at lower temperatures (20–30 °C) for the manufacturing of cheese, but thermophilic cultures flourish at higher temperatures (40–45 °C) when producing yogurt. Additionally, pH values are vital for LAB grow and it acidify best in particular pH ranges. Furthermore, the microbial metabolism and fermentation dynamics are influenced by the oxygen availability, moisture content, and nutritional composition of milk, underscoring the complex interaction between environmental conditions and starter culture activity (Li et al., 2016).

4.3.3 Milk Composition

The composition of milk provides vital nutrients for microbial development and metabolism, acting as the main substrate for dairy fermentation. The composition of milk can vary greatly; variations in fat, protein, lactose concentration, and mineral content can all have a significant impact on starter culture activity and product quality. For example, the composition of casein affects the proteolytic activity of some starter cultures throughout the cheese-making process, which influences the development of texture and flavor. Similar fluctuations in lactose concentration can affect the kinetics of acidification during the manufacturing of yogurt, resulting in variations of viscosity and tartness. Moreover, bioactive substances like immunoglobulins and lactoferrin can also alter microbial metabolism and enhance the functional qualities of dairy products that have undergone fermentation (Mohammadi et al., 2012).

4.3.4 Processing Methods

The methods used to ferment dairy products have a significant impact on the properties of the final product and the activity of the starter culture. Every stage of the production process affects the results of the fermentation process, from inoculation techniques and fermentation operation to post-fermentation treatments. For example, in yogurt production, texture development and acidification kinetics are controlled by the incubation temperature and the density of starter cultures injected. Similar to this, when manufacturing cheese, microbial activity, proteolysis, and flavor development are influenced by the curd-cutting method, salting schedule, and ripening environment. Furthermore, post-fermentation processes like aging, smoking, and brining enhance the richness of flavor and lengthen the shelf life of dairy products that have undergone fermentation, highlighting the significance of processing methods in dairy fermentation (Sodini et al., 2002).

4.3.5 Interactions with Indigenous Microbiota

Apart from starter cultures, fermentation results can also be influenced by indigenous microbiota found in raw milk and dairy processing conditions. The kinetics of fermentation and the properties of the final product can be affected by these native microorganisms through cooperation, competition, or hostility toward starting cultures. Furthermore, the dynamics and composition of the native microbiota are influenced by a number of variables, including geography, agricultural methods, and processing cleanliness. Therefore, in order to guarantee constant product quality and safety in dairy fermentation, it is essential to regulate the native microbiota. Techniques including sanitation, selective culture enrichment, and pasteurization assist reduce the chance of contamination and guarantee the dominance of targeted microbial consortia (Blaya et al., 2018).

4.3.6 Technological Developments

Dairy fermentation has been transformed by advances in bioprocessing and fermentation technology, which also present new chances to improve product quality and starter culture activity. Automated bioreactors and continuous fermentation platforms are examples of controlled fermentation systems that allow for exact control over fermentation conditions, maximizing the kinetics of microbial growth and uniformity of the final product. Furthermore, improvements in fermentation efficiency and product functioning may be possible with strain development breakthroughs like genetically modified cultures with increased metabolic capacities or stress tolerance. To guarantee the sustainability and safety of fermented dairy products, however, technological developments must be accompanied with ethical and regulatory frameworks (de Vuyst, 2000).

4.3.7 Quality Control and Assurance

Strong quality control and assurance procedures are required at every stage of the production process to guarantee the safety and quality of fermented dairy products. Every stage is carefully examined to maintain quality standards, from sensory evaluation and shelf-life studies to microbial enumeration and testing on raw milk. Producers can keep an eye on metabolite profiles, product attributes, and starter culture activity by using analytical techniques like gas chromatography-mass spectrometry (GC-MS), high-performance liquid chromatography (HPLC), and PCR-based microbial identification. Moreover, adherence to Good Manufacturing Practices (GMP) and Hazard Analysis and Critical Control Points (HACCP) principles mitigates the risk of contamination and ensures compliance with regulatory requirements (Hutkins, 2006).

4.4 Use of Selected Starter Culture, Adjunct Starter Culture, and Probiotic Starter Culture

Starter cultures play crucial role in dairy fermentation and more specially cheese production because they influence the flavour, texture, and nutritional value of the cheese. It usually falls into three basic groups that are frequently used: selected starter culture, adjunct starter culture and probiotic starter culture. Each has a distinct function during the fermentation process, producing a wide variety of delectable foods.

4.4.1 Selected Starter Culture

Carefully chosen strains of microorganisms, usually yeast or bacteria, are called selected starter cultures and are added to the food matrix to start fermentation. These cultures are chosen according to their capacity to yield the required final product flavors, scents, and textures. Particularly when it comes to dairy products such as cheese and yogurt, certain starting cultures are essential. For example, LAB like *Lactobacillus casei* and *Lactococcus lactis* are frequently utilized as chosen starting cultures in the cheese-making process. After consuming the lactose in milk, these bacteria produce lactic acid, which acidifies the milk and starts the curdling process that is required for the creation of cheese (Vinicius De Melo Pereira et al., 2020).

There are many benefits to employ specific starting cultures. They provide product quality consistency, enabling manufacturers to duplicate particular flavors and textures from batch to batch. Additionally, they promote food safety by eliminating dangerous microbes and fostering an atmosphere that supports advantageous fermentation. Notwithstanding, there are drawbacks to depending only on chosen starter cultures, including the possibility of losing microbial diversity and the requirement for consistent upkeep of culture stocks (Bassi et al., 2015).

4.4.2 Adjunct Starter Culture

Adjunct starter cultures improve the fermentation process and add more flavor complexity to chosen starter cultures. Adjunct cultures, as opposed to carefully selected starter cultures, are frequently native microflora that are already present in the raw materials or fermentation environment. Their distinct enzymes, metabolites, and microbial diversity enhance the finished product's sensory attributes (Irlinger et al., 2017).

Using adjunct starting cultures encourages a closer relationship to the region's terroir and culinary history while also celebrating the diversity of microorganisms. But it also adds an element of randomness and unpredictability to the fermentation process, necessitating the ability of trained craftspeople to adjust and react to shifting environmental circumstances (Nieto-Arribas et al., 2009).

4.4.3 Probiotic Culture Starter

Consuming sufficient amounts of probiotic starter cultures can have positive effects on one's health as they include live bacteria. Probiotic cultures give priority to the development of intestinal health and general well-being, in contrast to chosen and auxiliary cultures, which are primarily focused on flavor and fermentation. Fermented foods and beverages often contain probiotic strains of LAB, such as *Lactobacillus* and *Bifidobacterium*. A strong dose of probiotic bacteria is delivered to the digestive system by foods like yogurt, kefir, kimchi, and kombucha, which act as carriers for probiotic cultures. In addition to bolstering immunity, enhancing digestion, and easing some gastrointestinal diseases, these probiotics support a balanced microbial community in the gut (Yerlikaya, 2014).

Although probiotic starter cultures provide many health advantages, it is important to carefully examine aspects like viability, stability, and regulatory compliance when including them into fermented foods. For probiotic microbes to survive throughout the production process and the shelf life of the finished product, proper handling and storage are crucial (Rakib et al., 2017). To sum up, the utilization of carefully chosen starter cultures, supplement starter cultures, and probiotic starter cultures embodies a well-balanced combination of science, customs, and advancements in the field of fermentation. Whether they are producing rustic breads, artisanal cheeses, or probiotic beverages that boost health, these cultures enhance our culinary experiences and feed our bodies and souls. Blaiotta et al. (2017) standardized a process for production of "Italico" cheese from milk. Similarly, Gomes and Malcata (1998) reported development of probiotic processed cheese using goat milk.

4.5 Starter Stimulants and Their Properties

Starter stimulants are essential for speeding up and maximizing the fermentation process in the context of food production and fermentation. These materials, which are frequently organic compounds sourced from different sources, promote the development and activity of helpful microorganisms that are in charge of fermentation, such yeast and bacteria. Starter stimulants encourage microbial proliferation and improve the synthesis of desired end products by offering the required nutrients, energy sources, and ambient conditions. Some popular starter culture stimulants and their characteristics are hereunder:

4.5.1 Sugars

Fermentative bacteria use sugars as their main source of energy to power their metabolic processes and enable fast growth and fermentation. The sugars glucose, sucrose, maltose, and fructose are frequently utilized as starting stimulants. LAB and yeast can easily break down these carbohydrates to produce ethanol, lactic acid, and other fermentation byproducts. Furthermore, sugars help fermented foods and beverages acquire the flavors, scents, and textures that are desired (Vénica et al., 2018).

4.5.2 Yeast Extracts

Derived from autolyzed yeast cells, yeast extracts are a rich source of vitamins, minerals, amino acids, and nucleotides. They provide vital nutrients needed for protein synthesis, cell division, and metabolism, and act as strong promoters of yeast growth and fermentation. The kinetics of fermentation are improved by yeast extracts, which also raises the general yield and caliber of fermented goods. Furthermore, through the fermentation process, they release savory and umami chemicals that enhance the variety of flavors and fragrances (Chazaud & Larpent, 1980).

4.5.3 Organic Acids

Certain microbial populations are stimulated by organic acids, which also have antibacterial qualities. Examples of these are lactic acid, citric acid, and acetic acid. These acids produce an acidic environment that encourages the growth of species that can withstand acidity while suppressing the growth of unwanted bacteria. Apart from their role as preservatives, organic acids also add to the distinct sour and tanginess of fermented foods and drinks (Bressani et al., 2020).

4.5.4 Minerals and Trace Elements

Cofactors and activators for the enzymatic reactions involved in fermentation, minerals and trace elements such as iron, manganese, zinc, and magnesium are vital. These micronutrients have an impact on microbial development and metabolic pathways because they are necessary for the creation of enzymes, DNA, RNA, and cellular components. Minerals and trace elements work as stimulants for robust fermentation and the generation of high-quality end products by guaranteeing appropriate nutrient availability (Vedamuthu, 2013).

4.5.5 Prebiotics

Prebiotics are indigestible fibers and carbohydrates that specifically encourage the growth and function of probiotics, the good bacteria in the gut. When fermentation occurs, probiotic bacteria like Lactobacillus and Bifidobacterium are stimulated by prebiotics like inulin, oligosaccharides, and resistant starches. The probiotic microbes in fermented foods and beverages benefit from the colonization and growth of these prebiotics, which increases their health-promoting qualities (Hati et al., 2013).

4.5.6 Enzymes

Enzymes function as biological catalysts to quicken chemical reactions as it is unaffected by the environment. To convert complicated substrates into simpler molecules appropriate for microbial fermentation, certain enzymes—such as lipase, amylase, and protease—are employed as starting stimulants. Enzymes stimulate microbial growth and metabolic activity by making fermentable sugars and amino acids more accessible through the breakdown of starches, proteins, and lipids (Bezie & Regasa, 2019).

4.6 Conclusion

All things considered, the investigation of starter cultures offers a thorough comprehension of their importance in the field of fermentation and food production. After giving the idea a brief overview, we looked at how starters were categorized and acknowledged the variety of uses for them. It is possible to gain insight into the complex equilibrium between environmental factors and nutritional needs that is essential for microbial growth and metabolic activity by knowing the parameters that influence starter activity. Complicated methods for attaining specific flavor profiles, textures, and nutritional benefits in fermented foods and beverages were made evident through the use of adjunct starters, probiotic microorganisms, and specially chosen starters. Every type of starter culture adds something different to the fermentation process, from the accuracy of carefully chosen cultures to the complexity of adjunct cultures and the health-promoting qualities of probiotics. Additionally, the starter stimulant debate clarified the function of different chemicals in enhancing microbial growth and fermentation efficiency. As vital stimulants, sugars, yeast extracts, organic acids, minerals, prebiotics, and enzymes supply the nutrients, energy, and environmental cues required for microbial growth and product development. Our knowledge of starter cultures and their characteristics will continue to improve as long as innovation and research are conducted in the dynamic field of fermentation science. Food scientists and craftspeople will continue to push the limits of flavor, nutrition, and sustainability in fermented foods and beverages by utilizing the potential of stimulants and starter cultures.

References

Bassi, D., Puglisi, E., & Cocconcelli, P. S. (2015). Comparing natural and selected starter cultures in meat and cheese fermentations. *Current Opinion in Food Science, 2*, 118–122.

Beresford, T. P., Fitzsimons, N. A., Brennan, N. L., & Cogan, T. M. (2001). Recent advances in cheese microbiology. *International Dairy Journal, 11*(4–7), 259–274.

Bezie, A., & Regasa, H. (2019). The role of starter culture and enzymes/rennet for fermented dairy products manufacture—A review. *Nutrition and Food Science International Journal, 9*, 21–27.

Bintsis, T. (2018). Lactic acid bacteria as starter cultures: An update in their metabolism and genetics. *AIMS Microbiology, 4*(4), 665–684.

Bissonnette, F., Labrie, S., Deveau, H., Lamoureux, M., & Moineau, S. (2000). Characterization of mesophilic mixed starter cultures used for the manufacture of aged Cheddar cheese. *Journal of Dairy Science, 83*, 620–627.

Blaiotta, G., Murru, N., Di Cerbo, A., Succi, M., Coppola, R., & Aponte, M. (2017). Commercially standardized process for probiotic "Italico" cheese production. *LWT-Food Science and Technology, 79*, 601–608.

Blaya, J., Barzideh, Z., & LaPointe, G. (2018). Symposium review: Interaction of starter cultures and nonstarter lactic acid bacteria in the cheese environment. *Journal of Dairy Science, 101*(4), 3611–3629.

Bottari, B., Agrimonti, C., Gatti, M., Neviani, E., & Marmiroli, N. (2013). Development of a multiplex real time PCR to detect thermophilic lactic acid bacteria in natural whey starters. *International Journal of Food Microbiology, 160*, 290–297.

Bottazzi, V. (1993). *Microbiologia lattiero-casearia*. Edagricole.

Bressani, A. P. P., Martinez, S. J., Sarmento, A. B. I., Borém, F. M., & Schwan, R. F. (2020). Organic acids produced during fermentation and sensory perception in specialty coffee using yeast starter culture. *Food Research International, 128*, 108773.

Carafa, I., Stocco, G., Franceschi, P., Summer, A., Tuohy, K. M., Bittante, G., & Franciosi, E. (2019). Evaluation of autochthonous lactic acid bacteria as starter and non-starter cultures for the production of Traditional Mountain cheese. *Food Research International, 115*, 209–218.

Chazaud, M. T., & Larpent, J. P. (1980). Stimulation of the activity of starter cultures by yeast extract. *Revue Laitière Française, 383*, 19–24.

De Filippis, F., La Storia, A., Stellato, G., Gatti, M., & Ercolini, D. (2014). A selected core microbiome drives the early stages of three popular Italian cheese manufactures. *PLoS One, 9*, e89680.

De Vuyst, L. (2000). Technology aspects related to the application of functional starter cultures. *Food Technology and Biotechnology, 38*(2), 105–112.

Derkx, P. M. F., Janzen, T., Sørensen, K. I., Christensen, J. E., Stuer-Lauridsen, B., & Johansen, E. (2014). The art of strain improvement of industrial lactic acid bacteria without the use of recombinant DNA technology. *Microbial Cell Factories, 13*(Suppl. 1), S5.

Deveau, H., Garneau, J. E., & Moineau, S. (2010). CRISPR/Cas system and its role in phage-bacteria interactions. *Annual Review of Microbiology, 64*, 475–493.

Gatti, M., Trivisano, C., Fabrizi, E., Neviani, E., & Gardini, F. (2004). Biodiversity among Lactobacillus helveticus strains isolated from different natural whey starter cultures as revealed by classification trees. *Applied and Environmental Microbiology, 70*, 182–190.

Gemechu, T. (2015). Review on lactic acid bacteria function in milk fermentation and preservation. *African Journal of Food Science, 9*(4), 170–175.

Glättli, H. (1990). Starter and starter production for Swiss type cheeses. *Proceedings of the second cheese symposium, Teagasc, Dublin*, pp. 23–30.

Gomes, A. M., & Malcata, F. X. (1998). Development of probiotic cheese manufactured from goat milk: Response surface analysis via technological manipulation. *Journal of Dairy Science, 81*(6), 1492–1507.

Hati, S., Mandal, S., & Prajapati, J. B. (2013). Novel starters for value added fermented dairy products. *Current Research in Nutrition and Food Science Journal, 1*(1), 83–91.

Hutkins, R. W. (2006). Starter cultures. In *Microbiology and technology of fermented foods* (Vol. 32, p. 67). Wiley.

Irlinger, F., Hélinck, S., & Jany, J. L. (2017). Secondary and adjunct cultures. In *Cheese* (pp. 273–300). Academic Press.

Leisner, J. J. (2005). Weeds, heat and pure cultures—On the differential success of new technologies in the Danish and American creamery industries in the 1890s. *Food Policy, 30*, 419–433.

Li, P., Lin, W., Liu, X., Wang, X., & Luo, L. (2016). Environmental factors affecting microbiota dynamics during traditional solid-state fermentation of Chinese Daqu starter. *Frontiers in Microbiology, 7*, 1237.

Limsowtin, G.K.Y., Powell, I.B. and Parente, E. (1996). Types of starters. In Dairy Starter Cultures, Cogan, T.M. and Accolas, J.-P., eds. VCH Publishers, New York. pp. 101–129.

Mohammadi, R., Sohrabvandi, S., & Mohammad Mortazavian, A. (2012). The starter culture characteristics of probiotic microorganisms in fermented milks. *Engineering in Life Sciences, 12*(4), 399–409.

Nieto-Arribas, P., Poveda, J. M., Seseña, S., Palop, L., & Cabezas, L. (2009). Technological characterization of Lactobacillus isolates from traditional Manchego cheese for potential use as adjunct starter cultures. *Food Control, 20*(12), 1092–1098.

O'Toole, D. K. (2004). The origin of single strain starter culture usage for commercial Cheddar cheesemaking. *International Journal of Dairy Technology, 57*, 53–55.

Parente, E. (2006). Diversity and dynamics of microbial communities in natural and mixed starter cultures. *Australian Journal of Dairy Technology, 61*, 44–51.
Parente, E., Cogan, T. M., & Powell, I. B. (2017). Starter cultures: General aspects. In *Cheese* (pp. 201–226). Academic Press.
Pedersen, M. B., Iversen, S. L., Sørensen, K. I., & Johansen, E. (2005). The long and winding road from the research laboratory to industrial applications of lactic acid bacteria. *FEMS Microbiology Reviews, 29*(3), 611–624.
Rakib, M. R. H., Kabir, A., & Amanullah, S. M. (2017). Starter cultures used in the production of probiotic dairy products and their potential applications: A review. *Chemical and Biomolecular Engineering, 2*(2), 83–89.
Rossetti, L., Fornasari, M. E., Gatti, M., Lazzi, C., Neviani, E., & Giraffa, G. (2008). Grana Padano cheese whey starters: Microbial composition and strain distribution. *International Journal of Food Microbiology, 127*, 168–171.
Santarelli, M., Gatti, M., Lazzi, C., Bernini, V., Zapparoli, G. A., Zapparoli, G. A., & Neviani, E. (2008). Whey starter for Grana Padano cheese: Effect of technological parameters on viability and composition of the microbial community. *Journal of Dairy Science, 91*, 883–891.
Sodini, I., Lucas, A., Oliveira, M. N. D., Remeuf, F., & Corrieu, G. (2002). Effect of milk base and starter culture on acidification, texture, and probiotic cell counts in fermented milk processing. *Journal of Dairy Science, 85*(10), 2479–2488.
Stadhouders, J., & Leenders, G. J. M. (1984). Spontaneously developed mixed-strain cheese starters: Their behaviour toward phages and their use in the Dutch cheese industry. *Netherlands Milk and Dairy Journal, 38*, 157–181.
Tamang, J. P., Watanabe, K., & Holzapfel, W. H. (2016). Diversity of microorganisms in global fermented foods and beverages. *Frontiers in Microbiology, 7*, 181961.
Vedamuthu, E. R. (2013). Starter cultures for yogurt and fermented milks. In *Manufacturing yogurt and fermented milks* (pp. 115–148). Wiley.
Vénica, C. I., Wolf, I. V., Suárez, V. B., Bergamini, C. V., & Perotti, M. C. (2018). Effect of the carbohydrates composition on physicochemical parameters and metabolic activity of starter culture in yogurts. *LWT, 94*, 163–171.
Vinicius De Melo Pereira, G., De Carvalho Neto, D. P., Junqueira, A. C. D. O., Karp, S. G., Letti, L. A., Magalhães Júnior, A. I., & Soccol, C. R. (2020). A review of selection criteria for starter culture development in the food fermentation industry. *Food Reviews International, 36*(2), 135–167.
Yerlikaya, O. (2014). Starter cultures used in probiotic dairy product preparation and popular probiotic dairy drinks. *Food Science and Technology, 34*, 221–229.

Chapter 5
Coagulation of Milk

Subhadip Manik , Akshay Ramani, Rakendhu Saji, Chandrakanta Sen, and Reshab Majumder

5.1 Introduction

The fundamental objective behind cheese manufacturing is to employ the fermentation and coagulation to turn milk into a stable product. Rennet is typically added to milk in order to form gel by destabilising the casein micelles. After coagulation process, rennet curd is produced and whey is released when the casein network contracts. During this process, components such as whey proteins, lactose, mineral and water are lost into the whey while fat droplets are entrapped within casein or cheese matrix (Fox & McSweeney, 2017). Increasing cheese yields has been the subject of extensive research for economic purposes. In recent decades, a number of strategies have been studied to improve cheese's ability to retain serum protein (Hinrichs, 2001). Ultrafiltration of milk is employed frequently in order to increase the concentration of serum proteins in cheese, and enhance the efficiency of cheese plants. Cheese yield is substantially affected due to the utilization of concentrated milk (Mistry & Maubois, 2017). Additionally, whey proteins from cheese whey can be retrieved and added directly to the cheese milk via thermal treatment (Schenkel et al., 2011). Analogously, aggregated form of whey proteins gets entrapped into casein matrix and flourish higher yield (Hinrichs, 2001; Perreault et al., 2017). At last, thermal processing is employed to milk prior to cheese making (Singh & Waungana, 2001). Heat treatment is a common practice to ensure microbiological safety of milk. Despite this, it has been shown that traditional pasteurisation

S. Manik (✉)
Dairy Technology Division, ICAR-National Dairy Research Institute, Karnal, Haryana, India

A. Ramani · R. Saji · C. Sen
Dairy Chemistry Division, ICAR-National Dairy Research Institute, Karnal, Haryana, India

R. Majumder
Jharkhand Milk Federation, Ranchi, Jharkhand, India

S. Mansha Rafiq (ed.), *Technological Advances and Trends in Cheese Making*,
https://doi.org/10.1007/978-3-031-94476-5_5

techniques such as heat treatment of 72 °C/15 s, denatured ~3.2% of serum proteins. The degree of denaturation increasing with increasing of heat treatment intensity (Guinee et al., 1998). Denatured serum proteins have a tendency to self-aggregate or form complexes with κ-casein during the heat treatment process, which is mostly due to disulfide interactions. Whey proteins are then incorporated to cheese after undergoing denatured (Kethireddipalli & Hill, 2015). This chapter discusses rennet types and their coagulation in milk, different factors that are widely affected milk coagulation, role of calcium in milk coagulation and rheological and textural changes during coagulation.

5.2 Type of Rennet

The term 'rennet' originally referred to the milk-clotting enzyme derived from the stomach of a calf, containing chymosin (also known as rennin), an active digestive enzyme (Harboe et al., 2010; Mohanty & Behare, 2017). It is naturally found in the stomach lining of young mammals, including calves. The abomasum is sliced and left in a saline solution (pH range of 5.0–5.5) for several days in order to macerate and remove rennet. The maceration fluid is then filtered after being clarified with alum. However, till today, most of rennet used in commercial cheese production is derived from microbial sources or genetically engineered microorganisms. This is particularly important for vegetarians, as animal-derived rennet may not align with their dietary choices (Troch et al., 2017).

Three main types of rennet used during cheese-making are following:

Animal Rennet: It is derived from the stomach lining of calves, lambs, or other young mammals. It contains chymosin and other enzymes naturally present in the animal's stomach.

Microbial Rennet: It is produced from fungi or bacteria, such as *Rhizomucor miehei* or *Bacillus subtilis*, which are genetically modified to produce chymosin. It is often preferred by vegetarians and is widely used in cheese production.

Plant Rennet: It is extracted from plants such as fig leaves, thistles, or nettles. While less common, vegetable rennet can be used in cheese-making and is suitable for both vegetarians and vegans.

Each type of rennet can affect the texture, flavor, and production process of cheese differently. However, they all serve the same fundamental purpose of coagulating milk to form curds, which is essential for cheese production.

Animal Rennet

Animal rennet is an extract from the inner mucosa of the fourth stomach chamber (abomasum) of a calf or other young animals. Rennet can be derived from cows, buffaloes, camels, sheep, and goats. Bovine (cow/buffalo) rennets were initially used in cheese-making due to their unique properties, including milk clotting index (MCI) at specific pH and temperature levels. Additionally, their thermolability ensured that whey products were free of active coagulant residues. Camel rennet (an acid protease) is yet another suitable milk coagulant which has been recently reported for the production of cheese with similar sensory characteristics to that by bovine rennet (Fguiri et al., 2021). Food in Tunisia and Saudi Arabia, the agriculture organisation of the United Nations promoted the use of camel milk and rennet in cheese production, with variations in starter culture and calcium content, to improve the nutritional status of African populations. Lamb and goat rennets are similar to calf rennet, but are most effective for clotting milk from their own species (El Zubeir & Jabreel, 2008; Crabbe, 2004). Rennet paste is commonly used in the production of Southern Italian and protected designation of origin (PDO) cheeses (Foltmann, 1992). The paste is made by macerating and drying the stomach of recently suckled calves, lambs, or kids. This preparation, which contains or adds lipase, gives the cheese its distinct flavour. Adult bovine rennets are not recommended for cheese making due to their high pH sensitivity and general proteolytic activity (PA) (Ben Amira et al., 2017).

5.3 Composition of Rennet

Milk-clotting enzymes are typically classified as proteolytic activity, which function best at acidic pH levels. The ratio of chymosin and pepsin in animal rennet varies based on the age and diet of the slaughtered animal. The ratio of chymosin to pepsin in calf rennet can reach 95% in unweaned bovine calves. Camel rennet had a significantly higher MCI than bovine rennet, approximately sevenfold. Despite sharing 85% sequence identity, camel chymosin has a 70% higher milk-clotting activity (MCA) towards bovine milk compared to bovine chymosin (Murlidhar et al., 2017; Liu et al., 2021). These enzymes cleave the Phe_{105}-Met_{106} bond in bovine κ-casein. Chymosin is highly specific to cleaving the caseinomacropeptide from κ-casein. This disrupts the casein micelles, leading to milk clotting (Corredig & Salvatore, 2016). Bovine rennet contains three forms of chymosin (A, B, and C) with different amino acid sequences. A fourth form of prochymosin, chymosin D, has been found to have different electrophoretic mobility than the other forms (Andren, 2002). Pepsin has lower specificity than chymosin and can hydrolyze bonds with Leu, Phe, Tyr, or Val residues. Calf rennet and adult bovine rennet both contain gastricsin, an aspartic protease that accounts for 1–5% of total enzymes (Harboe et al., 2010). Lamb and kid rennet contain lipases and proteases, which contribute to the organoleptic properties of cheese. During commercial manufacture of bovine rennet, zymogenic forms of proteases are converted to active enzymes in an activation step at acidic pH. Whereas, zymogens of goat and lamb do not require this step as they are converted to active enzymes automatically under the pH conditions of abomasa (Addis et al., 2008).

Animal rennet has high specificity for κ-casein. Rennet can hydrolyze α- and β-caseins at multiple sites, depending on its source. Lamb rennet is highly specific to α-casein, particularly α1, and causes extensive lipolysis. Kid rennet causes high proteolysis in cheese made from milk of the same species (Garg & Johri, 1994). Furthermore, unlike other animal rennets, camel rennet activity increases with age. κ-casein hydrolysis resembles that of other animal rennets. Using camel milk for cheese production resulted in faster hydrolysis of α- than β-casein. Utilization of cow milk instead of camel milk significantly reduced α-casein hydrolysis. Furthermore, flocculation time was shorter compared to other rennets. Fat recovery and cheese yield are also similar to commercial bovine rennet. Chymosin originated from camel shows unique properties and make it suitable for use in cheese production (Mbye et al., 2021). Table 5.1 summarises the characteristics of various animal rennets.

5.3.1 Genetically Modified Chymosin

Many countries now use recombinant chymosin to manufacture cheese due to the limited availability of calf stomachs and ethical concerns about animal slaughter. Furthermore, cheese consumption does not raise ethical concerns about animal rennet because chymosin is produced by genetically modified microorganisms rather than young animals. Currently, fermentation-produced chymosin (FPC) accounts ~80% of the global market. FPC production involves isolating mRNA and synthesising cDNA using reverse transcriptase (Neveu et al., 2010). cDNA is commonly used to clone eukaryotic genes in prokaryotes due to its lack of introns, unlike eukaryotic genes. Purified DNA is inserted into plasmids, which are then introduced into microbes. Yeast, bacteria, and fungi are used as hosts to express recombinant enzymes and produce chymosin on a large scale (Demain & Vaishnav, 2009; Ward, 2012). The recombinant enzyme, caprine chymosin, can cleave κ-casein and MCA at acidic pH, making it a suitable alternative milk coagulant for cheese making. For the first time, active chymosin from buffalo (*Bubalus arnee bubalis*) was expressed in *Pichia pastoris* and showed higher affinity for κ-casein than conventional

Table 5.1 Characteristics of different animal rennet

Type of rennet	Milk specificity	Molecular weight of chymosin	Casein specificity	Sensory attributes	References
Bovine	Bovine milk	35–40 kDa	κ	Soft, firm cheese	Beigomi et al. (2014)
Camel	Bovine and camel milk	40 kDa	κ	Soft, firm cheese	Mbye et al. (2021)
Lamb	Sheep milk	36 kDa	κ, α_1	Hard, elastic cheese with piquant taste and pungent flavor	Simpson et al. (2012)
Kid	Goat milk	36 kDa	κ, α & β	Semi hard cheese with distinctive flavor	Huppertz et al. (2006) and Liu et al. (2021)

chymosin (Vallejo et al., 2008). This method of producing recombinant chymosin eliminates the need for *in-vitro* activation, providing an additional source of rennet for cheese production. FPCs are derived from genetically modified microorganisms (GMOs), but are not living organisms (Blair & Regenstein, 2015; Mills, 2019).

Several studies have been used plants as hosts to express chymosin. Tobacco and potato leaves, as well as *Brassica napus* and Flax (*Linum usitatissimum*) seeds, have been used as hosts for chymosin expression. Tobacco and potato leave express chymosin at 0.1–0.5% of total soluble protein, while transgenic plant seeds accumulate at least 0.5% (w/w) of total seed protein (Murlidhar et al., 2017). While plant seeds are being researched for recombinant protein production, there are still technical challenges that need to be addressed before they can be considered a viable option for large-scale production. Recent advancements have resulted in the development of plant-derived proteins with pharmaceutical applications, including antibodies, vaccines, human blood products, and growth regulators. Large-scale production of recombinant chymosin using plants as bioreactors has yet to be investigated. The absence of bovine pepsin, which is present at variable levels in traditional calf rennet, improves cheese yield. FPC produces cheese with comparable yield and quality to calf rennet. In 1990, the FDA approved the first bioengineered enzyme, FPC expressed in *E. coli*, as generally recognised as safe (GRAS) (Nicosia et al., 2022). In *E. coli*, recombinant proteins are typically synthesised as intracellular inclusion bodies, resulting in a significant increase in process costs. *E. coli* is not widely accepted as safe for human consumption. Currently, *Aspergillus niger var. awamori* and *Kluyveromyces lactis* are used to produce bovine chymosin B, which is sold under the brands Chymax (Chr. Hansen, Denmark) and Maxiren (DSM Food Specialties, Denmark) (Annapure & Gaur, 2022).

Chr. Hansen recently released Chy-Max M, a second-generation FPC made from fermentation-produced camel chymosin. This novel coagulant has unique features, including greater milk coagulation to general proteolysis ratio in bovine milk compared to chymosin (Murlidhar et al., 2017). Recombinant chymosin from different animal source has been studied to enhance cheese production and MCI (Mohsin et al., 2023). Compared to FPC, recombinant lamb chymosin produced in *E. coli* showed superior quality and coagulation properties. The temperature instability above 45 °C may benefit the production of hard cheeses. Caprine (*Capra hircus*) preprochymosin was cloned, characterised, and expressed in *K. lactis* and *S. cerevisiae* (Vega-Hernández et al., 2004). After activation at acidic pH, both yeast transformants' culture supernatants demonstrated κ-casein and MCA cleavage, indicating caprine chymosin as an alternative milk coagulant in cheese manufacturing. Using active chymosin from buffalo (*Bubalus arnee bubalis*) in *Pichia pastoris* for the first time showed better κ-casein affinity than previous approaches. The recombinant chymosin produced does not need in vitro activation, providing another rennet source for cheese production.

5.3.2 Microbial Rennet

Microbial rennets derived from microbial fermentation, have emerged as valuable alternatives to traditional animal rennets in cheese making. Consumer acceptance of cheeses crafted with microbial rennets remains favourable, as such products are commonly categorized as "vegetarian" cheeses. Microorganisms often develop additional cellular proteases that exhibit MCA. These sources of milk coagulant are attractive due to their low production costs, biochemical variety, and facile genetic modification. However, microbial coagulants are associated with high proteolytic activity that leads to low yield and bitter taste of cheese produced. Microbial rennets are derived from bacteria and fungi. Bacterial rennets are derived from bacteria (*Bacillus subtilis*), while fungal rennets are acquired from fungi, (*Rhizomucor miehei* and *Rhizomucor pusillus*). Fungal sources are ideal for producing milk-clotting proteases. Several studies have shown that bacterial species can produce milk-clotting proteases. Bacterial species such as *Bacillus subtilis natto*, *B. amyloliquefaciens* JNU002, *B. amyloliquefaciens* D4, *B. subtalis*, *B. licheniformis*, *Nocardiopsis sp*, *Enterococcus faecalis*, and *Myxococcus xanthus* strain 422.85 are being studied for their possible milk-clotting properties (Ding et al., 2012; Feijoo-Siota et al., 2014). Fungal rennet is produced through surface or submerged culture fermentation, followed by filtration to remove any non-proteolytic components. These enzymes range in molecular weight from 30 to 45 kDa and have one or two conserved aspartic acid residues at their active site. *Rhizomucor miehei* generates proteases with similar structure to bovine chymosin (Wang et al., 2021). These proteases have been produced under various settings, including media composition and reactor structure.

Large-scale production of microbial rennet (MR) has been produced using *R. miehei*, *R. pusillus*, and *Cryphonectria parasitica*. Fungal proteases' high thermostability allows them to remain intact throughout cheese maturation, resulting in changes in texture and flavour (Ardö et al., 2017; Ramani et al., 2023). They are suitable for making Emmental and Italian-style cheese due to their high processing temperatures. *Rhizomucor* species heat stability has been reduced using novel means, including chemical usage and genetic changes (Trindade et al., 2016). Methionine oxidation with hydrogen peroxide or dye-sensitized photo-oxidation decreased the thermal stability of MR *from Rhizomucor* species but did not significantly reduce their MCA. A mutagenesis technique has produced fungal MR with lower thermostability (Garg & Johri, 1994). *R. pusillus* single and double amino acid mutations for MR showed decreased thermostability while maintaining enzymatic activity and MCI (Murlidhar et al., 2017). Mutations at specific amino acid sites in *R. miehei* resulted in greater MCI and increased cheese output. Acylation of MR from *R. pusillus* with different acid anhydrides improved its MCA and MCI, both independently and in combination. Commercially available microbial (Table 5.2) rennets include fromase, hannilase, marzyme, meito rennet super (MRS), milase XQL/TQL/premium, enzymaks, reniplus, microclerici (*R. miehei*), suparen and thermolase (*C. parasitica*), and meito microbial rennet (*R. pusillus/*

Table 5.2 Commercial fungal rennet

Microorganisms	Trade name
Rhizomucormiehei	Hannilase, Rennilase, Fromase, Miki, Marzyme
Rhizomucorpusillus	Noury, Meito, Emporase, Novadel
Endothiaparasitica	Suparen, Surecurd
Cryphonecteriaparasitica	Suparen, Surecurd

miehei) (Feijoo-Siota et al., 2014). Milk coagulants can be arranged in the order of increasing specificity: *C. parasitica* transforms into *R. miehei*, which produces bovine pepsin and calf chymosin, respectively (Feijoo-Siota et al., 2014).

5.3.3 *Plant Rennet*

Plant-based rennet can be used in cheese production and shows potentiality towards replacement of animal derived rennet. The first evidence of plant-based milk coagulants to prepare cheese was during 5000 BC. The use of plant coagulants in cheese production improves nutritional intake and increases acceptance among vegetarians. Due to high PA, many of them have extremely low MCI, which results in cheese with pasty body that is unsuitable for commercial manufacturing of cheese. It also shows extensive curd digestion, off flavour development, and poor taste (Sousa et al., 2001). In West African and Southern European regions, plant coagulants were often employed in traditional cheese making (Ben Amira et al., 2017; Nyamakwere et al., 2021; Rout et al., 2023). The largest cheese variants, made using plant coagulant (aqueous extracts of *Cynara sp.*), are found in Spain and Portugal. Despite having a high PA, *Cynara sp.* has found success in the market as a rennin replacement. *Cynara sp* derived cheese varieties comprise Serra, Serpa, Los Pedroches, La Serena, Torta del Casar, Los Ibores, and Flor de Guía (milk of ewe and cow) (Shah et al., 2014). In rural Sudan, dairy farmers make white soft cheese from *Solanum dubium* berries using goat and sheep milk. Traditional cheese making in West African countries like Nigeria and Benin uses leaf extracts from *Calotropis procera* (Sodom apple), resulting in a soft, creamy texture and a fine flavour that can become bitter or piquant with ripening (Di Pierro, 2013; Murlidhar et al., 2017). Cheese production has been explored using papaya, pineapple, fig, rice seeds, and sodom apple (procerain) in various regions. Novel plant-based milk-clotting enzymes are being sought to meet the growing global demand for diverse, high-quality cheese.

Furthermore, the coagulant derived from the flowers of cardoon (*Cynara cardunculus*) has been identified as a superior alternative to traditional rennet in specific cheese types, including Serra cheese from sheep's milk, as well as Edam and Roquefort cheeses. However, its high proteolytic activity may lead to challenges such as yield loss, flavor defects (including acidity and bitterness), and poor texture in the final cheese product.

5.4 Rennet Coagulation

Proteolytic enzymes perform enzymatic coagulation in order to turn liquid milk into gel. The coagulant used for milk coagulation was formerly calf rennet. Rennet contains two active fractions: pepsin and chymosin, which together account for 95% of the enzyme activity. The primary function of chymosin is to hydrolyse κ-casein. It has two functions, firstly it exhibits strong affinity for κ-casein that is found at the edge of casein micelles, which causes micelle destabilisation during the coagulation phase and secondly it has a mild broad proteolysis activity on different casein fractions that mostly happens when cheese ripens. Rennet's proteolytic activity is significantly impacted by environmental variables that impact the substrate's condition as well as its surroundings. The ideal pH for rennet coagulant activity on milk is about 5.5 (Troch et al., 2017).

The ideal temperature range for rennet activity is 40–42 °C; activity decreases below 20 °C, and activity completely stops over 65 °C. The amount of Ca^{2+} ions present in milk is a major factor in regulating rennet activity, especially the ratio of calcium to nitrogen. "Slow" milk clotting happens at a ratio of 0.20, while normal clotting is observed at a ratio of 0.23. The coagulation properties of milk can be changed during temperature alteration that occurs before rennet is added. When milk cools, casein solubilization takes place, increasing micelle stability and thus lowering rennet activity. On the other hand, mild heating (≤60 °C) causes alterations in salt equilibrium, raises the size of insoluble phosphate colloidal forms, and increases micelle size, all of which decrease stability of milk proteins and promote rennet activity (Dejmek & Walstra, 2004; Fox & McSweeney, 2017; Troch et al., 2017).

In addition, even though rennet is widely used, there are other coagulants that come from vegetable, animal, or microbiological sources. Animal-derived coagulants, like trypsin and chymotrypsin, are less preferred since they have a big effect on production procedures and the quality of the final product. Compared to chymosin, swine pepsin has a more acidic profile and is most active in acidic environments, declining sharply above pH 6.3. At the pH of fresh milk, coagulation does not take place. Compared to porcine pepsin, bovine pepsin exhibits reduced pH dependence and clotting behaviour that is very comparable to rennet. Furthermore, chicken pepsin can be used as a coagulant. Additionally, a variety of plant body parts, including cleavers, artichokes, thistles, figs, papayas, and pineapples, can be macerated to produce coagulant preparations. However, the resulting coagulant activity varies greatly, depending on variables including plant maturity, conditions during collecting, and conditions during storage. As a result, the application of these protease coagulants is restricted to specific producing regions (Broome & Limsowtin, 1998).

Coagulation usually occurs in two stages: an enzymatic phase in which chymosin mediates the degradation of κ-casein, and a non-enzymatic phase in which the disintegrated micelles aggregate (Frederiksen et al., 2011). Micelle destabilisation yields from the hydrolysis of C-terminal region in κ-casein, which is catalysed by chymosin during the enzymatic phase. Para-κ-casein (N-terminal portion of κ-casein) is produced via enzyme hydrolysis and stays attached to the micelles, whereas glycomacropeptide (the C-terminal portion) is generally lost into the whey stream. Glycomacropeptide greatly lowers the electrostatic potential of casein micelles, which weakens the electrostatic repulsions that exist between them as it its strongly charged. Hydrolysis also removes steric repulsion from the hydration layer, which is one of casein's stabilising factors. In the second stage of coagulation, these para-casein molecules cluster because they lack stabilising components (Walstra et al., 2006). As soon as the electrostatic, steric, and hydration-related repulsive forces that maintain colloidal stability are sufficiently reduced, proximal or neighbouring micelles begin to assemble. These conditions are typically attained with a degree of κ-casein hydrolysis of approximately 80% at pH 6.6 (Troch et al., 2017). The coagulum, also called the gel, is a three-dimensional network made up of the destabilised micelles that ensnares fat and whey globules. Due to the intrinsic instability of this gel structure, the liquid phase of the curd is expelled when the micelle contracts. Synaeresis is the term used to describe this procedure that separates the fat and casein-containing curd from the whey, which also contains soluble proteins minerals and lactose (Lapointe-Vignola, 2002).

5.5 Role of Added Calcium in Cheese Making

The addition of calcium plays a crucial role in cheese making, influencing various stages of the process and ultimately impacting the quality and characteristics of the final product. Calcium contributes to the coagulation of milk proteins, which is a fundamental step in cheese making. During coagulation, calcium ions interact with casein proteins, particularly κ-casein, leading to the formation of a gel-like structure known as curd. This process is facilitated by rennet enzymes, which cleave κ-casein molecules and release calcium ions, promoting the aggregation of casein micelles and the formation of a stable curd matrix (Janhøj & Qvist, 2010).

Furthermore, calcium influences the texture, body, and firmness of cheese. Adequate calcium levels contribute to the development of a firm and cohesive curd, essential for proper whey expulsion and curd formation. Insufficient calcium may result in weak or fragile curds that are prone to breaking or disintegrating during processing, leading to lower cheese yield and compromised texture. Additionally, calcium affects the interactions between casein proteins and other components in milk, such as fat globules and whey proteins, influencing the overall structure and mouthfeel of the cheese (da Silva et al., 2020; Deshwal et al., 2023).

Moreover, calcium influences the ripening and flavor development of cheese. Calcium ions play a crucial role in enzymatic reactions during cheese ripening, influencing the activity of proteolytic enzymes and contributing to the breakdown of proteins into smaller peptides and amino acids. This proteolysis process is essential for the development of cheese flavor and texture, as it releases volatile compounds responsible for the characteristic aroma and taste of cheese. Additionally, calcium ions may interact with flavor compounds generated during ripening, influencing their distribution and perception in the final product. Therefore, the addition of calcium during cheese making not only affects the physical properties of cheese but also plays a significant role in flavor development and maturation (McSweeney, 2007).

Hence, the addition of calcium is essential for various aspects of cheese making, including coagulation, texture development, and flavor formation. Adequate calcium levels contribute to the formation of a stable curd structure, ensuring proper whey expulsion and curd formation. Furthermore, calcium influences the texture, body, and firmness of cheese, impacting its overall quality and consumer acceptance.

5.6 Effect of Coagulation on Rheology and Texture

5.6.1 Rheological Attributes

The rheological characteristics of rennet-induced gels, including their small and large deformation behaviour, were significantly influenced by the processing factors, such as gelation pH, temperature, and MSNF (milk solids-not-fat). The addition of rennet increased bond number and strength with time, resulting in an increase in yield stress (δy) and dynamic moduli (storage modulus-G′ and loss modulus-G″). On the other hand, the yield strain (γy) reduced over time, most likely because of shorter texture caused by network rearrangements. The power law exponents of rennet-induced gels were found to be similar for both small and large deformation properties. This observation could indicate the kind and composition of bonds found in rennet gels. Reduced pH led to higher G′, G″, and δy values because protein strands showed more fusing at lower pH levels because casein micelle instability was increased and particle repulsion was decreased. Lower pH levels caused a decrease in gelation time because they reduced electrostatic repulsion and increased rennet activity. Loss Tangent increased as pH dropped, suggesting that there are more rearrangements and flexibility at lower pH values. Higher temperatures resulted in higher G′, G″, and δy values, which were explained by increased casein micelle fusion and accelerated rennet activity. On the other hand, δ_y showed a maximum temperature beyond which it declined, probably because bond breakdown happened faster as the temperature increased. Temperature-dependent decrease in gelation time was caused by increased rennet activity and a faster rate of thermal collisions between casein micelles (Mishra et al., 2005).

5.6.2 *Textural Attributes*

Gel structure is mostly shaped by casein micelles (CM). In the progression of CM coagulation, the initially established CM network experiences structural changes that cause gel shrinkage in later stages of the process due to the restructuring of interparticle contacts (Risso et al., 2007). These rearrangements of interparticle interactions determine the level of gel compactness. These rearrangements are incomplete when gel formation happens quickly, resulting in less compact gels. On the other hand, slower gelation processes result in a denser network with ever smaller holes by disrupting preexisting contacts and creating new ones (Ingrassia et al., 2013).

Gel permeability is associated with liquid release during deformation and water-holding capacity (WHC), while gel hardness or firmness and WHC are closely related to protein network connection (Pascua et al., 2013). Greater pore diameters in whey protein gels result in lower WHC and higher gel permeability (Verheul & Roefs, 1998; Galante et al., 2017). Optimising gel firmness at cutting is essential in cheese manufacturing to reduce the amount of protein and fat that is lost from curd particles into the whey. An excessive amount of coagulum softness causes extensive fracturing, which increases the amount of fat and protein lost in the whey. On the other hand, excessively rigid coagulum could present difficulties when cutting with traditional tools (Fox et al., 2017).

5.7 Factors Affecting Rennet Coagulation

The composition of milk, particularly the amounts of casein, calcium, and pH, has a significant impact on every aspect of curd development, including enzymatic coagulation, curd hardness, and synaeresis of whey (Fox & McSweeney, 2004).

The composition and yield of cheese are greatly affected by the composition of milk. The performance of milk coagulation is significantly influenced by the quantity of fat and casein. The amount of fat in the dry matter of the cheese product is determined by the fat to casein ratio. Additionally, the water content of the finished product is impacted by this ratio's influence on synaeresis (Walstra et al., 2006). The amount of casein in milk also affects the rate at which curd develops; the concentration of casein enhances gel strength. However, there is only a slight increase in the curd's setting time when the casein level increases (Lapointe-Vignola, 2002). Furthermore, a critical factor in milk coagulation is the size of casein micelles; smaller micelles result in the development of firmer curds (Glantz et al., 2010). Genetic lactoprotein polymorphisms can affect milk coagulation characteristics (MCPs) in addition to the overall casein content. For every protein in milk, there are several genetic variations that have been found, and correlations between these variations and coagulation properties have been demonstrated. According to various research, κ-casein variant B improves coagulation (Walsh et al., 1998; Patil et al.,

2003), on the other hand, variation E adversely affects coagulation (Wedholm et al., 2006). Nevertheless, these investigations fail to specify whether the correlation between protein genotype and MCPs might be attributed to unique structural characteristics of individual protein variants or modifications in protein composition (Bonfatti et al., 2010).

The κ-casein genotype affects the amount of casein in milk as well as its proportion in relation to other caseins (Wedholm et al., 2006; Heck et al., 2009). Additionally, the genotypes of β-lactoglobulin and β-casein influence the protein content of milk (Heck et al., 2009). Particularly, β-lactoglobulin variant B expresses itself at a lower level than variant A, which leads to a larger concentration of casein (Mayer et al., 1997; Hallén et al., 2008) and improved coagulation properties (Caroli et al., 2009). The β-casein variant B had the quickest clotting time of all the variants found and more favorable properties that aid in successful coagulation (Buchberger & Dovč, 2000).

The maximal hardness, the rate at which gel firms, and the clotting time are all strongly influenced by milk pH. At pH values between 6.0 and 6.7, coagulation proceeds more quickly, producing a gel with greater strength and stiffness (Remeuf et al., 1991). However, Troch et al. (2017) reported that the ideal pH for rennet activity is roughly 5.1–5.5, with rennet's effectiveness increasing as pH decreases.

It is important to recognise the importance of titratable acidity at every stage of the coagulation process. This process includes rennet reactivity and the rate at which para-casein micelles aggregate. Titratable acidity also affects the rate of synaeresis and establishes whether milk is suitable for cheese production (De Marchi et al., 2009). The amount of lactic acid that results from bacterial activity during milk collection and transportation, as well as the buffering capacity of the milk represented the titratable acidity of fresh milk. The standard unit of measurement for titratable acidity is the degree of Dornic (°D), where one Dornic degree is equivalent to 0.01% lactic acid. Titratable acidity in normal milk falls between 16 and 18 °D while late lactation or mastitis milk had less than 15 °D. Contrarily, milk that has only begun to lactate registers between 19 and 20 °D. Titratable acidity of milk above 20 °D can cause coagulation during sterilisation, while over 22 °D can cause coagulation during pasteurisation. So that, these situations present technological hurdles for processing of milk (Colinet et al., 2013). The amount of calcium in milk has an important effect on the gel's stiffness and coagulation time. The aqueous phase's calcium to phosphate and citrate ratio increases the rate of coagulation, and a significant relationship between gel strength and calcium content is detected (Remeuf et al., 1991).

Temperature also play a crucial role on milk. The enzyme is still active at temperatures below 10 °C, however gelation does not take place. Coagulation moves slowly between 10 and 20 °C, and clotting gradually occurs between 30 and 42 °C. Over 42 °C, coagulation decreases and stops completely around 55 °C. Temperature has a very small impact on the enzymatic action process and it is correlated with the enzyme's coefficient of diffusion to its site of action (Walstra et al., 2006). Additionally, pre-cooling of milk for a few days before to processing has an impact on gelation because colloidal calcium phosphate dissolves at low temperatures and disrupts micelles. Partially reversing these effects requires bringing the milk back to renneting temperature (Troch et al., 2017).

Another factor influencing the coagulation characteristics of milk is the somatic cell count (SCC) (Vallas et al., 2010). Reduced SCC levels have been associated to improved milk coagulation properties. Protein incorporation into curd may be difficult in milk with increased SCC as per Politis and Ng-Kwai-Hang (1988). They also established a negative relationship between total solids, fat, and protein content of cheese and its SCC levels.

5.8 Conclusion

In summary, the coagulation process, which involves the enzymatic conversion of liquid milk into a gel-like structure, is a crucial step in the production of cheese. Casein micelles aggregate as a result of this process, which is mostly carried out by rennet enzymes such chymosin, and a cohesive curd is formed. A number of variables, such as pH, temperature, calcium concentration, and genetic variations in milk proteins, affect the rate and degree of coagulation, which in turn shapes the properties of the finished cheese.

Cheese producers must optimise coagulation parameters in order to get the appropriate cheese flavour, texture, and functioning. The coagulum's rheological characteristics, such as gel hardness and water-holding ability, are important factors in regulating the yield and quality of cheese. Developing ways to control and manipulate the coagulation process in order to make cheeses with consistent quality and desirable qualities requires an understanding of the enzymatic mechanisms behind milk coagulation as well as the factors influencing coagulation properties. Clarifying the molecular mechanisms of rennet action, studying novel coagulants, and examining the effects of genetic and environmental factors on milk coagulation properties are potential areas of future research in this topic.

References

Addis, M., Piredda, G., & Pirisi, A. (2008). The use of lamb rennet paste in traditional sheep milk cheese production. *Small Ruminant Research, 79*(1), 2–10.

Andren, A. (2002). Rennets and coagulants. In *Encyclopedia of dairy sciences* (Vol. 1, pp. 283–286). Academic Press.

Annapure, U. S., & Gaur, S. S. (2022). Commercial enzymes in dairy processing. In *Value-addition in food products and processing through enzyme technology* (pp. 205–219). Academic Press.

Ardö, Y., McSweeney, P. L., Magboul, A. A., Upadhyay, V. K., & Fox, P. F. (2017). Biochemistry of cheese ripening: Proteolysis. In *Cheese* (pp. 445–482). Academic Press.

Beigomi, M., Mohammadifar, M. A., Hashemi, M., Rohani, M. G., Senthil, K., & Valizadeh, M. (2014). Biochemical and rheological characterization of a protease from fruits of Withania coagulans with a milk-clotting activity. *Food Science and Biotechnology, 23*, 1805–1813.

Ben Amira, A., Besbes, S., Attia, H., & Blecker, C. (2017). Milk-clotting properties of plant rennets and their enzymatic, rheological, and sensory role in cheese making: A review. *International Journal of Food Properties, 20*(sup1), S76–S93.

Blair, R., & Regenstein, J. M. (2015). *Genetic modification and food quality: A down to earth analysis*. Wiley.

Bonfatti, V., Di Martino, G., Cecchinato, A., Degano, L., & Carnier, P. (2010). Effects of β-κ-casein (CSN2-CSN3) haplotypes, β-lactoglobulin (BLG) genotypes, and detailed protein composition on coagulation properties of individual milk of Simmental cows. *Journal of Dairy Science, 93*(8), 3809–3817.

Broome, M. C., & Limsowtin, G. K. Y. (1998). Milk coagulants. *Australian Journal of Dairy Technology, 53*(3), 188–190.

Buchberger, J., & Dovč, P. (2000). Lactoprotein genetic variants in cattle and cheese making ability. *Food Technology and Biotechnology, 38*(2), 1–10.

Caroli, A. M., Chessa, S., & Erhardt, G. J. (2009). Invited review: Milk protein polymorphisms in cattle: Effect on animal breeding and human nutrition. *Journal of Dairy Science, 92*(11), 5335–5352.

Colinet, F., Troch, T., Vanden Bossche, S., Soyeurt, H., Gengler, N., Abbas, O., et al. (2013). Etude de la variabilité des aptitudes à la transformation laitière en Région wallonne basée sur l'utilisation de la spectrométrie infrarouge. In *18ième Carrefour des Productions animales: Nouvelles approches pour une optimisation de nos élevages laitiers*. Centre wallon de Recherches agronomiques.

Corredig, M., & Salvatore, E. (2016). Enzymatic coagulation of milk. In *Advanced dairy chemistry: Volume 1B: Proteins: Applied aspects* (pp. 287–307). Springer.

Crabbe, M. J. C. (2004). Rennets: General and molecular aspects. In *Cheese: Chemistry, physics and microbiology* (Vol. 1, pp. 19–45). Elsevier.

da Silva, V. B., da Bahia, U. F., Barreto Bezerra, L. S. M., de Oliveira Alves, B. P., & dos Santos, B. C. (2020). Rheology and texture applied to goat and sheep milk cheeses: Principles, applications and future trends. In *Avanços em Ciência e Tecnologia de Alimentos* (Vol. 2, pp. 409–436). Editora Científica Digital.

De Marchi, M., Fagan, C. C., O'Donnell, C. P., Cecchinato, A., Dal Zotto, R., Cassandro, M., et al. (2009). Prediction of coagulation properties, titratable acidity, and pH of bovine milk using mid-infrared spectroscopy. *Journal of Dairy Science, 92*(1), 423–432.

Dejmek, P., & Walstra, P. (2004). The syneresis of rennet-coagulated curd. In *Cheese: Chemistry, physics and microbiology* (Vol. 1, pp. 71–103). Elsevier.

Demain, A. L., & Vaishnav, P. (2009). Production of recombinant proteins by microbes and higher organisms. *Biotechnology Advances, 27*(3), 297–306.

Deshwal, G. K., Gómez-Mascaraque, L. G., Fenelon, M., & Huppertz, T. (2023). A review on the effect of calcium sequestering salts on casein micelles: From model milk protein systems to processed cheese. *Molecules, 28*(5), 2085.

Di Pierro, G. (2013). Enzymatic characterization of two plant coagulants: Cynara cardunculus L. and Ficus carica L. Doctoral Disseration. University of Padua.

Ding, Z., Wang, W., Wang, B., Ouyang, A., Xiao, S., Wang, Y., et al. (2012). Production and characterization of milk-clotting enzyme from Bacillus amyloliquefaciens JNU002 by submerged fermentation. *European Food Research and Technology, 234*, 415–421.

El Zubeir, I. E., & Jabreel, S. O. (2008). Fresh cheese from camel milk coagulated with Camifloc. *International Journal of Dairy Technology, 61*(1), 90–95.

Feijoo-Siota, L., Blasco, L., Luis Rodriguez-Rama, J., Barros-Velázquez, J., de Miguel, T., Sánchez-Pérez, A., Villa, G., & T. (2014). Recent patents on microbial proteases for the dairy industry. *Recent Advances in DNA & Gene Sequences (Formerly Recent Patents on DNA & Gene Sequences), 8*(1), 44–55.

Fguiri, I., Atigui, M., Sboui, A., Samira, A., Marzougui, C., Dbara, M., et al. (2021). Camel milk-clotting using plant extracts as a substitute to commercial rennet. *Journal of Chemistry, 2021*, 1–8.

Foltmann, B. (1992). Chymosin: A short review on foetal and neonatal gastric proteases. *Scandinavian Journal of Clinical and Laboratory Investigation, 52*(sup210), 65–79.

Fox, P.F., Guinee, T.P., Cogan, T.M., McSweeney, P.L.H. (2017). Starter Cultures. In: Fundamentals of Cheese Science. Springer, Boston, MA. https://doi.org/10.1007/978-1-4899-7681-9_6

Fox, P. F., & McSweeney, P. L. H. (2004). *Cheese: Chemistry, physics and microbiology. Vol. 1. General aspects*. Elsevier Academic Press.

Fox, P. F., & McSweeney, P. L. H. (2017). Cheese: An overview. In P. L. H. McSweeney, P. F. Fox, P. D. Cotter, & D. W. Everett (Eds.), *Cheese: Chemistry, physics and microbiology; general aspects* (4th ed., pp. 5–21). Elsevier.

Frederiksen, P. D., Andersen, K. K., Hammershøj, M., Poulsen, H. D., Sørensen, J., Bakman, M., et al. (2011). Composition and effect of blending of noncoagulating, poorly coagulating, and well-coagulating bovine milk from individual Danish Holstein cows. *Journal of Dairy Science, 94*(10), 4787–4799.

Galante, M., Boeris, V., Álvarez, E., & Risso, P. (2017). Microstructural and textural properties of rennet-induced milk protein gel: Effect of guar gum. *International Journal of Food Properties, 20*(sup3), S2569–S2578.

Garg, S. K., & Johri, B. N. (1994). Rennet: Current trends and future research. *Food Reviews International, 10*(3), 313–355.

Glantz, M., Devold, T. G., Vegarud, G. E., Månsson, H. L., Stålhammar, H., & Paulsson, M. (2010). Importance of casein micelle size and milk composition for milk gelation. *Journal of Dairy Science, 93*(4), 1444–1451.

Guinee, T. P., Fenelon, M. A., Mulholland, E. O., O'Kennedy, B. T., O'Brien, N., & Reville, W. J. (1998). The influence of milk pasteurization temperature and pH at curd milling on the composition, texture and maturation of reduced fat cheddar cheese. *International Journal of Dairy Technology, 51*, 1–10.

Hallén, E., Wedholm, A., Andrén, A., & Lundén, A. (2008). Effect of β-casein, κ-casein and β-lactoglobulin genotypes on concentration of milk protein variants. *Journal of Animal Breeding and Genetics, 125*(2), 119–129.

Harboe, M., Broe, M. L., & Qvist, K. B. (2010). The production, action and application of rennet and coagulants. In *Technology of cheesemaking* (pp. 98–129). Wiley.

Heck, J. M. L., Schennink, A., Van Valenberg, H. J. F., Bovenhuis, H., Visker, M. H. P. W., Van Arendonk, J. A. M., & Van Hooijdonk, A. C. M. (2009). Effects of milk protein variants on the protein composition of bovine milk. *Journal of Dairy Science, 92*(3), 1192–1202.

Hinrichs, J. (2001). Incorporation of whey proteins in cheese. *International Dairy Journal, 11*, 495–503.

Huppertz, T., Upadhyay, V. K., Kelly, A. L., & Tamime, A. Y. (2006). Constituents and properties of milk from different species. In *Brined cheeses* (pp. 1–42). Wiley.

Ingrassia, R., Costa, J. P., Hidalgo, M. E., Canales, M. M., Castellini, H., Riquelme, B., & Risso, P. (2013). Application of a digital image procedure to evaluate microstructure of caseinate and soy protein acid gels. *LWT - Food Science and Technology, 53*(1), 120–127.

Janhøj, T., & Qvist, K. B. (2010). The formation of cheese curd. In *Technology of cheesemaking* (Vol. 14, p. 130). Wiley.

Kethireddipalli, P., & Hill, A. R. (2015). Rennet coagulation and cheesemaking properties of thermally processed milk: Overview and recent developments. *Journal of Agricultural and Food Chemistry, 63*, 9389–9403.

Lapointe-Vignola, C. (2002). *Science et technologie du lait: transformation du lait*. Presses inter Polytechnique.

Liu, X., Wu, Y., Guan, R., Jia, G., Ma, Y., & Zhang, Y. (2021). Advances in research on calf rennet substitutes and their effects on cheese quality. *Food Research International, 149*, 110704.

Mayer, H. K., Ortner, M., Tschager, E., & Ginzinger, W. (1997). Composite milk protein phenotypes in relation to composition and cheesemaking properties of milk. *International Dairy Journal, 7*(5), 305–310.

Mbye, M., Mohamed, H., Raziq, A., & Kamal-Eldin, A. (2021). The effects of camel chymosin and Withania coagulans extract on camel and bovine milk cheeses. *Scientific Reports, 11*(1), 13573.
McSweeney, P. L. H. (2007). Cheese manufacture and ripening and their influence on cheese flavour. In *Improving the flavour of cheese* (Woodhead publishing series in food science, technology and nutrition) (pp. 1–25). Woodhead.
Mills, A. (2019). *Genetically engineered foods*. Scientific e-Resources.
Mishra, R., Govindasamy-Lucey, S., & Lucey, J. A. (2005). Rheological properties of rennet-induced gels during the coagulation and cutting process: Impact of processing conditions. *Journal of Texture Studies, 36*(2), 190–212.
Mistry, V. V., & Maubois, J.-L. (2017). Application of membrane separation technology to cheese production. In P. L. H. McSweeney, P. F. Fox, P. D. Cotter, & D. W. Everett (Eds.), *Cheese: Chemistry, physics and microbiology; cheese technology and major cheese groups* (4th ed., pp. 677–697). Elsevier.
Mohanty, A. K., & Behare, P. (2017). Enzymes in dairy processing. In *Microbial enzyme technology in food applications* (pp. 347–359). CRC Press.
Mohsin, A. Z., Norsah, E., Marzlan, A. A., Abd Rahim, M. H., & Hussin, A. S. M. (2023). Exploring the applications of plant-based coagulants in cheese production: A review. *International Dairy Journal, 108*, 105792.
Murlidhar, M., Anusha, R., & Bindhu, O. S. (2017). Plant-based coagulants in cheese making. In *Dairy engineering* (pp. 3–35). Apple Academic Press.
Neveu, C., Charvet, C. L., Fauvin, A., Cortet, J., Beech, R. N., & Cabaret, J. (2010). Genetic diversity of levamisole receptor subunits in parasitic nematode species and abbreviated transcripts associated with resistance. *Pharmacogenetics and Genomics, 20*(7), 414–425.
Nicosia, F. D., Puglisi, I., Pino, A., Caggia, C., & Randazzo, C. L. (2022). Plant milk-clotting enzymes for cheesemaking. *Foods, 11*(6), 871.
Nyamakwere, F., Esposito, G., Dzama, K., & Raffrenato, E. (2021). A review of artisanal cheese making: An African perspective. *South African Journal of Animal Science, 51*(3), 296–309.
Pascua, Y., Koç, H., & Foegeding, E. A. (2013). Food structure: Roles of mechanical properties and oral processing in determining sensory texture of soft materials. *Current Opinion in Colloid & Interface Science, 18*(4), 324–333.
Patil, M. R., Borkhatriya, V. N., Boghra, V. R., & Sharma, R. S. (2003). Effect of bovine milk κ-casein genetic polymorphs on curd characteristics during cheddar cheese manufacture. *Journal of Food Science and Technology, 40*(6), 582–586.
Perreault, V., Remillard, N., Chabot, D., Morin, P., Pouliot, Y., & Britten, M. (2017). Effect of denatured whey protein concentrate and its fractions on cheese composition and rheological properties. *Journal of Dairy Science, 100*, 5139–5152.
Politis, I., & Ng-Kwai-Hang, K. F. (1988). Effects of somatic cell count and milk composition on cheese composition and cheese making efficiency. *Journal of Dairy Science, 71*(7), 1711–1719.
Ramani, A., Manik, S., Hazra, T., Barak, S., & Mudgil, D. (2023). Whey: As a fermentation substrate for the production of exopolysaccharides. In *Whey valorization: Innovations, technological advancements and sustainable exploitation* (pp. 167–186). Singapore.
Remeuf, F., Cossin, V., Dervin, C., Lenoir, J., & Tomassone, R. (1991). Relations entre les caractères physico-chimiques des laits et leur aptitude fromagère. *Le Lait, 71*(4), 397–421.
Risso, P. H., Relling, V. M., Armesto, M. S., Pires, M. S., & Gatti, C. A. (2007). Effect of size, proteic composition, and heat treatment on the colloidal stability of proteolyzed bovine casein micelles. *Colloid and Polymer Science, 285*, 809–817.
Rout, P., Ramani, A., Bhattacharjee, S., & Rajput, V. D. (2023). Micro-and nanoparticle of chitosan for vitamin encapsulation: A nutshell overview. In *Nano-biofortification for human and environmental health* (pp. 187–210). Springer.
Schenkel, P., Samudrala, R., & Hinrichs, J. (2011). Fat-reduced semi-hard cheese enriched with a microparticulated whey protein concentrate: Impact on cheese-making properties and rheological characteristics. *Milchwissenschaft, 66*, 43–47.

Shah, M. A., Mir, S. A., & Paray, M. A. (2014). Plant proteases as milk-clotting enzymes in cheesemaking: A review. *Dairy Science & Technology, 94*, 5–16.

Simpson, B. K., Rui, X., & Klomklao, S. (2012). Enzymes in food processing. In *Food biochemistry and food processing* (pp. 181–206). Wiley.

Singh, H., & Waungana, A. (2001). Influence of heat treatment of milk on cheesemaking properties. *International Dairy Journal, 11*, 543–551.

Sousa, M. J., Ardö, Y., & McSweeney, P. L. H. (2001). Advances in the study of proteolysis during cheese ripening. *International Dairy Journal, 11*(4–7), 327–345.

Trindade, L. V., Desagiacomo, C., Polizeli, M. D. L. T. D. M., Damasio, A. R. D. L., Lima, A. M. F., Gomes, E., & Bonilla-Rodriguez, G. O. (2016). Biochemical characterization, thermal stability, and partial sequence of a novel exo-polygalacturonase from the thermophilic fungus Rhizomucor pusillus A13.36 obtained by submerged cultivation. *BioMed Research International, 2016*, 8653583.

Troch, T., Lefébure, É., Baeten, V., Colinet, F., Gengler, N., & Sindic, M. (2017). *Cow milk coagulation: Process description, variation factors and evaluation methodologies* (pp. 21–287). A review.

Vallas, M., Bovenhuis, H., Kaart, T., Pärna, K., Kiiman, H., & Pärna, E. (2010). Genetic parameters for milk coagulation properties in Estonian Holstein cows. *Journal of Dairy Science, 93*(8), 3789–3796.

Vallejo, J. A., Ageitos, J. M., Poza, M., & Villa, T. G. (2008). Cloning and expression of buffalo active chymosin in Pichia pastoris. *Journal of Agricultural and Food Chemistry, 56*(22), 10606–10610.

Vega-Hernández, M. C., Gómez-Coello, A., Villar, J., & Claverie-Martín, F. (2004). Molecular cloning and expression in yeast of caprine prochymosin. *Journal of Biotechnology, 114*(1–2), 69–79.

Verheul, M., & Roefs, S. P. F. M. (1998). Structure of whey protein gels, studied by permeability, scanning electron microscopy and rheology. *Food Hydrocolloids, 12*(1), 17–24.

Walsh, C. D., Guinee, T. P., Reville, W. D., Harrington, D., Murphy, J. J., O'Kennedy, B. T., & FitzGerald, R. J. (1998). Influence of κ-casein genetic variant on rennet gel microstructure, cheddar cheesemaking properties and casein micelle size. *International Dairy Journal, 8*(8), 707–714.

Walstra, P., Wouters, J. T., & Geurts, T. J. (2006). *Dairy science and technology*. Taylor & Francis Group.

Wang, S., Zhang, P., Xue, Y., Yan, Q., Li, X., & Jiang, Z. (2021). Characterization of a novel aspartic protease from Rhizomucor miehei expressed in Aspergillus niger and its application in production of ACE-inhibitory peptides. *Food, 10*(12), 2949.

Ward, O. P. (2012). Production of recombinant proteins by filamentous fungi. *Biotechnology Advances, 30*(5), 1119–1139.

Wedholm, A., Larsen, L. B., Lindmark-Månsson, H., Karlsson, A. H., & Andrén, A. (2006). Effect of protein composition on the cheese-making properties of milk from individual dairy cows. *Journal of Dairy Science, 89*(9), 3296–3305.

Chapter 6
Accelerated Cheese Ripening

Ronit Mandal, Partha Pratim Debnath, Payal Karmakar, Rashmi Kumari, Anindita Debnath, and Jui Lodh

6.1 Introduction

Cheese is a fermented dairy product having appreciable varieties depending upon the flavor and forms available throughout the world. A diverse group of microorganisms including starter and adjunct cultures are used for fermentation of raw milk for cheese making, known as cheese flora. These cheese florae are considered as the major contributors towards the formation of characteristic sensory and textural attributes of the several varieties of cheese due to their complex interactivity with milk components like carbohydrates, proteins and fats. This phenomenon is termed as 'ripening', a foremost technological processing step in cheese making (Forde & Fitzgerald, 2000). Usually, the ripening period of rennet coagulated cheeses extended from 2 weeks (e.g. mozzarella) to 2 years or more (e.g. Extra-mature Cheddar or Parmigiano-Reggiano) (Khattab et al., 2019). In ripening process, with time, microbiological and biochemical changes occurred in curd mass are responsible for the formation of characteristic flavor and texture of the particular cheese variety. The microbiological differences involve in the changes to the cheese microflora, including the growth, death and lysis of the starter cells, progress of an adventitious non-starter cheese microorganisms or in case of other varieties, the growth of a secondary cheese microflora, e.g. in Swiss cheese variety: *Propionibacterium*

R. Mandal · P. Karmakar
Agriculture and Food Engineering Department, Indian Institute of Technology Kharagpur, Kharagpur, India

P. P. Debnath
Faculty of Dairy Technology, Department of Dairy Technology, West Bengal University of Animal and Fishery Sciences, Nadia, WB, India

R. Kumari · A. Debnath · J. Lodh (✉)
Department of Dairy Technology, Sanjay Gandhi Institute of Dairy Technology, Bihar Animal Sciences University, Patna, Bihar, India

S. Mansha Rafiq (ed.), *Technological Advances and Trends in Cheese Making*,
https://doi.org/10.1007/978-3-031-94476-5_6

freudenreichii subsp. *shermanii;* in mold mould-ripened varieties: molds; in smear-ripened cheeses: complex Gram-positive bacterial microflora (Beresford & Williams, 2004).

During proteolysis, hydrolysis of casein micelles produce changes in pH and water-binding ability of the coagulum, may cause migration and precipitation of calcium phosphate from the system which soften cheese texture (O'Callaghan & Guinee, 2004). Degree of ripening determines the extent of changes taken place in chemical composition which regulate the progress of typical cheese aroma and flavor. Determination of extent of ripening and its related flavor compounds are used for ascertaining the quality of cheese (Forde & Fitzgerald, 2000). Several technological incentives are found to be useful to boost the rate of cheese ripening and to minimize the costs which might significantly influence the overall quality of cheese (Cocolin et al., 2018).

The biochemical reactions during ripening can be classified into two occurrences i.e., primary and secondary. Primary event involves the biochemical transformation of residual lactose, lactate and citrate which is collectively known as 'glycolysis'; metabolism of milk proteins involving mainly different casein fractions, termed as proteolysis and metabolism of fats or lipolysis (McSweeney, 2011). The secondary biochemical events are involved in the biochemical transformation of fatty acids and of amino acids lead to the production of many volatile flavor compounds (Curtin & McSweeney, 2004; McSweeney & Fox, 2004; Upadhyay et al., 2004). Enzymes, take part in the ripening process appear from various origins including milk lipoprotein lipase, rennet enzyme "chymosin" or may be produced from bacterial activity. In this context, non-starter lactic acid bacteria (NSLAB) and secondary or adjunct cheese cultures are among the prime contributors (Clark et al., 2009). The biochemistry of cheese ripening is an immense area of research which has been reviewed thoroughly (McSweeney & Sousa, 2000; Sousa et al., 2001; Yvon & Rijnen, 2001; Smit et al., 2002; Collins et al., 2004; McSweeney & Fox, 2004).

6.2 Ripening Agents

Ripening agents are the main causative agents for ripening in cheese, influencing the nature, rate and extent of the process. Ripening agents participate in three primary reactions namely glycolysis, lipolysis and proteolysis.

Ripening agents used in the process of ripening include

(i) Native milk enzymes
(ii) Non-starter bacteria
(iii) Rennet or rennet substitute (e.g. Pepsin, chymosin, microbial proteinases etc.)
(iv) Secondary starter enzymes (e.g. enzymes from fungi, smear bacteria and/or gas-producing bacteria)
(v) Enzymes from starter bacteria

Ripening agents in cheese manufacturing include coagulants such as rennet or rennet replacements, enzymes obtained from milk, starter or non-starter microorganisms and/or secondary starter microorganisms. Chymosin, a coagulant found in milk-fed calves, kids, and lambs, destabilizes the colloidal configuration of milk protein mainly casein by disrupting the peptide bond between Phe105-Met106 in κ-casein. This activity is more susceptible than the other bonds found in milk proteins (Sousa et al., 2001). While the majority of coagulants utilized are lost through whey due to their water solubility, a minor fraction (about 4–6%) is retained in curd. The level of retention depends on parameters such as coagulant type, temperature of cooking of curd, pH of whey at draining, and the curd's moisture level (Upadhyay et al., 2004). During ripening, the leftover rennet is important for the initiation of proteolysis of casein and participate in the breakdown of other casein fractions namely αs1-, αs2-, or β-caseins. The action of chymosin on β-casein is intensely inhibited by sodium chloride concentrations exceeding 5% and fully suppressed by 10% NaCl. Other coagulation agents used in cheese production include pepsin, which also aids in ripening process of cheese. However, its proteolytic action exceeds that of chymosin. In milk-fed young calves or ruminants, the ratio of chymosin to pepsin can be as high as 9:1; however, the ratio can be reduced to 1:1 if the ruminant is fed nutrients other than milk. The most often utilized coagulant is from *R. miehei*, which is highly specific to chymosin. *C. parasitica* proteinase degrades ovine casein more effectively than protease from *R. miehei* (Trujillo et al., 2000). Enzymes obtained from lactic acid bacteria viz. mesophilic *Lactococci* and *Leuconostocs*, thermophilic *Lactobacilli*, and *Streptococcus thermophiles*, participate in secondary proteolysis during cheese ripening.

These microorganisms are utilized in cheese manufacturing as starter cultures for acid production or as secondary starters for development of flavor and texture. The main function of starter culture is acid production and lowering of pH. Secondary starter cultures also contribute to tasks such as developing flavor, modifying ripening, or changing the color and appearance of the cheese. Lactic acid bacteria are not highly proteolytic, but they have a wide range of proteinase/peptidase systems that catalyze the formation of various peptides and amino acids. Secondary starter organisms play a vital role in proteolysis and flavor development in cheese. The most common secondary starters used in cheese are *Lactobacillus* spp. (*L helveticus*, *L casei*), yeasts (*Kluyveromyces lactis*, *Saccharomyces* spp., *Debaryomyces hansenii*), molds (*P camemberti*, *Geotrichum candidum*, *Penicillium roqueforti*), *Propionibacterium* spp., *Micrococcus* spp., *Enterococcus* spp. and *Brevibacterium linens*. These secondary starter organisms have a wide range of enzyme systems, able to catabolize both milk fat and casein-derived peptides resulting in ripening indices and flavor development. The ripening process is very prompt in surface-ripened and/or mold-ripened cheeses due to the presence of intricate bacterial or fungal flora on their surface.

6.3 Accelerating Cheese Ripening

Ripening or curing of cheese is a gradual and an expensive process. Ripening of cheese is not inevitable and/or manageable (Sihufe et al., 2010). Several indigenous and modern approaches are practiced to accelerate the rate of biochemical changes during ripening of cheese. The process to hasten the ripening or curing of cheese is known as accelerated cheese ripening.

Approaches used for controlled and accelerated cheese ripening are given below:

(i) Modified starters
(ii) Exogenous enzymes
(iii) Exploitation of cheese slurries
(iv) Adjunct microbial cultures
(v) Attenuated starters
(vi) Elevated temperatures (Source: Azarnia et al., 2006).

Each approach has coupled with its own merit and demerit. Utilization of modified starters are easy to apply. This process is able to retain the natural enzyme proportion, but alteration of starters, either by physical and/or genetically is a complex phenomenon, in a strict sense.

Usage of exogenous enzymes are comparatively cheap, have specific activity on given substrate with a preference of flavor options while the process of addition of exogenous enzyme is limited to the preference of advantageous enzymes. However, this approach is associated with the threat of over-ripening, hurdles with uniform incorporation, legitimate limit of addition etc.

Enzyme modified cheese (EMC) and cheese with high- moisture have been employed successfully for making cheese with tailored and constant flavor. EMC flavors are designed with authentic combination of proteolytic and lipolytic enzymes which may be incorporated to cheese emulsions as natural substrates. EMCs can be utilized to any food where cheesy flavor is essential to impart (Wilkinson et al., 2011). However, EMCs are not able to develop characteristic flavor and/or texture associated with the natural cheeses.

Addition of adjunct cultures together with NSLAB may have potential ability to speed up ripening, but their application has been restricted due to the accessibility of specific strains.

Ripening at elevated temperature may restrict the progress of undesirable microbial contaminants; however, can be safely applied when cheese is prepared with pasteurized milk under good manufacturing conditions (Azarnia et al., 2006). It is one of the most effective, least complicated and most economical method for accelerated cheese ripening (Sihufe et al., 2010). Use of high pasteurization temperature is undoubtedly speed up cheese ripening by modifying enzymatic activities, progressing rennet or acid coagulation of milk and thereby increase the yield of cheese (Trujillo et al., 2000). Ripening at elevated temperature i.e. 16 °C, was stated to boost the ripening of Dutch-type cheeses though rigorous hygienic measures have to be followed to check the fast growth of LAB, NSLAB and other microorganisms. Otherwise, intoxication or infection may occur which deteriorate the cheese quality (Sihufe et al., 2010). However, use of elevated ripening temperature may increase the activity of decarboxylases of LAB leading to the production of biogenic amines which have negative health consequences on cheese consumers.

6.4 Novel Detection Techniques for Cheese Flavor

Sensory quality attributes of cheeses are amongst the important criteria for acceptability by consumers as well as for their industrial production. The most crucial technological stage during the production of cheese is ripening, which involves a series of biochemical processes carried out by a wide range of microorganisms and contributing to cheese's perceived sensory qualities—its aroma, flavor. Microbial fermentation in cheese includes a sequence of enzymatic and chemical reactions during this ripening phase of cheese to produce the final texture and flavor characteristics which cater to the overall sensory experience of consumer. Sensorial disparities may exist in different cheese varieties or even same varieties even though both industrial and artisanal cheese making involve a ripening time because of the different ways that cheese matures in controlled and uncontrolled environments. The food industry is required by the growing customer consciousness of food safety and quality to create and preserve a product's unique identity and attributes. Evaluation of the cheese quality attributes therefore becomes necessary. These aroma, flavor and other quality attributes of cheeses are assessed by different descriptive, instrumental, and computational techniques, with these methods involving varying level of complexity, destructive/non-destructive nature, and ease of operation (Khattab et al., 2019; Lei & Sun, 2019).

6.4.1 Electronic Noses (e-Noses)

Electronic noses or e-noses are used as a cutting-edge analytical sensory instrument to identify flavors in the food products. E-noses comprise a biomimetic device made up of an intricate network of electronic and chemical multi-sensor array, with information analysis and processing unit comprising of an artificial neural network (ANN), digital pattern-recognition algorithms software, along with reference-libraries that can replicate the way the human olfactory system works to detect both simple and complex aromas (Wilson & Baietto, 2009; Khattab et al., 2019). The way e-nose sensors work by looking for interactions between the materials that coat the sensor and the gaseous molecules, which changes the electrical current that flows through the sensor. A transducer transforms this into an equivalent electrical signal that is recorded. There are various types of electrochemical sensors, such as conducting polymer gas sensors, metal-oxide gas sensors, and quartz crystal microbalance sensors (Wilson & Baietto, 2009). The e-nose sensor digital outputs are further analysed and interpreted by different techniques to provide operators with meaningful information. Firstly, by graphical analysis, such as using offset polar plots, polar plots, bar charts, etc., multivariate data analysis by canonical discriminant analysis (CDA), PCA, cluster analysis (CA), and featured within (FW) and lastly by ANN and radial basis function (RBF) algorithms (Wilson & Baietto, 2009).

E-noses have the capability to detect and recognize various VOCs that form a part of cheese flavor. When chemicals are detected, the sensors react by producing electrical signals that are digitized into multidimensional matrices depending on the concentration of the compounds (Khattab et al., 2019). Pattern recognition analysis of the e-nose is useful in building an olfactory fingerprint of product samples. In a study, e-nose analysis was accurately distinguished between various types of cheese that came from different sources or at different ages (Gursoy et al., 2009). In Danish Blue cheeses, the proliferation of *Penicillium roqueforti* performs as a key factor during ripening, affecting their quality, flavor, and taste (And & Nielsen, 2005; Khattab et al., 2019). Therefore, by modeling the e-nose fingerprints, this technology may be immediately used to the quality assessment of such cheese products as well as ripening monitoring. And and Nielsen (2005) used generated multivariate models from e-nose fingerprints to predict the age at which cheese from the same or different manufacturing units will ripen. E-nose technique has demonstrated its adaptability in monitoring cheese sensory qualities without requiring sophisticated panels or costly laboratory equipment, allowing quick decision-making or process control during cheese manufacture.

6.4.2 Electronic Tongues (e-Tongues)

Similar to e-nose, electronic-tongue or e-tongue has potentiometric or voltametric sensor array intended to identify non-volatile compounds in mostly liquid matrices, replicating human taste perception. A variety of taste sensations, such as sweet, salty, sour, bitter, umami, spicy, and metallic can be detected by e-tongue (Schlossareck & Ross, 2019). Studies are scarce for cheese flavor detection using e-tongue since the sensors are primarily used with liquid and semi-liquid samples milk, yogurt, and protein hydrolysates (Valente et al., 2018; Lipkowitz et al., 2018). Cheese flavors were tested and analysed using e-tongue after ultrasonic treatment following with PLS regression and PCA by Cong et al. (2015). In a study, Schlossareck and Ross (2019) used e-tongue to distinguish *paneer* cheese samples containing varying capsaicin content. In another study by Lipkowitz et al. (2018), taste and flavor of aged full-fat Cheddar cheeses with varying protein-fat ratios was analyzed for more than 12 months by consumer and trained panels. Statistical analysis was used to correlate sensory qualities with electronic tongue data. PCA was employed for discrimination of cheese based on aging time and protein-fat ratio and PLS regression was performed to correlate sensory with instrumental attributes. The e-tongue was even more sensitive to slight changes in metallic, bitter, salty, sweet, sour, spicy, and umami tastes (Lipkowitz et al., 2018). Cheeses obtained from raw and pasteurized milk was accurately distinguished and classified by using e-tongue (Valente et al., 2018). In another study, e-tongue was utilized to study cheese flavors in traditional Chinese cheese and it was correlated with cheese microflora (Zhang et al., 2021). The researchers were able to distinguish between fresh cheese and 7-day aged cheese. Another study compared samples surface-ripened with five strains of *koji* mold, genus Aspergillus, to commercial Camembert cheese to analyze the taste attributes of *koji* cheese using e-tongue (Hayashida et al., 2023). *Koji* cheese had less sourness, but more astringency, bitterness, saltiness, and umami tastes compared to Camembert.

6.4.3 Spectroscopic Techniques

Intrinsic fluorescence of molecules in food matrices can be used by fluorescence spectroscopy to ascertain the distribution of molecules in the sample. Therefore, Fluorescence spectroscopy provides a sensitive way to map the spatial distribution of fluorescent flavor compounds in cheese analysis. The product of the fluorescent compound concentration (C) and fluorescence quantum yield (Φ) yields the fluorescence intensity (I_f) emitted by a sample ($I_f = \Phi. C$). Fluorescence lamps selectively excite fluorescent molecules inside the sample at particular wavelengths, which are recorded by detectors to create spatial maps of fluorescence intensity. Naturally milk and cheese have aromatic amino acids, vitamin A, tryptophan which show fluorescence. Because of the variations in chemical composition, samples of different cheese varieties have distinct fluorescence spectra (Khattab et al., 2019). This feature allows for the differentiation of different cheeses and the use of fluorescence fingerprints to assess authenticity. Fluorescence spectroscopy techniques such as conventional front-face fluorescence (FF) spectroscopy, synchronous fluorescence (SF) spectroscopy have been used for the prediction of cheese physicochemical properties (Abbas et al., 2012), discrimination of plant-oil-adulterated cheeses (Dankowska et al., 2015), ash and protein prediction in French blue cheeses (Abbas et al., 2012), prediction of moisture, fat and protein in Tilsit cheese (Ozbekova & Kulmyrzaev, 2017). Applying chemometric data analysis on SF and FF spectroscopy has also allowed monitoring changes in protein during maturation, prediction of cheese maturation index, proteolysis, free amino acid content (Karoui et al., 2004; Kokawa et al., 2015). In a study, Kraggerud et al. (2014) used FF spectral data to predict cheese sensory quality while ripening.

Raman spectroscopy technique allows determination of material molecular composition by seeing how monochromatic light scatters due to molecular vibration in a sample. Raman spectroscopy can offer important insights into the chemical composition, structure, and quality of cheese samples. Raman spectroscopy was used for qualitative and quantitative analysis of non-volatile compounds in cheese (Kilcawley, 2017), for instance, successful classification of adulterated cheese (de Sá Oliveira et al., 2016), clotting time (Nicolau et al., 2015). The Raman spectra in the 3050–2770 cm^{-1} band, originating from lipids, has allowed primary distinctions among the artisanal cheeses based on quantity of lipids (Costa et al., 2022). Yaman et al. (2022) found bands in the region 986 cm^{-1}, 1070 and 1077 cm^{-1}, pertaining to bending vibrations of CH, COH, CN groups in proteins. This allows changes in proteins, lipids to be monitored using Raman signals along with quantitative analysis of generated volatiles/non-volatiles during ripening.

Nuclear magnetic resonance (NMR) gives precise knowledge regarding the molecular dynamics and structure of the food products. It works by exposing nuclei to a magnetic field and radiofrequency pulses. When nuclei with a non-zero nuclear spin is subjected to a magnetic field, such as H^+ (protons), align themselves with the field. This allows for the detection of distinctive signals released during their relaxation. NMR has been utilized in metabolite profiling in cheeses in many studies (Scano et al., 2019; Chen et al., 2021; Balthazar et al., 2021). Ruyssen et al. (2013) studied sensory characteristics of Gouda cheese by ^{1}H NMR. Piras et al. (2013) studied NMR based metabolomics for the ripening process in Fiore Sardo cheese prepared by autochthonous adjunct cultures. Scano et al. (2019) used ^{1}H NMR spectroscopy to examine Fossa cheese (an Italian type cheese) ripening. Higher quantities of free amino acids (FAAs) after proteolysis and butyric acid and diacylglycerols were detected after lipolysis. Extended ripening resulted increase in rumenic acid (18:2 cis-9, trans-11) and glycerol levels (Scano et al., 2019). In Chen et al. (2021) studied metabolites in aqueous extracts of Cheddar cheese by NMR in terms of H^+ ratio of citrulline and arginine to assess ripening where they observed tyrosine, tyramine, lysine to be highly correlated with mature cheese profile. Balthazar et al. (2021) observed aromatic compounds like tyrosine, phenylalanine, histidine, benzoic acid, and formic acid in Prato cheese after metabolic profiling using NMR. Conte et al. (2020) identified 9Z-octadecenoic acid as a biomarker in Parmigiano Reggiano cheese.

In the food sector, infrared (IR) spectroscopy has become fastest, most effective non-destructive method for process and product quality monitoring and control. The IR techniques analyze the molecular composition of products by interaction with IR (780–25,000 nm). NIR spectroscopy operates in wavelengths in the near-IR range (780–2500 nm) and measures the rotations and vibrations of chemical bonds in molecules. It is frequently used for the analysis of proteins, lipids, and carbohydrates in food items. NIR is studied in reflectance and transmittance mode, where mainly the former is used in cheese analysis. Blazquez et al. (2006) used NIR spectroscopy (750–2498 nm) to study sensory properties of processed cheese. González-Martín et al. (2011) predicted sensory properties of cheese such as rancid flavor and pungency, retro-nasal sensation by NIR spectroscopy. In another study (González-Martín et al., 2014), acetaldehyde, ethanol, 1-propanol, 2-butanol, 2-pentanol, 3-methyl-1-butanol, 2-butanone, 2-pentanone, 2-heptanone and 2-nonanone were identified in cheese. MIR spectroscopy operates in the mid-IR range (2500–25,000 nm) and works on the principle of infrared absorption by molecular vibrations to offer comprehensive details about the functional groups in the molecules. It can be used in the prediction of formation of fat- and protein-derived flavor compounds in cheese.

Fourier transform infrared spectroscopy (FTIR) provides an improvement in spectrum resolution and sensitivity with excellent precision and accuracy. FTIR spectroscopy has been applied in cheese analysis by many researchers. In a study, Subramanian et al. (2011) used FTIR to predict levels of organic acids in aged cheese by detection in the region 1800–900 cm^{-1}. Fagan et al. (2007) used FTIR to study maturity and sensory properties of Camembert cheese. In another study, Koca et al. (2007) predicted levels of acetic, butyric, and propionic acid in Swiss cheese. Subramanian et al. (2009) used FTIR spectra (1450 to 1350 and 1200 to 990 cm^{-1}) to distinguish Cheddar cheese based on flavor derived from esters of fatty acids, organic acids, and amino acids.

6.4.4 DNA Sequencing, Multi-omics and Cheesomics

Cheese microbial communities change over time and depend on the variety of cheese. These microorganisms include starters, adjunct cultures, environmental, processing plant microorganisms which have a significant impact on flavor, quality of cheeses. The field of next-generation sequencing (NGS) and bioinformatics techniques have led to increasing research on the composition of the cheese microbiota and ripening. NGS techniques such as metagenomics and rRNA gene amplicon sequencing along with meta-transcriptomics, metaproteomics, metabolomics, meta-omics, multi-omics platforms (cheesomics) have helped enhance understanding of cheese microbiota and its functioning and predict cheese attributes like flavor, quality, texture (Afshari et al., 2020). Total DNA sequencing or polymerase chain reaction (PCR) amplified DNA of microbiota from cheese samples help understanding cheese microbiology. Microbial community profiling (MCP) uses molecular methods like PCR, particular primers that target 16S rRNA gene, temperature gradient gel electrophoresis (TGGE), differential gradient gel electrophoresis (DGGE), single-strand conformation polymorphism (SSCP) etc. generated DNA fingerprints for dominant microbial populations in cheese (Afshari et al., 2020). These analyses helped in profiling of microbial community in different types of cheese which are summarized elsewhere (Afshari et al., 2020).

Microbial cheese ripening is a complex phenomenon, which is not able to be explained just by profiling of the microbial community in cheese samples. Meta-omics approach has been introduced for going beyond microbial analysis to accurately predict functions of cheese microbiota in the overall ripening process. Meta-genomics comprise the study of genomes, and it involves analyzing genetic sequences to determine functions that are genome-encoded. Meta-genomics studies carried out to predict microbial species, are responsible for a particular flavor in different cheese (De Filippis et al., 2016; Escobar-Zepeda et al., 2016). Transcriptomics have been studied for cheese microbiota like *Lactobacillus rhamnosus*, *Propionibacterium freudenreichii* to identify possible metabolic pathways and generation of cheese flavor and volatiles (Falentin et al., 2010; Lazzi et al., 2014). Proteomics has also been used for studying proteolysis and cheese flavor formation. Metabolomic analysis of cheese is the study and quantification of metabolites formed by various pathways in cheese analysed using chromatographic techniques coupled with MS detectors. It has used to detect the formation of various metabolites during Mozzarella cheese maturation and can serve as a chemical fingerprint to distinguish cheese varieties (Mazzei & Piccolo, 2012; Pisano et al., 2016). Metabolomic analysis of nonvolatile taste compounds have also been carried out using liquid chromatographic systems (Le Boucher et al., 2013). Combining of multiple 'omics' and cheesomics will help studies on determining influence of cheese microbial community on the specific flavor compounds, selection of specific cultures, and identify novel bioactive markers beneficial to human health.

Metagenomics, metatranscriptomics, and biochemical analyses were used to investigate surface-ripened cheese ecosystem over 1 month ripening (Dugat-Bony et al., 2015). First ripening was dominated by *Lactobacillus lactis* and *Kluyveromyces lactis*, followed by *Geotrichum candidum*, *Debaromyces hansenii*, *Corynebacterium casei* and *Hafnia alvei*. Lactate dehydrogenase transcript increases indicated the lactose consumption by *L. casei,* and the lactic acid was consumed by *G. candidum* leading to the decomposition of proteins and fats. De Filippis et al. (2016) used meta-transcriptomics and metabolomics to examine the effects of ripening conditions (lower (16 °C) and higher (20 °C)) temperatures on the *pasta filata* cheese microbiota functional diversity. Higher temperature maturation showed expression of genes related to proteases, amino acid metabolism, fatty acid oxidation, etc., which correlated with metabolic data that showed cheese flavor compounds developed more readily at higher temperatures during ripening.

6.4.5 Imaging Techniques

Imaging techniques are becoming popular due to their non-destructive nature and minimum sample preparation requirements. Some of the imaging techniques used in the food industry include hyperspectral imaging (HSI), multispectral imaging (MSI), computer vision (CV), magnetic resonance imaging (MRI), computed tomography (CT).

Samples can be thoroughly chemically analyzed using hyperspectral imaging, which simultaneously capture spectral and spatial data of the products forming a 4-dimensional *hypercube* (two spatial dimension and one continuous spectral signal at different wavelengths, λ, at interval 10 nm) (Feng & Sun, 2012; Khattab et al., 2019). Usually, a hyperspectral camera is involved in scaning the images of cheese samples, producing a sequence of photos at $\boldsymbol{n}$ different wavelengths, and hyperspectral data cube $\boldsymbol{X}$ is acquired. Every pixel in the image possesses a spectrum $\boldsymbol{x}_i$, representing the intensity of light reflected from the sample at that wavelength, such that,

$$X = [x_1, x_2, \ldots, x_n]$$

HSI may be employed for mapping the spatial distribution of flavor components across the surface of cheese samples, analysed using the spectral data. PCA and linear discriminant analysis (LDA) methods can be used to evaluate the hyperspectral data and extract useful information pertaining to the chemical composition of cheese samples. PCA helps finding spectral signatures connected to chemical components and minimize the dimensionality of the hyperspectral data. Using PCA scores, the spatial distribution of these components (proteins, fats) can be modeled as

$$Y = XW$$

Where, Y = matrix of PCA scores; W = matrix of PCA loadings.

HSI has been earlier for prediction of fat content and hardness of semi-hard cheeses with high accuracy (Darnay et al., 2017), hardness of Swiss cheeses (Vásquez et al., 2018), starch content in adulterated samples of fresh cheeses (Baretto et al., 2018). HSI spectra in the near infrared regions have also been used for discrimination of traditional cheeses (da Silva Medeiros et al., 2024), maturity modeling for long-ripened hard cheeses (Priyashantha et al., 2020, 2021), monitoring proteolysis and lipolysis in semi-hard Italian cheeses (Malegori et al., 2021). MSI method has been for classification of Fourme d'Ambert and Bleu d'Auvergne blue cheeses with high accuracy (Kulmyrzaev et al., 2008).

Other imaging techniques like CV has been used for grading and quality inspection of foods in the food industries. CV captures regular images of products using a camera and an illumination source and used for classification according to the physical attributes of cheese (Khattab et al., 2019). Other advanced imaging techniques like CT, MRI, X-ray imaging can acquire 3D images of the samples and can be used for evaluation of quality of cheese. Due to their working principles, they have been used in the cheese industry for monitoring of physical attributes like oiling off, composition analysis, microstructure, eye formation during cheese maturation with great accuracy (Wang & Sun, 2004; Everard et al., 2007; Jeliński et al., 2007; Guggisberg et al., 2013; Schuetz et al., 2013, 2016). With suitable modifications and development of algorithms such techniques can be used for monitoring formation of volatiles during cheese ripening and thus inform about cheese flavor characteristics.

6.4.6 Ultrasound Based Sensing Techniques

As in the case of imaging techniques, ultrasound based instruments derive their working from their physical nature of sensing obstacles, cracks, and crevices in this propagation field. They offer low-cost and non-destructive cheese quality evaluation techniques and have been used for monitoring of eyes and cracks in various cheese types (Eskelinen et al., 2007; Conde et al., 2008; Nassar et al., 2010). While not being directly able to detect flavor in cheese due to their mechanisms, ultrasonic based techniques can be used to complement tests related to sensory analysis of cheese and validation of cheese maturity.

6.5 Application of Biotechnology to Promote Cheese Maturation and Flavor

Bio-technology includes the usage of living organisms or their derivatives, for instance, cells and enzymes, for various industrial applications, specifically in the field of agriculture, food production, as well as, medicine (Bentahar et al., 2023; Clark & Pazdernik, 2016). In the dairy sector, bio-technology has been utilized not only in dairy production sector, but also in dairy processing sector. Animal cloning, DNA fingerprinting, recombinant vaccine's development are some of the areas where bio-technology has been successfully explored. In the milk processing sector too, bio-technology has been applied for production of various enzymes, bio-preservatives, probiotics, functional foods, nutraceutical and many others (Sundarraj et al., 2018). Some of the advantages of usage of bio-technology in various dairy products are listed in Table 6.1.

Similar to the above stated advantages, various bio-technological approaches have also been successfully applied in cheese, which would be discussed here in details. Usage of exogenous enzymes (protease, lipases and ß-Galactosidase), incorporation of cheese slurry (for instance enzyme modified cheese), utilization of various microbial agents for example, lactic acid bacteria, yeasts and molds, attenuated and adjunct cultures are some of the methodologies which has been utilized to fasten the cheese ripening, thereby improving the organoleptic parameters, namely, flavor, aroma, body and texture.

Table 6.1 Applications of bio-technology in dairy products

S. no	Biological agent	Area of application	Advantages reported	References
1	Probiotic microbes			
	Lactobacillus acidophilus	Fermented dairy products	Exerts antibiotic impact on gut pathogens	Ahmed et al. (2010)
	Lactobacillus casei	Actimel, gefilus, vifit, yakult, kefir, provolone and parmesan cheese	Enhancement of flavor profile along with body and texture	Yerlikaya (2014)
	Lactobacillus gasseri	Fermented milk products	Stimulates macrophages in intestine	Uzuner (2009) and Amiri et al. (2021)
2	Enzymes			
	ß-galactosidase	Milk and other dairy products	Hydrolyses lactose, thereby making the product suitable for lactose intolerant persons	Nivetha and Mohanasrinivasan (2017)
	ß-galactosidase	Ice-cream, condensed milk, whey spreads	Improvement in body and texture by inhibition of lactose crystallization	Nivetha and Mohanasrinivasan (2017)
	Lipase and protease	Ghee	Flavor enhancement	Debnath and Ghosh (2018)

6.5.1 *Enzymes*

Cheese ripening or maturity includes three foremost reactions, for instance, lipolysis, proteolysis along with glycolysis. The aforesaid reaction plays the most vital part for the formation of desirable organoleptic attributes in cheese ripening. Lipase and protease enzyme which are required for lipolysis and proteolysis are present in milk endogenously. They are also derived from starter culture, attenuated and adjunct cultures. But, sometimes due to various processing conditions, reduction in activity of endogenously derived enzymes is observed (Ardö, 2021). Thus, in order to fasten ripening and facilitate flavor, body and texture development, exogenous enzymes are being added during cheese manufacture. These enzymes can be directly incorporated in milk or in blocks of cheese or during the dry salting step. They can be also added along with starter culture or rennet (Azarnia et al., 2006).

6.5.1.1 Lipase

The principal role of an enzyme is to act as a bio-catalyst in any reaction. Here, too the main reason behind the external addition of lipase enzyme is to accelerate the flavor production in cheese ripening, by fat hydrolysis. This particular phenomenon is explained by the underlying picture (Fig. 6.1).

Various researches have been conducted for examining the impact of exogenous lipase in flavor improvement, some of which would be discussed here. The research group of Kendirci used four varying lipase enzymes with different degrees of incubation to attain the desirable aroma of white ripened cheese (Kendirci et al., 2020). The lipolytic action resulted in 12.5–81.9 times increase in free fatty acid as compared to untreated sample. Similarly, concentration of volatile compounds got elevated by 30.1–86.9 times after lipolysis. Formation of undesirable volatile compounds such as n-aldehydes was not reported in the above mentioned study (Kendirci et al., 2020). In another study, usage of microbial lipase at the rate of 200 U in Swiss cheese has reduced the period of ripening from 90 to 60 days, thereby reducing the final cost of manufacture of the product (Rani & Jagtap, 2019). Hardness increased, whereas brittleness declined during this period. Negative impact on sensory analysis and physico-chemical properties were not reported here (Rani & Jagtap, 2019).

Lipase enzymes are substrate specific as being discussed herewith. Protein engineered lipase of *Pleurotus citrinopileatus* has been shown as an effective substitute

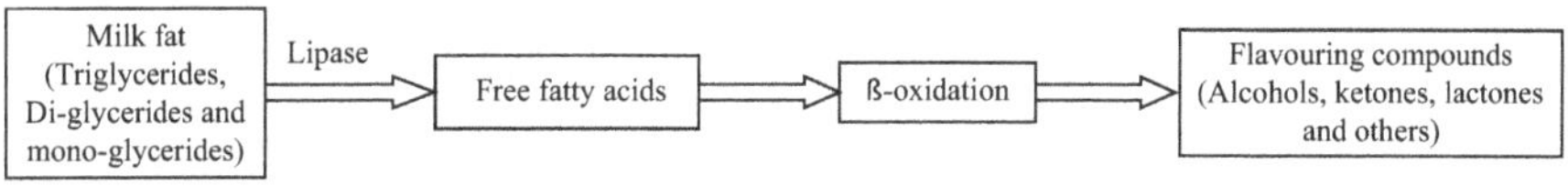

Fig. 6.1 Production of Flavoring compounds by hydrolysis of milk fat by lipase enzyme. (Source: Verma et al., 2022)

of pre-gastric esterases for manufacturing Feta type cheeses (Broel et al., 2022). Here, protein engineering was carried out to alter substrate specificity as the said lipase is more specific to long chain fatty acids (LCFA). Hydrolysis of LCFA may yield off flavor. As a result, by protein engineering, researchers changed the specificity of lipase from long chain to medium chain, which resulted in superior sensory attributes of cheese (Broel et al., 2022). In another research work, done by Zhang along with his group, lipolytic efficacy of various lipase enzymes ("Lipase AY 30 SD, lipase DF 15, lipase F3G, lipase MER and Palatase 20000L"), so as to hasten cheese ripening was analyzed (Zhang et al., 2023). Hydrolysis in almost all the cases resulted in formation of acids, esters and about 27 novel volatile compounds. However, degree of production of the aforesaid components varies according to the type of enzyme used. For instance, volatile chain fatty acids were produced in a greater amount by lipase AY, DF and MER, whereas, creamy sensation was more perceived in the case of lipase F3G (Zhang et al., 2023). Similarly, another study was carried out by Gao and co-workers to determine the effect of particular lipase enzymes ("Lipase MER, lipase AY 30G and Palatase") in the flavor development of cheese. In contrast to Palatase enzyme, greater amount of fatty acids having short chain length was formed with the help of lipase MER, which plays the pivotal role in flavor formation. However, concentration of esters was more after hydrolysis by lipase AY 30G, which made it suitable if fruity cheese flavor is desired (Gao et al., 2022).

Consumers now a days are very conscious regarding their calorie intake, thus dietetic or low calorie products, for instance, low fat cheeses have become extremely popular in recent times. Flavor enhancement of low fat cheeses has been carried out by exogenous lipase addition. Zhao and group used lipase of *Bacillus licheniformis* NCU CS-5 in low fat cheese for improving its flavor profile. The said lipase enzyme was highly specific towards short along with medium chain length fatty acids. After lipolysis, elevated production of butyric, hexanoic, caprylic and decanoic acid was reported (Zhao et al., 2022). Immobilized lipases has also been utilized for augmenting flavor in low fat cheeses. In a research study carried out by Guan and co-workers, α-lactalbumin nanotubes has been utilized in the form of immobilization carriers of lipase enzymes. Hydrolytic efficacy of lipase enzyme got improved by 68% after immobilization as compared to free lipase enzyme. As a consequence, generation of free fatty acids got enhanced by 1.5 times after immobilization, which considerably improved the flavor profile of cheese having lower fat content (Guan et al., 2021).

6.5.1.2 Protease

Among the three bio-chemical reactions (proteolysis, lipolysis along with glycolysis) take place in cheese ripening, proteolysis plays the most crucial part in achieving the cheese's desirable organoleptic attributes (Karaca & Güven, 2018). Protease enzyme breakdown the caseins into amino acids which eventually undergoes decarboxylation, transamination, oxidative deamination and degradation reactions, as a consequence amines and α keto acids gets formed, followed by aldehydes, alcohol and acids production (McSweeney & Sousa, 2000). The aforesaid end products lead to matured cheese's delicate aroma.

El-Salam along with his group used bacterial protease enzyme to fasten domiati cheese's ripening. Amount of free amino acids, for instance, glutamic acid, proline, lysine as well as many others increased significantly just after 2 months of ripening. Protease incorporated Domiati cheese was more acceptable to sensory panelists in contrast to untreated cheese (El-Salam et al., 2019). Ripening of Domiati cheese was also accelerated by usage of neutrase enzyme (Taha et al., 2020). Here, apart from increasing the ripening rate, the main focus was to increase the anti-oxidant, anti-cancer and angiotensin converting enzyme (ACE) inhibitory activity with the help of proteolysis. Neutrase enzyme led to significant increase in release of bioactive peptides, which eventually resulted in the enhancement of aforesaid properties. As a result, health promoting attributes and functional properties of cheese got increased (Taha et al., 2020). Carboxypeptidase enzyme has been utilised to speed up the hard cheese's ripening, for example, Reggianito cheese. Increased amount of amino acids, in specific, lysine, glutamic acid, leucine along with valine, was reported after proteolysis. Sensory evaluation has not been done in the above study (Ceruti et al., 2016).

Faster ripening rate is being reported in white cheeses after incorporation of neutrase enzyme, which is derived from *Bacillus subtilis*. Free amino acid increased at a significant rate after ripening in contrast to the untreated cheese. Glu, Lys, Pro and Asp were the prominent amino acids found after ripening. Duration of ripening decreased here because of enhanced proteolysis (Karaca & Güven, 2018). Another research group experimented various combination of protease enzyme (exopeptidase and endopeptidase) in white cheese in order to find the ideal enzymatic combination, which could have the best balance between the proteolysis rate and degree of bitterness. They concluded mixture of Neutrase and Flavopro937 enzyme as the best combination. Bitterness is basically exhibited because of hydrophobic peptides. Combination of exo and endo-peptidase helps to counter this and produces various components such as phenylacetaldehyde, acetophenone and many others (Bas et al., 2019). Many other researches are being carried out in the same aspect, which are summarized in Table 6.2:

6.5.2 Micro-organisms

Apart from usage of exogenous enzymes, various microbes play a pre-dominant role in accelerating the ripening of cheese. Although, microbes are already utilized for production of cheese as starter culture, which also plays a very crucial part in flavor formation during ripening. But for further enhancement of ripening attenuated and adjunct cultures are incorporated, which would be discussed in the following paragraphs.

Table 6.2 Various other researches which has been carried out to speed up the maturation of cheese by exogenous enzyme addition

S. no	Research study	Inference	References
1	Usage of lipase and protease enzyme in combination in Ras cheese, Egyptian hard cheese	Acceleration of ripening was observed just after a maturation period of 30 to 45 days. Desirable organoleptic characteristics have been reported	Shahin et al. (2023)
2	Expediting the ripening of Cheddar cheese with *Penicillium candidum* PCA1/TT031 protease by using response surface methodology (RSM)	No significant variation was observed in sensory analysis. However, differences in amino acids were reported between control and treated sample. RSM proved to be a suitable technique for the optimizing the ripening acceleration process	Alhelli et al. (2021)
3	Impact of liposome encapsulated flavourzyme on Iranian brined cheese's proteolysis rate	Highest degree of proteolysis along with maximum sensory acceptance was attained by incorporation of exogenous enzyme with a maturation period of 30 days. In addition, for accomplishing the aforesaid, cheese curd was dipped within saturated brine solution for a duration of 8 h	Jahadi et al. (2016)
4	Efficacy of 13 fungal enzymes for carrying out lipolysis, proteolysis and glycolysis of Egyptian Ras cheese in the ripening period	Each fungal enzymes used here gave positive results for proteolysis whereas all of them exhibited negative results for glycolysis except the enzyme derived from *Aspergillus nidulans*. In contrast, the same species displayed negative correlation with lipolysis, while the rest of them exhibited positive co-relation	El-Fadaly et al. (2015)
5	Preparation of enzyme modified white cheese by external lipase and protease enzyme addition. The prepared product was used as a flavoring ingredient in other products	Concentration of free amino acids and free fatty acids got increased, which resulted in reduction of off-flavors and produced more caramel, sweaty and nutty flavors. The prepared product could be used as a flavoring ingredient where cheese flavor is desirable	Ali et al. (2019)
6	Manufacturing of imitation cheese from soy and cow milk's blend, in the presence of protease along with lipase enzyme for accelerating the ripening and produce superior sensory attributes	Increase in amino acids and free fatty acids was reported. In addition, formation of volatile compounds was observed. Intensity of bitter, pungent and other off odours got decreased, while pleasant and sweet sensory attributes were reported. Varieties of flavored cheeses could be produced by the combination of soy and cow milk	Ali et al. (2017)

Probiotic adjunct cultures of *Lactobacillus* species were utilized for alleviating the physicochemical along with organoleptic characteristics of Ras cheese, a hard cheese variety from Egypt. Increased amount of soluble nitrogen, soluble tyrosine and tryptophan as well as total volatile fatty acids have been reported after ripening in the treated sample in contrast to the untreated one. Fungal growth got diminished and growth of Coliform got restricted because of usage of probiotic strains. In addition, overall acceptability of treated sample got enhanced after ripening (Abd-Elmonem et al., 2022). Starter XT-312 which has been heat treated is being successfully used as a supplement in Dutch-type cheese along with *Lactobacillus helveticus LH-B01* for accelerating the proteolysis, by increased release of bioactive components (Garbowska et al., 2021). Nisin Z producer "*Lactococcus lactis* subsp. lactis CBM 21" has been utilised in the form of adjunct culture for manufacturing of "Squacquerone cheese", an Italian cow milk cheese. The study revealed activity of nisin was prominent throughout the storage period. However, no damaging impact was reported on the conventional *Streptococcus thermophilus* starter, which was used for the said cheese production. Inhibition of coliforms and *Pseudomonas* species was observed after the inclusion of adjunct culture. Proteolysis was also not affected by the CBM 21 species, instead it improved the organoleptic characteristics of Squacquerone cheese (Siroli et al., 2020). In another study, *Lactobacillus paracasei 4321,* has been utilized in the form of adjunct culture in Caciotta type cheese, an Italian semi-hard cheese for improving its flavor within reduced ripening period. Higher amount of volatile constituents was formed in the treated sample because of the interaction between the primary starter organisms (*Streptococcus thermophilus* and *Lactobacillus delbrueckii*) and adjunct culture. Any significant impact on the activity of *Streptococcus thermophilus* was not observed, however, growth of *Lactobacillus delbrueckii* was reduced in the presence of adjunct starter. This resulted in prolonged acid production which leads to over-acidification of cheese (Bancalari et al., 2020). Usage of *Lactobacillus Brevis* KX572376 in the form of adjunct starter in ultra-filtered Feta Cheese increased its anti-oxidant activity, thereby improving its functional property (Yousefi et al., 2021).

Freeze shocked *Lactobacillus* species was utilised as an adjunct starter at the rate of 1% to accelerate the rate of ripening in Edam cheese. Significant difference was not recorded in overall composition in comparison to that of control, except that of low pH readings in treated sample. Increased production of free amino acids along with cumulative volatile fatty acids elevated proteolysis and lipolysis's rate. As a consequence, flavor and texture improved after treatment, thereby resulting in the reduction of ripening duration from 60 to 45 days (Abou El-Nour et al., 2023). Maturation of gouda and danbo cheese has also been accelerated in pilot scale by using a novel mesophilic, thermotolerant, autolytic *Lactococcus lactis* strain, named ERC. The research group stated that efficacy of this *Lactococcus lactis* ERC strain, was superior to that of market available Flora Danica starter. In addition, it was reported that adjunct starter will not be required for enhancing the ripening rate, if ERC culture is being used as the primary starter. The ripening period of Danbo cheese was reduced from 175 days to 91 days by usage of this ERC culture (Zhao et al., 2023). Adjunct starter of *Aspergillus* species has also been utilized to increase the rate of lipolysis of semi-hard cheeses by generating elevated amount of free fatty acids along with volatile components (Chintagavongse et al., 2022). Strains of *Bifidobacterium* species derived from human origin has been applied as a possible probiotic adjunct starter for manufacturing heat processed sheep milk cheese. Significant impact was not reported on primary starter's activity, organoleptic parameters and on cheese's gross composition along with its pH. However, a prominent increase in the rate of proteolysis was recorded in the above trial (Langa et al., 2020). Ripening rate of Edam cheese has been accelerated by usage of ultrasonically processed *Lactobacilli* species as an adjunct culture. Amount of soluble and non-protein nitrogen increased, as a result proteolysis got fastened, which reduced the ripening period from 60 to 45 days. Both organoleptic taste and body and texture of Edam cheese improved because of the usage of adjunct culture (Ahmed et al., 2023). Maturation of probiotic cheddar cheese is also being accelerated by usage of probiotic *Pediococcus acidilactici AS185* in the form of adjunct starter. Amount of free amino acids as well as short chain fatty acids got elevated, as a result sensory acceptance of cheese got improved (Wang et al., 2023). Various other research works are being done in this regard, which can be gone through in the given references (Yang et al., 2024; Giménez et al., 2024; Calasso et al., 2020).

6.6 Conclusion

Cheese ripening involves several biochemical alterations that are responsible for characteristic flavor and, body and texture formation. However, the process is time consuming and difficult to be carried out under suitable controlled condition for longer duration. On the other hand, the knowledge on cheese ripening, mainly secondary biochemical alterations, is still at infantile state. Though, several emerging advanced technologies specially based on genetic engineering have pulled the attention of manufacture to cop up the difficulties associated with ripening process, more research was still needed in this arena. The diversity in characteristics, processing (mainly ripening) complexities and at the same time the increasing popularity of ripened varieties of cheese and interest among the producers open a new area for future research.

References

Abbas, K., Karoui, R., & Aït-Kaddour, A. (2012). Application of synchronous fluorescence spectroscopy for the determination of some chemical parameters in PDO French blue cheeses. *European Food Research and Technology, 234*, 457–465.

Abd-Elmonem, M. A., Tammam, A. A., El-Desoki, W. I., Zohri, A. N. A., & Moneeb, A. H. (2022). Improving the properties of the Egyptian hard cheese (ras type) with adding some probiotic Lactobacillus spp. as adjunct cultures. *Assiut Journal of Agricultural Sciences, 53*(1), 12–30.

Abou El-Nour, A. M., Khalil, R. A. M., Ahmed, H. A. A., El-Sisi, A. S., Ibrahim, H., & A. (2023). Impact of using freeze-shocked adjunct cultures on Edam cheese quality during ripening. *Ismailia Journal of Dairy Science and Technology, 10*(1), 1–9.

Afshari, R., Pillidge, C. J., Dias, D. A., Osborn, A. M., & Gill, H. (2020). Cheesomics: The future pathway to understanding cheese flavour and quality. *Critical Reviews in Food Science and Nutrition, 60*(1), 33–47.

Ahmed, Z., Wang, Y., Cheng, Q., & Imran, M. (2010). Lactobacillus acidophilus bacteriocin, from production to their application: An overview. *African Journal of Biotechnology, 9*(20), 2843–2850.

Ahmed, H. A. A., Khalil, R. A. M., Abou El Nour, A. M., El-Sisi, A. S., & Aida, H. (2023). Ultrasonication of lactobacilli cultures on Edam cheese quality during ripening. *Journal of Food and Nutrition Research, 11*(7), 474–482.

Alhelli, A. M., Mohammed, N. K., Khalil, E. S., & Hussin, A. S. M. (2021). Optimizing the acceleration of Cheddar cheese ripening using response surface methodology by microbial protease without altering its quality features. *AMB Express, 11*, 1–13.

Ali, B., Khan, K. Y., Majeed, H., Xu, L., Wu, F., Tao, H., & Xu, X. (2017). Imitation of soymilk–cow's milk mixed enzyme modified cheese: Their composition, proteolysis, lipolysis and sensory properties. *Journal of Food Science and Technology, 54*, 1273–1285.

Ali, B., Khan, K. Y., Majeed, H., Xu, L., Bakry, A. M., Raza, H., Shoaib, M., Wu, F., & Xu, X. (2019). Production of ingredient type flavoured white enzyme modified cheese. *Journal of Food Science and Technology, 56*, 1683–1695.

Amiri, S., Aghamirzaei, M., Mostashari, P., Sarbazi, M., Tizchang, S., & Madahi, H. (2021). The impact of biotechnology on dairy industry. In *Microbial biotechnology in food and health* (pp. 53–79). Academic Press.

And, J. T., & Nielsen, P. V. (2005). Electronic nose technology in quality assessment: Monitoring the ripening process of Danish blue cheese. *Journal of Food Science, 70*(1), E44–E49.

Ardö, Y. (2021). Enzymes in cheese ripening. In *Agents of change: Enzymes in milk and dairy products* (pp. 363–395). Springer.

Azarnia, S., Robert, N., & Lee, B. (2006). Biotechnological methods to accelerate Cheddar cheese ripening. *Critical Reviews in Biotechnology, 26*(3), 121–143.

Balthazar, C. F., Guimarães, J. T., Rocha, R. S., Neto, R. P. C., Esmerino, E. A., Silva, H. L. A., Alves Filho, E. G., Pimentel, T. C., Rodrigues, S., Brito, E. S., Tavares, M. I. B., Silva, M. C., Raices, R. S. L., Freitas, M. Q., Granato, D., & Cruz, A. G. (2021). Metabolic profiling of probiotic low-sodium Prato cheese with flavour enhancers: Usefulness of NMR spectroscopy and chemometric tools. *International Dairy Journal, 119*, 104992.

Bancalari, E., Montanari, C., Levante, A., Alinovi, M., Neviani, E., Gardini, F., & Gatti, M. (2020). Lactobacillus paracasei 4341 as adjunct culture to enhance flavor in short ripened Caciotta-type cheese. *Food Research International, 135*, 109284.

Barreto, A., Cruz-Tirado, J.P., Siche, R. and Quevedo, R., 2018. Determination of starch content in adulterated fresh cheese using hyperspectral imaging. *Food Bioscience, 21*, pp.14–19.

Bas, D., Kendirci, P., Salum, P., Govce, G., & Erbay, Z. (2019). Production of enzyme-modified cheese (EMC) with ripened white cheese flavour: I-effects of proteolytic enzymes and determination of their appropriate combination. *Food and Bioproducts Processing, 117*, 287–301.

Bentahar, S., Abada, R., & Nadia, P. Y. (2023). Biotechnology: Definitions, types and main applications. *Ymer, 22*(4), 563–575.

Beresford, T., & Williams, A. (2004). The microbiology of cheese ripening. In *Cheese: Chemistry, physics and microbiology* (Vol. 1, pp. 287–317). Academic Press.

Blazquez, C., Downey, G., O'Callaghan, D., Howard, V., Delahunty, C., Sheehan, E., Everard, C., & O'Donnell, C. P. (2006). Modelling of sensory and instrumental texture parameters in processed cheese by near infrared reflectance spectroscopy. *Journal of Dairy Research, 73*(1), 58–69.

Broel, N., Sowa, M. A., Manhard, J., Siegl, A., Weichhard, E., Zorn, H., Li, B., & Gand, M. (2022). Altering the chain length specificity of a lipase from *Pleurotus citrinopileatus* for the application in cheese making. *Foods, 11*(17), 2608.

Calasso, M., Minervini, F., De Filippis, F., Ercolini, D., De Angelis, M., & Gobbetti, M. (2020). Attenuated Lactococcus lactis and surface bacteria as tools for conditioning the microbiota and driving the ripening of semisoft Caciotta cheese. *Applied and Environmental Microbiology, 86*(5), e02165–e02119.

Ceruti, R. J., Pirola, M. B., Ramos, E., Robert, L., Rubiolo, A. C., & Sihufe, G. A. (2016). Use of an exogenous carboxypeptidase to accelerate proteolysis in Reggianito cheese. *Czech Journal of Food Science, 34*(5), 445–455.

Chen, Y., MacNaughtan, W., Jones, P., Yang, Q., Williams, H., & Foster, T. (2021). Selection of potential molecular markers for cheese ripening and quality prediction by NMR spectroscopy. *LWT, 136*, 110306.

Chintagavongse, N., Takiguchi, H., Ming-Hsuan, C., Tamano, K., Hayakawa, T., Wakamatsu, J. I., Mitani, T., & Kumura, H. (2022). A study of lipolysis induced by adjuncts from edible *Aspergillus* sp. solid culture products on ripened semi-hard cheese. *Journal of the Science of Food and Agriculture, 102*(10), 4355–4362.

Clark, D. P., & Pazdernik, N. J. (2016). *Basics of biotechnology* (pp. 1–31). Academic Cell.

Clark, S., Costello, M., Drake, M. A., & Bodyfelt, F. (Eds.). (2009). *The sensory evaluation of dairy products* (Vol. 251). Springer. https://doi.org/10.1007/978-0387-77408-4

Cocolin, L., Mataragas, M., Bourdichon, F., Doulgeraki, A., Pilet, M. F., Jagadeesan, B., Kalliopi, R., & Phister, T. (2018). Next generation microbiological risk assessment meta-omics: The next need for integration. *International Journal of Food Microbiology, 287*, 10–17. https://doi.org/10.1016/j.ijfoodmicro.2017.11.008

Collins, Y. F., McSweeney, P. L. H., & Wilkinson, M. G. (2004). Lipolysis and catabolism of fatty acids in cheese. In P. F. Fox, P. L. H. McSweeney, T. M. Cogan, & T. P. Guinee (Eds.), *Cheese: Chemistry, physics and microbiology, Vol 1: General aspects* (3rd ed., pp. 373–389). Elsevier.

Conde, T., Mulet, A., Clemente, G., & Benedito, J. (2008). Detection of internal cracks in Manchego cheese using the acoustic impulse-response technique and ultrasounds. *Journal of Dairy Science, 91*(3), 918–927.

Cong, Y. J., Yi, H., & Zheng, F. P. (2015). Discriminating the taste of cheese under different ultrasonic treatment durations by using electronic tongue. *Food Science, 36*, 114–118.

Conte, P., Cinquanta, L., Meo, P. L., Mazza, F., Micalizzi, A., & Corona, O. (2020). Fast field cycling NMR relaxometry as a tool to monitor Parmigiano Reggiano cheese ripening. *Food Research International, 139*, 109845.

Costa, J. R., Pereira, D. A., de Paula, I. L., de Abreu, L. R., Pinto, S. M., Edwards, H. G., Stephani, R., & de Oliveira, L. F. (2022). The taste of a champion: Characterization of artisanal cheeses from the Minas Gerais region (Brazil) by Raman spectroscopy and microstructural analysis. *Journal of Food Composition and Analysis, 112*, 104704.

Curtin, A. C., & McSweeney, P. L. H. (2004). Catabolism of amino acids in cheese during ripening. In P. F. Fox, P. L. H. McSweeney, T. M. Cogan, & T. P. Guinee (Eds.), *Cheese: Chemistry, physics and microbiology, Vol 1: General aspects* (3rd ed., pp. 435–454). Elsevier.

da Silva Medeiros, M. L., de Carvalho, L. M., Madruga, M. S., Rodríguez-Pulido, F. J., Heredia, F. J., & Barbin, D. F. (2024). Comparison of hyperspectral imaging and spectrometers for prediction of cheeses composition. *Food Research International, 183*, 114242.

Dankowska, A., Małecka, M., & Kowalewski, W. (2015). Detection of plant oil addition to cheese by synchronous fluorescence spectroscopy. *Dairy Science and Technology, 95*, 413–424.

Darnay, L., Králik, F., Oros, G., Koncz, Á., & Firtha, F. (2017). Monitoring the effect of transglutaminase in semi-hard cheese during ripening by hyperspectral imaging. *Journal of Food Engineering, 196*, 123–129.

De Filippis, F., Parente, E., & Ercolini, D. (2016). Metagenomics insights into food fermentations. *Microbial Biotechnology, 10*(1), 91–102.

de Sá Oliveira, K., de Souza Callegaro, L., Stephani, R., Almeida, M. R., & de Oliveira, L. F. C. (2016). Analysis of spreadable cheese by Raman spectroscopy and chemometric tools. *Food Chemistry, 194*, 441–446.

Debnath, P. P., & Ghosh, B. C. (2018). Enzyme modified ghee flavour: Application and shelf life studies. *International Journal of Pure and Applied Bioscience, 6*, 822–826.

Dugat-Bony, E., Straub, C., Teissandier, A., Onésime, D., Loux, V., Monnet, C., Irlinger, F., Landaud, S., Leclercq-Perlat, M. N., Bento, P., & Fraud, S. (2015). Overview of a surface-ripened cheese community functioning by meta-omics analyses. *PLoS One, 10*(4), e0124360.

El-Fadaly, H. A., El-Kadi, S. M., Hamad, M. N., & Habib, A. A. (2015). Role of fungal enzymes in the biochemistry of Egyptian Ras cheese during ripening period. *Open Access Library Journal, 2*(8), 1–13.

El-Salam, B. A. E. Y. A., Ahmed, M. S., Yasser, M. M., & Taha, S. S. E. D. (2019). Soft white cheese ripening using bacterial protease enzyme. *Acta Scientiarum Polonorum Technologia Alimentaria, 18*(4), 385–397.

Escobar-Zepeda, A., Sanchez-Flores, A., & Baruch, M. Q. (2016). Metagenomic analysis of a Mexican ripened cheese reveals a unique complex microbiota. *Food Microbiology, 57*, 116–127.

Eskelinen, J. J., Alavuotunki, A. P., Haeggström, E., & Alatossava, T. (2007). Preliminary study of ultrasonic structural quality control of Swiss-type cheese. *Journal of Dairy Science, 90*(9), 4071–4077.

Everard, C. D., O'callaghan, D. J., Fagan, C. C., O'Donnell, C. P., Castillo, M., & Payne, F. A. (2007). Computer vision and color measurement techniques for inline monitoring of cheese curd syneresis. *Journal of Dairy Science, 90*(7), 3162–3170.

Fagan, C. C., O'Donnell, C. P., O'Callaghan, D. J., Downey, G., Sheehan, E. M., Delahunty, C. M., Everard, C., Guinee, T. P., & Howard, V. (2007). Application of mid-infrared spectroscopy to the prediction of maturity and sensory texture attributes of cheddar cheese. *Journal of Food Science, 72*(3), E130–E137.

Falentin, H., Deutsch, S. M., Jan, G., Loux, V., Thierry, A., Parayre, S., Maillard, M. B., Dherbecourt, J., Cousin, F., Jardin, J., Siguier, P., Couloux, A., Barbe, V., Vacherie, B., Wincker, P., Gibrat, J.-F., Gaillardin, C., & Lortal, S. (2010). The complete genome of *Propionibacterium freudenreichii* CIRM-BIA1T, a hardy Actinobacterium with food and probiotic applications. *PLoS One, 5*(7), e11748.

Feng, Y. Z., & Sun, D. W. (2012). Application of hyperspectral imaging in food safety inspection and control: A review. *Critical Reviews in Food Science and Nutrition, 52*(11), 1039–1058.

Forde, A., & Fitzgerald, G. F. (2000). Biotechnological approaches to the understanding and improvement of mature cheese flavour. *Current Opinion in Biotechnology, 11*(5), 484–489. https://doi.org/10.1016/S0958-1669(00)00130-0

Gao, P., Su, Y., Zhang, W., Pang, X., Xie, N., Zhang, M., & Zhang, S. (2022). Chemical and flavor characteristics of enzyme-modified cheese made by two-stage processing. *Gels, 8*(3), 160.

Garbowska, M., Berthold-Pluta, A., Stasiak-Różańska, L., & Pluta, A. (2021). The impact of the adjunct heat-treated starter culture and Lb. helveticus LH-B01 on the proteolysis and ACE inhibitory activity in Dutch-type cheese model during ripening. *Animals, 11*(9), 2699.

Giménez, P., Peralta, G. H., Wolf, I. V., Hynes, E. R., & Bergamini, C. V. (2024). Adjunct cultures of autochthonous lactobacillus strains to diversify and improve cheese maturation. *International Journal of Dairy Technology, 78*. https://doi.org/10.1111/1471-0307.13111

González-Martín, M. I., Severiano-Pérez, P., Revilla, I., Vivar-Quintana, A. M., Hernández-Hierro, J. M., González-Pérez, C., & Lobos-Ortega, I. A. (2011). Prediction of sensory attributes of cheese by near-infrared spectroscopy. *Food Chemistry, 127*(1), 256–263.

González-Martín, I., Hernández-Hierro, J. M., González-Pérez, C., Revilla, I., Vivar-Quintana, A., & Ortega, I. L. (2014). Potential of near infrared spectroscopy for the analysis of volatile components in cheeses. *LWT-Food Science and Technology, 55*(2), 666–673.

Guan, T., Liu, B., Wang, R., Huang, Y., Luo, J., & Li, Y. (2021). The enhanced fatty acids flavor release for low-fat cheeses by carrier immobilized lipases on O/W Pickering emulsions. *Food Hydrocolloids, 116*, 106651.

Guggisberg, D., Fröhlich-Wyder, M. T., Irmler, S., Greco, M., Wechsler, D., & Schuetz, P. (2013). Eye formation in semi-hard cheese: X-ray computed tomography as a non-invasive tool for assessing the influence of adjunct lactic acid bacteria. *Dairy Science and Technology, 93*, 135–149.

Gursoy, O., Somervuo, P., & Alatossava, T. (2009). Preliminary study of ion mobility based electronic nose MGD-1 for discrimination of hard cheeses. *Journal of Food Engineering, 92*(2), 202–207.

Hayashida, S., Hagi, T., Kobayashi, M., Kusumoto, K. I., Ohmori, H., Tomita, S., Suzuki, S., Yamashita, H., Sato, K., Miura, T., & Nomura, M. (2023). Comparison of taste characteristics between koji mold–ripened cheese and Camembert cheese using an electronic tongue system. *Journal of Dairy Science, 106*(10), 6701–6709.

Jahadi, M., Khosravi-Darani, K., Ehsani, M. R., Mozafari, M. R., Saboury, A. A., Zoghi, A., & Mohammadi, M. (2016). Modelling of proteolysis in Iranian brined cheese using proteinase-loaded nanoliposome. *International Journal of Dairy Technology, 69*(1), 57–62.

Jeliński, T., Du, C. J., Sun, D. W., & Fornal, J. (2007). Inspection of the distribution and amount of ingredients in pasteurized cheese by computer vision. *Journal of Food Engineering, 83*(1), 3–9.

Karaca, O. B., & Güven, M. (2018). Effects of proteolytic and lipolytic enzyme supplementations on lipolysis and proteolysis characteristics of white cheeses. *Foods, 7*(8), 125.

Karoui, R., Dufour, É., Pillonel, L., Picque, D., Cattenoz, T., & Bosset, J. O. (2004). Fluorescence and infrared spectroscopies: A tool for the determination of the geographic origin of Emmental cheeses manufactured during summer. *Le Lait, 84*(4), 359–374.

Kendirci, P., Salum, P., Bas, D., & Erbay, Z. (2020). Production of enzyme-modified cheese (EMC) with ripened white cheese flavour: II-effects of lipases. *Food and Bioproducts Processing, 122*, 230–244.

Khattab, A. R., Guirguis, H. A., Tawfik, S. M., & Farag, M. A. (2019). Cheese ripening: A review on modern technologies towards flavor enhancement, process acceleration and improved quality assessment. *Trends in Food Science and Technology, 88*, 343–360.

Kilcawley, K. N. (2017). Cheese flavour. In P. F. Fox, T. P. Guinee, T. M. Cogan, & P. L. H. McSweeney (Eds.), *Fundamentals of cheese science* (pp. 443–474). Springer.

Koca, N., Rodriguez-Saona, L. E., Harper, W. J., & Alvarez, V. B. (2007). Application of Fourier transform infrared spectroscopy for monitoring short-chain free fatty acids in Swiss cheese. *Journal of Dairy Science, 90*(8), 3596–3603.

Kokawa, M., Ikegami, S., Chiba, A., Koishihara, H., Trivittayasil, V., Tsuta, M., Fujita, K., & Sugiyama, J. (2015). Measuring cheese maturation with the fluorescence fingerprint. *Food Science and Technology Research, 21*(4), 549–555.

Kraggerud, H., Naes, T., & Abrahamsen, R. K. (2014). Prediction of sensory quality of cheese during ripening from chemical and spectroscopy measurements. *International Dairy Journal, 34*(1), 6–18.

Kulmyrzaev, A., Bertrand, D., & Dufour, É. (2008). Characterization of different blue cheeses using a custom-design multispectral imager. *Dairy Science and Technology, 88*(4–5), 537–548.

Langa, S., Peirotén, A., Gaya, P., Garde, S., Arqués, J. L., Nuñez, M., Medina, M., & Rodríguez-Mínguez, E. (2020). Human Bifidobacterium strains as adjunct cultures in Spanish sheep milk cheese. *Journal of Dairy Science, 103*(9), 7695–7706.

Lazzi, C., Turroni, S., Mancini, A., Sgarbi, E., Neviani, E., Brigidi, P., & Gatti, M. (2014). Transcriptomic clues to understand the growth of *Lactobacillus rhamnosus* in cheese. *BMC Microbiology, 14*, 1–14.

Le Boucher, C., Courant, F., Jeanson, S., Chéreau, S., Maillard, M. B., Royer, A. L., Thierry, A., Dervilly-Pinel, G., Le Bizec, B., & Lortal, S. (2013). First mass spectrometry metabolic fingerprinting of bacterial metabolism in a model cheese. *Food Chemistry, 141*(2), 1032–1040.

Lei, T., & Sun, D. W. (2019). Developments of nondestructive techniques for evaluating quality attributes of cheeses: A review. *Trends in Food Science and Technology, 88*, 527–542.

Lipkowitz, J. B., Ross, C. F., Diako, C., & Smith, D. M. (2018). Discriminating aging and protein-to-fat ratio in Cheddar cheese using sensory analysis and a potentiometric electronic tongue. *Journal of Dairy Science, 101*(3), 1990–2004.

Malegori, C., Oliveri, P., Mustorgi, E., Boggiani, M. A., Pastorini, G., & Casale, M. (2021). An in-depth study of cheese ripening by means of NIR hyperspectral imaging: Spatial mapping of dehydration, proteolysis and lipolysis. *Food Chemistry, 343*, 128547.

Mazzei, P., & Piccolo, A. (2012). 1H HRMAS-NMR metabolomic to assess quality and traceability of mozzarella cheese from Campania buffalo milk. *Food Chemistry, 132*(3), 1620–1627.

McSweeney, P. L. H. (2011). Cheese: Biochemistry of cheese ripening. In *Encyclopedia of dairy sciences* (2nd ed., pp. 667–674). https://doi.org/10.1016/B978-0-12374407-4.00080-7

McSweeney, P. L. H., & Fox, P. F. (2004). Metabolism of residual lactose and of lactate and citrate. In P. F. Fox, P. L. H. McSweeney, T. M. Cogan, & T. P. Guinee (Eds.), *Cheese: Chemistry, physics and microbiology, Vol 1: General aspects* (3rd ed., pp. 361–372). Elsevier.

McSweeney, P. L., & Sousa, M. J. (2000). Biochemical pathways for the production of flavour compounds in cheeses during ripening: A review. *Le Lait, 80*(3), 293–324.

Nassar, G., Lefbvre, F., Skaf, A., Carlier, J., Nongaillard, B., & Noêl, Y. (2010). Ultrasonic and acoustic investigation of cheese matrix at the beginning and the end of ripening period. *Journal of Food Engineering, 96*(1), 1–13.

Nicolau, N., Buffa, M., O'callaghan, D. J., Guamis, B., & Castillo, M. (2015). Estimation of clotting and cutting times in sheep cheese manufacture using NIR light backscatter. *Dairy Science and Technology, 95*, 495–507.

Nivetha, A., & Mohanasrinivasan, V. (2017). Mini review on role of β-galactosidase in lactose intolerance. *IOP Conference Series: Materials Science and Engineering, 263*(2), 022046.

O'Callaghan, D. J., & Guinee, T. P. (2004). Rheology and texture of cheese. In P. F. Fox, P. L. H. McSweeney, T. M. Cogan, & T. P. Guinee (Eds.), *Cheese: Chemistry, physics and microbiology, Vol 1: General aspects* (3rd ed., pp. 511–540). Elsevier.

Ozbekova, Z., & Kulmyrzaev, A. (2017). Fluorescence spectroscopy as a non destructive method to predict rheological characteristics of Tilsit cheese. *Journal of Food Engineering, 210*, 42–49.

Piras, C., Marincola, F. C., Savorani, F., Engelsen, S. B., Cosentino, S., Viale, S., & Pisano, M. B. (2013). A NMR metabolomics study of the ripening process of the Fiore Sardo cheese produced with autochthonous adjunct cultures. *Food Chemistry, 141*(3), 2137–2147.

Pisano, M. B., Scano, P., Murgia, A., Cosentino, S., & Caboni, P. (2016). Metabolomics and microbiological profile of Italian mozzarella cheese produced with buffalo and cow milk. *Food Chemistry, 192*, 618–624.

Priyashantha, H., Höjer, A., Saedén, K. H., Lundh, Å., Johansson, M., Bernes, G., Geladi, P., & Hetta, M. (2020). Use of near-infrared hyperspectral (NIR-HS) imaging to visualize and model the maturity of long-ripening hard cheeses. *Journal of Food Engineering, 264*, 109687.

Priyashantha, H., Höjer, A., Saedén, K. H., Lundh, Å., Johansson, M., Bernes, G., Geladi, P., & Hetta, M. (2021). Determining the end-date of long-ripening cheese maturation using NIR hyperspectral image modelling: A feasibility study. *Food Control, 130*, 108316.

Rani, S., & Jagtap, S. (2019). Acceleration of Swiss cheese ripening by microbial lipase without affecting its quality characteristics. *Journal of Food Science and Technology, 56*, 497–506.

Ruyssen, T., Janssens, M., Van Gasse, B., Van Laere, D., Van der Eecken, N., De Meerleer, M., Vermeiren, L., Van Hoorde, K., Martins, J. C., Uyttendaele, M., & De Vuyst, L. (2013). Characterisation of Gouda cheeses based on sensory, analytical and high-field 1H nuclear magnetic resonance spectroscopy determinations: Effect of adjunct cultures and brine composition on sodium-reduced Gouda cheese. *International Dairy Journal, 33*(2), 142–152.

Scano, P., Cagliani, L. R., & Consonni, R. (2019). ^{1}H NMR characterisation of the lipid fraction and the metabolite profiles of Fossa (pit) cheese. *International Dairy Journal, 90*, 39–44.

Schlossareck, C., & Ross, C. F. (2019). Electronic tongue and consumer sensory evaluation of spicy paneer cheese. *Journal of Food Science, 84*(6), 1563–1569.

Schuetz, P., Guggisberg, D., Jerjen, I., Fröhlich-Wyder, M. T., Hofmann, J., Wechsler, D., Flisch, A., Bisig, W., Sennhauser, U., & Bachmann, H. P. (2013). Quantitative comparison of the eye formation in cheese using radiography and computed tomography data. *International Dairy Journal, 31*(2), 150–155.

Schuetz, P., Guggisberg, D., Fröhlich-Wyder, M. T., & Wechsler, D. (2016). Software comparison for the analysis of cheese eyes in X-ray computed tomography. *International Dairy Journal, 63*, 62–69.

Shahin, A. A., Nasser, A. M., & Ragab, W. A. (2023). Effect of addition of some enzymes on the acceleration of Ras cheese manufactured with addition of goat's Milk. *Egyptian Journal of Food Science, 51*(1), 101–111.

Sihufe, G. A., Zorrilla, S. E., Perotti, M. C., Wolf, I. V., Zalazar, C. A., Sabbag, N. G., Costa, S. C., & Rubiolo, A. C. (2010). Acceleration of cheese ripening at elevated temperature. An estimation of the optimal ripening time of a traditional Argentinean hard cheese. *Food Chemistry, 119*, 101–107.

Siroli, L., Patrignani, F., D'Alessandro, M., Salvetti, E., Torriani, S., & Lanciotti, R. (2020). Suitability of the Nisin Z-producer *Lactococcus lactis* subsp. *lactis* CBM 21 to be used as an adjunct culture for Squacquerone cheese production. *Animals, 10*(5), 782.

Smit, G., van Hylckama Vlieg, J. E. T., Smit, B. A., Ayad, E. H. E., & Engels, W. J. M. (2002). Fermentative formation of flavour compounds by lactic acid bacteria. *Australian Journal of Dairy Technology, 7*, 61–68.

Sousa, M. J., Ardo, Y., & McSweeney, P. L. H. (2001). Advances in the study of proteolysis in cheese during ripening. *International Dairy Journal, 11*, 327–345.

Subramanian, A., Harper, W. J., & Rodriguez-Saona, L. E. (2009). Cheddar cheese classification based on flavor quality using a novel extraction method and Fourier transform infrared spectroscopy. *Journal of Dairy Science, 92*(1), 87–94.

Subramanian, A., Alvarez, V. B., Harper, W. J., & Rodriguez-Saona, L. E. (2011). Monitoring amino acids, organic acids, and ripening changes in Cheddar cheese using Fourier-transform infrared spectroscopy. *International Dairy Journal, 21*(6), 434–440.

Sundarraj, A. A., Rajathi, A. A., Vishaal, S. C., Rohit, D., Prakash, M. S., Sam, A. A., & Seihenbalg, S. S. (2018). Food biotechnology applications in dairy and dairy products. *Journal of Pharmacy Research, 12*(4), 520–525.

Taha, S., El Abd, M., De Gobba, C., Abdel-Hamid, M., Khalil, E., Hassan, F. U., & Fathy, D. (2020). The multifunctional activity of water-soluble peptides' extract of Domiati cheese during accelerated ripening by Neutrase. *Journal of Food Processing and Preservation, 44*(5), e14434.

Trujillo, A. J., Guamis, B., Laencina, J., & Lopez, M. B. (2000). Proteolytic activities of some milk clotting enzymes on ovine casein. *Food Chemistry, 71*, 449–457.

Upadhyay, V. H., McSweeney, P. L. H., Magboul, A. A. A., & Fox, P. F. (2004). Proteolysis in cheese during ripening. In P. F. Fox, P. L. H. McSweeney, T. M. Cogan, & T. P. Guinee (Eds.), *Cheese: Chemistry, physics and microbiology* (Vol. 1, pp. 391–433). Elsevier Academic Press.

Uzuner, E. (2009). *Use of rice milk in probiotic yogurt production*. Ege University. Scientific Research Projects.

Valente, N. I., Rudnitskaya, A., Oliveira, J. A., Gaspar, E. M., & Gomes, M. T. S. (2018). Cheeses made from raw and pasteurized cow's milk analysed by an electronic nose and an electronic tongue. *Sensors, 18*(8), 2415.

Vásquez, N., Magán, C., Oblitas, J., Chuquizuta, T., Avila-George, H., & Castro, W. (2018). Comparison between artificial neural network and partial least squares regression models for hardness modeling during the ripening process of Swiss-type cheese using spectral profiles. *Journal of Food Engineering, 219*, 8–15.

Verma, D. K., Al-Sahlany, S. T. G., Niamah, A. K., Thakur, M., Shah, N., Singh, S., Baranwal, D., Patel, A. R., Utama, G. L., & Aguilar, C. N. (2022). Recent trends in microbial flavour compounds: A review on chemistry, synthesis mechanism and their application in food. *Saudi Journal of Biological Sciences, 29*(3), 1565–1576.

Wang, H. H., & Sun, D. W. (2004). Evaluation of the oiling off property of cheese with computer vision: Influence of cooking conditions and sample dimensions. *Journal of Food Engineering, 61*(1), 57–66.

Wang, C., Gao, L., Gao, Y., Yang, G., Zhao, Z., Zhao, Y., Wang, J., & Li, S. (2023). Evaluation of *Pediococcus acidilactici* AS185 as an adjunct culture in probiotic cheddar cheese manufacture. *Food Science and Nutrition, 11*(3), 1572–1583.

Wilkinson, M. G., Doolan, I. A., & Kilcawley, K. N. (2011). Cheese: Enzyme-modified cheese. In *Encyclopedia of dairy sciences* (2nd ed., pp. 799–804). https://doi.org/10.1016/B978-0-12-374407-4.00094-7

Wilson, A. D., & Baietto, M. (2009). Applications and advances in electronic-nose technologies. *Sensors, 9*(7), 5099–5148.

Yaman, H., Aykas, D. P., Jiménez-Flores, R., & Rodriguez-Saona, L. E. (2022). Monitoring the ripening attributes of Turkish white cheese using miniaturized vibrational spectrometers. *Journal of Dairy Science, 105*(1), 40–55.

Yang, Y., Xia, Y., Li, C., Wang, G., Xiong, Z., Song, X., Zhang, H., Wang, M., & Ai, L. (2024). Metabolites, flavor profiles and ripening characteristics of Monascus-ripened cheese enhanced by Ligilactobacillus salivarius AR809 as adjunct culture. *Food Chemistry, 436*, 137759.
Yerlikaya, O. (2014). Starter cultures used in probiotic dairy product preparation and popular probiotic dairy drinks. *Food Science and Technology, 34*, 221–229.
Yousefi, L., Dovom, M. R. E., Najafi, M. B. H., & Mortazavian, A. M. (2021). Antioxidant activity of ultrafiltered-Feta cheese made with adjunct culture during ripening. *Journal of Food Measurement and Characterization, 15*(5), 4336–4342.
Yvon, M., & Rijnen, L. (2001). Cheese flavour formation by amino acid catabolism. *International Dairy Journal, 11*(4–7), 185–201.
Zhang, K., Jia, M., Guo, Z., Li, Y., Li, B., & Li, X. (2021). Evaluation of bacterial diversity of traditional cheese in Tarbagatay prefecture, China, and its correlation with cheese quality. *Food Science and Nutrition, 9*(6), 3155–3164.
Zhang, Y., Gao, P., Zhang, W., Zhu, H., Wang, C., Xie, N., & Zhang, S. (2023). Free fatty acid hydrolyzed with lipases and their effects on enzyme-modified cheese flavor. *Food Science of Animal Products, 1*(3), 9240031.
Zhao, J., Ma, M., Yan, X., Zhang, G., Xia, J., Zeng, G., Tian, W., Bao, X., Zeng, Z., Yu, P., & Gong, D. (2022). Expression and characterization of a novel lipase from Bacillus licheniformis NCU CS-5 for application in enhancing fatty acids flavor release for low-fat cheeses. *Food Chemistry, 368*, 130868.
Zhao, S., Dorau, R., Tømmerholt, L., Gu, L., Tadesse, B. T., Zhao, G., & Solem, C. (2023). Simple and better–Accelerated cheese ripening using a mesophilic starter based on a single strain with superior autolytic properties. *International Journal of Food Microbiology, 407*, 110398.

Chapter 7
Overview of the Manufacturing Processes

Shivam Kayasth and Gauri Thakur

7.1 Introduction

One common dairy product that is consumed is cheese. Cheese comes in a vast array of kinds, produced all over the world, including gouda, cheddar, feta, and camembert. The worldwide cheese market was projected to be around 83.4 billion dollars in 2022; by 2028 (IMARC Group et al., 2024), this amount is expected to rise to over 120 billion dollars. Cheese production holds a significant place in both traditional dairy and the emerging plant-based food sectors. Over the years, cheese-making methods have advanced from artisanal practices to highly mechanized industrial processes. This evolution has allowed for the creation of a wide variety of cheeses, catering to different consumer preferences and dietary needs. While traditional cheeses like stretched and whey cheeses remain popular, Duto rising interest in sustainability, health, and ethical consumption, there is a growing market for alternative goods such vegan and hybrid cheeses (Fox & McSweeney, 2017). The benefits of cheese consumption are becoming more widely known because of particular nutrients like zinc, calcium, phosphorus, vitamin B12, and vitamin A needed for strong bones and muscles and it is driving the trends in the cheese market. The Key factors propelling the worldwide cheese market are rising interest in foods high in protein, and shifting consumer tastes, which account for half of the cheese produced, and the food service industry's growth and the expansion of distribution networks.

S. Kayasth (✉)
Department of Food Science and Technology, National Institute of Food Technology Entrepreneurship and Management-Kundli (NIFTEM-K), Sonipat, Haryana, India

G. Thakur
Department of Food Engineering and Technology, Institute of Chemical Technology (ICT), Mumbai, India

S. Mansha Rafiq (ed.), *Technological Advances and Trends in Cheese Making*,
https://doi.org/10.1007/978-3-031-94476-5_7

Technological developments in the cheese-making industry have been essential in satisfying the increasing demand while preserving product quality. Modern advancements in cheese production combine innovation with tradition. Dairy-based cheese production has seen improvements through mechanization and better bacterial cultures (Parente et al., 2017), particularly in stretched cheeses, where stretching techniques have become more efficient. At the same time, plant-based cheese alternatives have made significant strides in areas like texture, meltability, and flavor, especially in vegan cheese production. Hybrid cheeses, which blend dairy and plant-based ingredients, represent a growing trend aimed at merging the best qualities of both.

By examining the manufacturing processes of key cheese varieties including vegan, whey, hybrid, stretched, and dry milk cheeses important insights can be gained into their composition, production standards, methods, and end uses. Additionally essential to preserving product quality, shelf life, and sensory appeal are packaging and storage. The combination of traditional dairy practices with new plant-based innovations reflects the evolving nature of the cheese industry, shaped by technological progress and changing consumer preferences.

7.2 Overview of Different Cheese Types

There are many different types of cheese produced in the world, and efforts have been made to categorize them into groups that are somewhat similar. Many of these strategies are covered, such as those that rely on the texture and content of cheese, milk coagulation techniques, and ripening indices. There are a lot of different kinds of cheese. Cheeses can be divided into groups according to a variety of criteria, such as animal milk, country or region of origin, texture, fat content, length of fermentation, and methods of manufacturing.

The most widely used and conventional strategy is based on moisture content; results are then further refined using fat content and curing/ripening procedures. This results in a variety of cheeses, such as aged cheeses (like gouda and cheddar), fresh cheeses (like ricotta and cottage cheese), blue cheeses (like gorgonzola and roquefort), hard cheeses (like parmesan and asiago), and many more. The criteria can be applied separately or in combination, and no single technique is always applicable (Table 7.1).

7.3 Traditional Cheese Manufacturing Processes

Cheese comes in a variety of varieties (Fox et al., 2017a). The variety affects the cheese's properties, composition, and processing. Cheese is frequently separated into two groups: acid or rennet cheeses and natural or processed cheeses. Acid is added to milk to cause the proteins to coagulate and create acid cheeses. Fresh

Table 7.1 Indian cheeses and their unique characteristic

Cheese name	Description
Chhurpi	A traditional cheese from the Himalayan region made from yak milk, available in soft and hard varieties. Rich in nutrition and often used in local dishes
Kalari	Known as the "Mozzarella of Kashmir," this dense cheese is made from cow or goat's milk and is typically fried, enjoyed as a street food delicacy
Chhena	A crumbly cheese from West Bengal, similar to paneer, used in various Bengali sweets like rasgulla and Sandesh
Bandel	A smoky-flavored cheese from West Bengal with a dry texture, often enjoyed on its own or added to dishes
Qudam	A salty and chewy cheese from the Kashmiri Gujjar tribes, known for its rubber-like texture and long shelf life
Topli Na paneer	Also known as basket cheese, this Parsi cheese is soft and salty, typically enjoyed raw or with crackers
Churu	An aromatic cheese from the Himalayan regions, characterized by its pungent flavor and blue mold
Kalimpong	Originating from Sikkim, this cheese has a tangy flavor similar to gouda and is often enjoyed with wine

cheeses like cream cheese and queso fondo are made by direct acidification. The majority of cheeses, including swiss and cheddar, coagulate the milk using rennet, an enzyme, together with starting cultures. In the industry, cheese made exclusively from milk is referred to as "natural cheese." To make process cheese, natural cheese is combined with other additives and baked to alter its texture, melt consistency, and extend its shelf life. The art of producing cheese is called caseiculture, or cheese production. Like many other food preservation procedures, the making of cheese enables the concentrated preservation of a food material's nutritional and financial value.

7.4 General Manufacturing Processes

The process used to produce cheese is largely similar, with certain adjustments made to obtain desired results in various kinds. Distinct forms of cheese have distinct processing sequences, temperatures, durations, and target pHs for certain processes, as well as differences in block formation, aging, and usage of brining or salting. A very rough summary of the stages involved in creating cheese is shown in the flow chart below.

Standardization of milk

↓

Pasteurization/Heat treatment of milk

↓

Milk cooling

↓

Innoculate Non-starter, starter bacteria and ripen it

↓

Add rennet for formation of curd

↓

Cutting & shaping of curd

↓

Whey drainage

↓

Texture curd

↓

Dry salting or brining

↓

Cheese fromation into blocks

↓

Storage & aging of cheese

↓

Packaging

The initial phase in the procedure is standardizing milk to maximize the protein-to-fat ratio and assure superior cheese with a good yield. This is followed by mild pasteurization (72–74 °C for 15 s) to reduce spoilage organisms and inactive heat-sensitive enzymes, creating a favorable environment for starter cultures. Bactofugation may be used to lower spore counts, particularly for cheeses like Gouda and Swiss. Calcium chloride can be added to enhance coagulation, after which the milk is cooled to about 32 °C and inoculated with a starter culture, allowing it to ripen for 30–40 min to develop flavor through fermentation. Coagulation, facilitated by rennin, separates the milk into curd and whey, typically taking around 30 min. Next, for better whey evacuation, the curd is chopped into little pieces. For cheddar, this is heated to 38 °C to improve curd hardness. The whey is drained through heat or mechanical means, and in some cases, specific cultures like penicillium are added before this step. The curd undergoes texturizing, or cheddaring, where it is stacked and flipped to expel additional whey and continue fermentation, achieving a pH of 5.1 to 5.5. After milling the curd into smaller pieces, it is salted either by dry salting or brining, which aids drainage and flavor development. After that, the salted curd is pressed into blocks, which shapes the cheese and releases additional whey. Afterwards that, the cheese is packaged to prevent microbes and moisture loss and kept in coolers till it reaches its proper aging time, which can take anywhere from months to years.

7.5 Composition of Different Types of Cheeses (Table 7.2)

Table 7.2 Composition of stretched, hybrid, dry milk, vegan cheeses

Component	Stretched cheeses	Whey cheeses	Hybrid cheeses	Dry milk cheeses	Vegan cheeses	References
Moisture content	40–60%	70–80%	30–50%	2–5%	50–65%	Díaz et al. (2022)
Fat content	20–30%	10–24%	20–40%	20–30%	15–25%	Fernández et al. (2023)
Protein content	20–30%	8–12%	20–30%	30–40%	10–20%	Kumar and Sharma (2023)
Salt content	1.5–2.5%	–	–	–	1.0–2.0%	

7.6 Stretched Cheese

7.6.1 Definition and Characteristics

Stretched cheese, often referred to as spun curd, pasta filata, or string cheese, is a type of cheese produced by elongating curds into thin, stretched strands. "Stretch is the term used to characterize melted cheese's capacity to form fibrous strands that lengthen when tugged or stretched."

Stretched cheeses are known for their unique elastic and fibrous texture, which results from the heating and stretching of curds, allowing for the formation of long strands when melted. They typically have a higher moisture content compared to other cheese varieties, contributing to a soft and creamy mouthfeel. The flavor profile of stretched cheeses ranges from mild to slightly tangy, with fresh varieties being milder and aged ones exhibiting more complex flavors. Visually, these cheeses often present a shiny, smooth exterior and may release a white liquid upon cutting. Their versatility makes them popular in various culinary applications, including pizzas, salads, and sandwiches, due to their excellent melting properties.

7.6.2 Types of Stretched Cheese

Stretched cheeses include a variety of unique types, each with its own distinct features. Mozzarella is a fresh cheese recognized for its mild flavor and shiny, moist exterior, often releasing a white liquid when cut. Provolone is a semi-hard Italian cheese characterized by its soft, elastic, and stringy texture. Caciocavallo is distinguished by its gourd-like or teardrop shape, typical of the Molise and Southern Italy regions. Another notable type is Burrata, which contains a creamy filling of mozzarella, making it an excellent choice for grilled cheese sandwiches that require light toasting. Scamorza, a cow's milk cheese, matures in its own whey for several hours, enhancing its taste. Halloumi resembles mozzarella in its stretching technique but has a firmer texture, softening slightly when heated, which makes it ideal for grilling or frying. Finally, Stracciatella di bufala, originating from the Apulia region of southern Italy, is crafted from buffalo milk and contains shreds of stretched white curd.

7.6.3 Manufacturing Process of Stretched Cheese

The production of stretched cheese, such as mozzarella, involves several meticulously controlled steps, each critical to developing the cheese's unique texture and flavor. The process begins with the selection of high-quality milk, typically from water buffalo or cows. The milk is pasteurized at a temperature of approximately 72 °C (161 °F) for 15 s to eliminate pathogenic bacteria while preserving the natural flavors and beneficial microorganisms essential for fermentation. Some artisanal methods may use lower temperatures, such as 63 °C (145 °F) for 30 min, for a gentler pasteurization process.

Milk Preparation

↓

Acidification and rennet addition

↓

Cutting the Curd

↓

Heating and Washing

↓

Stretching

↓

Shaping and Brining

↓

Cooling and Packaging

The milk is chilled to around 30–32 °C (86–90 °F) after pasteurization. Lactic acid bacteria-based starter cultures are introduced to the milk to initiate fermentation, bringing the milk's pH down to roughly 5.2. In order for flavor and texture to develop, this acidity is essential. Rennet is added at a concentration of roughly 0.02–0.05% of the milk volume after fermentation. Within 30 to 60 min, this enzyme coagulates the milk, creating a curd that resembles gel. After the curd sets, it is sliced into tiny pieces, usually measuring 1–2 cm (0.4–0.8 in.). This cutting allows for efficient whey drainage. The curds are then allowed to rest for about 5–10 min to firm up before the next step. The curds are gently heated to around 40–45 °C (104–113 °F) while being stirred to promote whey expulsion. This step can take 30–45 min. In some processes, hot water (around 85 °C or 185 °F) may be added to the curds to enhance the drainage of whey and prepare them for stretching. The hallmark of stretched cheese production occurs next. The curds are drained and placed in hot water (about 70 °C or 160 °F) or steam until they become pliable, usually for about 5–10 min. The curds are then kneaded and stretched by hand or machine. This process involves repeatedly pulling and folding the curd until it reaches a smooth, elastic consistency, typically taking 10–15 min. The desired outcome is long, stringy strands of cheese that can be shaped into various forms. After stretching, the cheese is shaped into balls, logs, or other desired forms. The shaped cheese is then immersed in a brine solution, typically containing 8–10% salt, for about 30 min to several hours. This step enhances flavor and helps with preservation by inhibiting microbial growth. Finally, the cheese is cooled quickly, either in a cold water bath or by air cooling, to stop the cooking process. It is then packaged in moisture-*retentive* materials, often with some brine, to maintain freshness and prevent drying out. Proper storage conditions, typically at 4 °C (39 °F), are crucial for preserving the cheese's flavor and texture.

7.6.4 Recent Advancements in Stretched Cheese Production

7.6.4.1 Automation and Mechanization

Automation and mechanization have revolutionized the cheese production process, enhancing efficiency, consistency, and safety. Advanced automated systems are now employed across multiple stages of stretched cheese manufacturing, including pasteurization, curd cutting, and stretching. These systems enable precise control of critical parameters such as temperature, pH, and timing, which are essential for achieving the desired quality and safety standards. Robotic technology has also been integrated into curd handling processes, minimizing human intervention and reducing the risk of contamination. Automated curd manipulators can ensure uniformity in curd size and texture, thereby improving the final product's consistency. Furthermore, the implementation of continuous monitoring systems allows for real-time adjustments, optimizing production efficiency and product quality (O'Brien & Kelly, 2020).

7.6.4.2 Improved Bacterial Cultures

The manufacturing of stretched cheeses has been greatly improved by developments in the utilization of bacterial cultures, allowing for the formation of distinctive flavors and improved texture. Here customized starter cultures are used, made up of particular lactic acid bacteria strains chosen for their favorable fermentation properties. The accurate identification of strains such Lactococcus lactis & Streptococcus thermophilus, respectively, which contribute to the ideal stretchability and flavor profiles, is made possible by molecular techniques, such as DNA sequencing. Moreover, the incorporation of functional cultures, such as *Lactobacillus rhamnosus* probiotics, improves gut health and increases cheese shelf life by squelching rotting germs. Innovations in fermentation control, such as real-time monitoring and automated systems, enable manufacturers to maintain optimal conditions throughout the fermentation process, resulting in consistent product quality. The exploration of indigenous bacterial strains also offers opportunities to create cheeses that reflect regional flavors, thereby appealing to consumers seeking authenticity in their food choices.

7.6.4.3 Innovations in Stretching Techniques

Innovations in stretching techniques have greatly improved the physical properties of stretched cheeses. Traditional methods are being complemented by novel approaches, including microwave-assisted stretching and enzymatic treatments. Microwave-assisted techniques facilitate rapid and uniform heating of curds, enhancing elasticity and texture while reducing processing *time* (O'Brien & Kelly, 2020). Enzymatic treatments have been developed to optimize the stretchability of curds without compromising flavor integrity. These methods allow for precise control over the physical properties of the cheese, enabling producers to tailor products to meet specific market demands. Additionally, the design of specialized machinery for automated stretching processes has provided greater control over critical parameters, such as temperature and stretching duration, which are crucial for achieving the desired final characteristics.

7.7 Vegan Cheese

7.7.1 Definition and Characteristics

It has been acknowledged that plant-based or vegan cheeses, which are produced from a variety of plant-derived materials like fruits, vegetables (leaves, stems, blossoms, and tubers), and fruit seeds, are healthy substitutes (Pua et al., 2022). It may defined as "a food product made from plant ingredients, designed to resemble the appearance, texture, and flavor of animal-based cheeses." Creating plant-based cheeses that match the physical, sensory, and nutritional qualities of dairy cheese presents a significant challenge due to the intricate structure and composition of cheese. It's also essential to remember that laws in many nations restrict the terms associated to dairy being used for plant-based goods, which could have an impact on how these products are labeled. The belief that natural plant-based lipids are healthier than the components in dairy products manufactured from cow's milk has led to a rise in interest in plant-based cheeses during the past 10 years (Walther et al., 2022). Moreover, plant-based cheese offers a good substitute for people who cannot tolerate dairy products. But with plant-based proteins, allergenicity is still a big concern since some immune responses can cause mild to severe allergic reactions.

7.7.2 Key Ingredients in Vegan Cheese

The main ingredients used in vegan cheese production include a mix of fats, proteins, and carbohydrates sourced from various plant materials. These components help replicate the creamy, rich texture and flavor associated with dairy cheese. Such as follows:

(1) Plant-based fats—Coconut oil, palm oil, and other plant-derived oils are commonly used to provide the smooth texture and richness that are typical of cheese. Fats play an important role in mouthfeel, meltability, and overall texture. (2) Starches and Gums—Starches such as tapioca, potato, and corn starch are used to achieve a gelatinous, stretchable texture. They help create the firm structure needed in products like shredded or sliced cheese. Gums like xanthan, carrageenan, and agar-agar are utilized to stabilize the texture and improve shelf life. (3) Proteins and fermentation—Plant-based proteins, including soy, pea, and nut proteins, contribute to the structure and texture of vegan cheese. Fermented ingredients such as nutritional yeast add depth of flavor, mimicking the umami and tanginess found in traditional cheeses. In some cases, vegan cheeses are aged to further develop these flavors (Ismail et al., 2020a, b) (Table 7.3).

7.7.3 Manufacturing Process

Plant based cheese is produced similarly to traditional cheesemaking but uses plant-based ingredients. The process begins with homogenizing water and plant oils to create a smooth emulsion. Starch is then added and heated to undergo gelatinization, forming the base structure. To add texture, plant proteins (such soy or pea) are added after cooling. Texturizing agents enhance the consistency, and acidulants—such as citric acid—are added to raise the pH and encourage the coagulation of proteins. To get a texture, mouthfeel, and melting quality similar to cheese, the mixture is then molded, chilled, and stored. A very fundamental flowchart of the stages involved in manufacturing cheese found is below (Kovačević et al., 2024).

Table 7.3 Key ingredients in vegan cheese

Ingredient	Sources
Carbohydrates	Starches from tapioca, potato, and corn
Plant proteins	Derived from legumes, nuts, and seeds
Vegetable oils	Extracted from coconut, cocoa, and palm
Salt	Sodium Phophate and sodium citrate
Texturizers	Agar, xanthan gum and alginic acid
Acidulants	Citric acid, acetic acid and lactic acid

Water

↓

Homogenization

↓

Heating and Mixing

↓

Starch Gelatinization

↓

Cooling and Homogenization

↓

Addition of Texturizing Agents

↓

Heating and Mixing (Second Phase)

↓

Acidification

↓

Forming and Storage

The process start with water addition. Water serves as the primary base and solvent for various ingredients. It is crucial in hydrating starches, proteins, and other dry ingredients that will form the final cheese matrix. Water content in plant-based cheese typically ranges from 35% to 74.5%, as reported by Bergsma (2017). Controlling the final product's texture & shelf life depends on the water balance. Adequate hydration helps starches swell and proteins disperse, facilitating further processing steps. The first major step involves homogenization of water with oil. This step uses high-shear mixing to break down oil droplets and create a stable oil-in-water emulsion, an essential step to achieve the creamy and smooth texture of the final cheese product. Plant derived fats like coconut, the sunflower, or palm oils are frequently utilized in plant-based cheese. These oils help achieve the proper mouthfeel and meltability by imitating the fat level of dairy cheese. Typically, 15–35% of the mass of the cheese is made up of fat (Bergsma, 2017). During this phase, emulsifiers like lecithin or mono-and diglycerides are added to stabilize the emulsion and stop the separation of water and oil. Emulsifiers also improve the smooth texture and keep oil from pooling during melting, as Chen et al. (2019) point out.

The emulsion is subjected to heating to facilitate ingredient dispersion. Heating increases the fluidity of the mixture and aids in dissolving the starches and proteins, ensuring they are evenly distributed throughout the matrix. Starch (from sources such as corn, potato, rice, or tapioca) is incorporated at this stage. Starches are key to forming the gel structure of plant-based cheese, contributing to the firmness, elasticity, and stability of the final product. The process of heating causes starch to gelatinize, which causes the granules to absorb water, swell, and finally burst to form a gel-like structure that gives the cheese its body and texture. As Bergsma (2017) described, starch can be used either in its modified form (chemically or enzymatically modified to improve water-holding and gelling capacity) or in its non-modified form. Modified starches (like pre-gelatinized starches) are often employed for better control over texture and water retention. During this process, starch typically makes up 10–24% of the cheese mixture (Bergsma, 2017). A gel network is created as a result of gelatinization, and when the starch granules expand, they absorb water and oil, giving the cheese its viscoelastic qualities and causing it to act more like traditional dairy cheese.

To cause a sol-gel transition, the mixture is chilled once the starches have gelatinized. The starch is able to solidify into a more crystalline form because to this process. The gel can solidify during cooling, which is required to achieve the ultimate structure of the cheese. Plant proteins, such as soy protein isolate or pea protein isolate, are now included. Proteins play a central role in the creation of the cheese matrix, similar to how casein works in dairy cheese. The proteins form a network that traps water and oil, contributing to the firm structure of the cheese. As explained by Wang et al. (2017), the size and aggregation of protein particles influence the cheese's rheological properties, such as its elasticity and texture. Homogenization at this stage ensures that the proteins, starches, and fats are fully integrated into a uniform matrix. The dispersion of protein aggregates improves texture and emulsification by forming a network that supports oil and water evenly. Texturizing agent such as hydrocolloids (carrageenan, guar gum, or xanthan gum), are added to cheese to enhance its mechanical qualities. These agents increase the elasticity, improve the mouthfeel, and provide structural integrity to the cheese. Grossmann and McClements (2021) emphasize that the role of texturizing agents in plant-based cheese is similar to the role of casein micelles in dairy cheese, which provides the cohesive, stretchy nature of the product. A second heating and mixing phase is performed to ensure that the proteins and texturizing agents are fully incorporated into the gel network. The temperature is carefully controlled to avoid overcooking or breaking down the emulsion. As Wang et al. (2017) notes, controlling protein aggregation and matrix formation is crucial at this step, as the degree of protein denaturation and aggregation affects the rheology, meltability, and overall texture of the final product.

To reduce pH and cause acidification, acidulants like citric acid are applied. Plant protein gelation characteristics are largely controlled by the acidification process. Proteins can create a tighter network and increase the final product's firmness and mouthfeel by lowering the pH. According to Bergsma (2017) and Grasso et al. (2021), plant-based cheeses should ideally have a pH of between 5.21 and 5.87. This pH range is crucial for maintaining the proper protein-protein interactions that lead to texture, stability, and flavor development. It is comparable to that of regular dairy cheese. Cheese's shelf life is also impacted by acidification since microbiological growth is inhibited at lower pH levels. The cheese mixture is put into molds or other forms after acidification.

7.7.4 Recent Advancements in Vegan Cheeses

7.7.4.1 Improved Texture and Meltability

Recent innovations in vegan cheese production have significantly enhanced the texture and meltability of these products, making them more comparable to traditional dairy cheeses. Research by Kovačević et al. (2024) emphasizes that the incorporation of advanced emulsification techniques and the use of hydrocolloids, such as xanthan gum and locust bean gum, improve the creaminess and meltability of vegan cheeses. For instance, the development of emulsified blends that utilize coconut oil and starches has been shown to mimic the textural properties of dairy cheese effectively (Dannenberg et al., 2021). Moreover, the optimization of processing conditions, including temperature and shear rates, has been crucial in achieving desirable melt *characteristics* (Wang et al., 2017).

7.7.4.2 Enhanced Nutritional Profiles

The nutritional profiles of vegan cheeses have also improved, with a growing focus on fortifying these products with essential nutrients. For example, the addition of pea protein isolate not only enhances protein content but also provides a favorable amino acid profile, contributing to muscle maintenance and repair (Brennan et al., 2019). Furthermore, fortification with vitamins such as B12 and D, along with minerals like calcium and phosphorus, helps address common deficiencies in plant-based diets (Borchers et al., 2020). Research shows that contemporary vegan cheeses can match or even surpass dairy counterparts in terms of nutritional value, particularly when functional ingredients like omega-3 fatty acids from algae oil are included.

7.7.4.3 Use of Novel Plant Protein

The exploration of novel plant proteins has been a significant advancement in the production of vegan cheeses. Proteins derived from sources such as chickpeas, fava beans, and lentils are gaining attention due to their unique functional properties and nutritional benefits. Research has demonstrated that these alternative protein sources can enhance the emulsification, gelling, and water-holding capacities of vegan cheese formulations. Additionally, the use of fermented plant proteins is emerging as a promising avenue, as fermentation can improve digestibility and bioavailability of nutrients while also imparting desirable flavor profiles (Berkel et al., 2021).

7.7.5 Sustainability, Health, and Ethical Considerations

7.7.5.1 Environmental Impact

By lowering greenhouse gas emissions and pollutants, cutting back on the consumption of foods derived from animals can greatly improve environmental sustainability (Clark et al., 2019; Willett et al., 2019). Roughly 34% of all greenhouse gas emissions are caused by the food system. Using pastures for livestock can restrict carbon sequestration, even while animal production can utilize marginal lands and transform byproducts into consumable resources (Salter, 2017; Hayek et al., 2021). Growing public consciousness about the effects of food systems on the environment and human health is fuelling the market for plant-based cheese substitutes.

7.7.5.2 Climate Change Mitigation

Traditional cheese, especially that made from cow's milk, has the potential to cause global warming primarily through milk production, which accounts for 65–98% of CO_2-equivalent emissions (Bava et al., 2018; Finnegan et al., 2018). According to Clune et al. cow's milk typically has a global warming potential of 1.39 kg CO_2-eq per liter, so that each kilogram of cheese produces roughly 8.86 kg CO_2-eq in total emissions. Although there is a dearth of life cycle assessment (LCA) data regarding plant-based cheese, initial research suggests that plant-based ingredients, like almond and coconut milk, have emissions that are substantially lower than those of cow's milk (0.42 kg CO_2-eq per kilogram) (Braun et al., 2016. When nuts are used as the foundation for cheese, emissions may be even lower than.

7.7.5.3 Health Considerations

Plant-based cheeses may be essential in encouraging healthier eating habits as interest in plant-based diets grows. Plant-based diets high in nutrients are associated with better health outcomes, such as decreased mortality rates from cardiovascular and all causes. On the other hand, diets heavy in sugar and refined carbohydrates do not produce the same health benefits. It is important to differentiate between healthy and unhealthy plant-based diets because, although vegetarian diets are generally linked to a lower incidence of ischemic heart disease and some cancers, some studies point to possible risks for conditions like stroke (Appleby et al., 2016). It's essential to assess plant-based diets according to their beneficial components in order to comprehend their overall effects.

7.8 Whey Cheeses

7.8.1 Definition and Characteristic

"Whey, a byproduct of cheese production that remains after curds have been separated from milk during traditional cheesemaking". Whey cheeses are produced from whey. Unlike regular cheeses that primarily use milk curds, whey cheeses utilize the leftover proteins in whey, including whey protein and other solids. These proteins coagulate when heated, forming a soft, creamy texture. Whey cheeses, such as Ricotta and Brunost, are known for their light texture and relatively higher moisture content, offering a low-fat alternative to traditional cheese. They also provide a sustainable solution by reducing waste in the dairy industry and making use of byproducts that would otherwise be discarded.

7.8.2 Composition and Standard

Whey cheeses typically contain higher levels of whey proteins compared to casein, which distinguishes them from traditional cheeses. The composition can vary depending on the type of whey cheese and production methods. In general, whey cheeses are composed of proteins (mainly whey proteins), fats, lactose, and minerals. For example, Ricotta, a well-known whey cheese, generally contains 8–12% protein, 10–24% fat, and about 3–5% lactose. The moisture content of whey cheeses tends to be higher, often around 70–80%, contributing to their softer texture.

7.8.3 Manufacturing Process

Whey cheeses are dairy products prepared using particular techniques that might be solid, semi-solid, or soft. These include heating whey to coagulate its proteins, sometimes with the addition of acid, and molding the concentrated result. The whey can be pre-concentrated using either approach prior to additional processing. Furthermore, dairy components such as milk or cream can be added either prior to or following coagulation or concentration.

Milk selection and standardization

↓

Milk Coagulation and Cheese Curd Formation

↓

Whey Collection

↓

Whey Clarification and Filtration

↓

Whey Concentration

↓

Coagulation of Whey Proteins

↓

Addition of Milk or Cream

↓

Curd Formation and Separation

↓

Molding and Shaping

↓

Ripening

↓

Salting and Packaging

The initial step involves the selection and standardization of milk. High-quality milk, primarily from cows but also from goats and sheep, is crucial for achieving desirable cheese characteristics. Key factors in milk selection include its composition, specifically the balance of fat and protein, which affects curd formation and texture (Fox et al., 2017a, b). Additionally, microbial quality is vital; the milk must be free from pathogens, sourced from healthy animals, and collected under hygienic conditions. Freshness is also important, as fresh milk preserves optimal flavor and nutritional value (Hernandez-Ledesma et al., 2018). After selection, milk standardization adjusts its composition to ensure consistent quality across batches. This may involve fat adjustment, such as adding cream or skimming, and protein modification through blending different milk sources or adding milk powder (Hernandez-Ledesma et al., 2018). Milk is coagulated to separate the curd (solid) from the whey (liquid). This is typically achieved using rennet (enzymatic coagulation) or acid (such as citric acid or lactic acid). The curds are produced at this step by the milk's casein proteins coagulating. The leftover liquid, referred to as whey, comprises water, minerals, lactose, and soluble proteins such β-lactoglobulin and α-lactalbumin (Fox et al., 2017a, b).

Once the curds are separated, the whey is drained and collected. Whey can be classified into two forms: sweet whey, obtained from rennet-coagulated cheese production (e.g., Cheddar or Parmesan), which has a higher pH (around 6.4–6.7); and acid whey, derived from acid-coagulated cheese production (e.g. cottage cheese), which has a lower pH (around 4.5–5.0) (Fox et al., 2017a, b). Before whey concentration, the whey undergoes clarification to remove residual curd particles, fat globules, and impurities. This process involves centrifugation, which removes suspended solids and leftover curd particles from the whey. Additionally, microfiltration or ultrafiltration may be utilized to eliminate smaller impurities, excess fat, or lactose, depending on the desired end product. After clarification, the whey is prepared for concentration or coagulation. In the whey cheese production process, whey can be concentrated or directly coagulated to form curd. Concentration reduces the water content and increases the protein concentration in the whey. This can be achieved through evaporation, which removes water under vacuum conditions, or membrane filtration, which uses techniques like ultrafiltration and microfiltration to concentrate proteins while reducing lactose content (Fox et al., 2017a, b).

Coagulation is required for the whey proteins to produce cheese curd. One way to cause coagulation is by heating the whey over 85 °C, which denatures the whey proteins (mostly β-lactoglobulin and α-lactalbumin) and causes them to clump together forming curds (Fox et al., 2017a, b). On the other hand, by decreasing the pH of the whey and forcing the proteins to congregate, acids like citric or lactic acid might encourage protein coagulation. In comparison to regular cheese curds, the resultant curd usually has a finer, smoother texture, a higher moisture content, and a softer texture. To improve the flavor, texture, and fat content of the finished product, milk or cream may be added to the whey either before or after concentration in certain recipes. This method is typical for variations such as Ricotta.

Curds are transferred into molds, where they are shaped and allowed to drain further. The molds define the texture and structure of the final cheese. For firmer whey cheeses, light pressing may be applied to expel additional liquid. Certain whey cheeses, like Ricotta Salata, undergo ripening, which can last from a few days to several months. During this process, microbial activity and enzyme action contribute to flavor development, moisture reduction, and texture firming (Fox et al., 2017a, b). Salting is performed by either adding salt directly to the curd or immersing it in a brine solution. Salt enhances flavor and acts as a preservative by inhibiting undesirable microbial growth. After salting, the cheese is packaged and stored under proper conditions to maintain quality.

7.8.4 Recent Advancements

7.8.4.1 Efficient Whey Utilization Techniques

Recent advancements in whey cheese production emphasize the efficient utilization of whey, a byproduct of cheese manufacturing, to minimize waste and improve economic viability. Techniques such as ultrafiltration and nanofiltration have gained prominence, allowing for the concentration of proteins and other valuable components from whey. This procedure lowers the amount of waste produced in addition to improving protein recovery (Zhang et al., 2023). Additionally, innovative fermentation processes have been developed to convert whey into valuable products such as bioactive peptides and functional foods, providing added value and further reducing waste (Khatiwada et al., 2022).

7.8.4.2 Enhanced Flavor and Texture Profiles

The flavor and texture of whey cheese have been significantly improved through various methods, including the use of enzymatic treatments and the incorporation of natural flavor enhancers. Enzymatic hydrolysis can improve the flavor profile by releasing free amino acids and peptides that contribute to umami and sweetness, as shown by recent studies (Kumar et al., 2023). Furthermore, the incorporation of various milk fat sources and emulsifiers has been shown to optimize the mouthfeel and creaminess of whey cheese, resulting in a more desirable texture (Patel & Vyas, 2023). The application of microencapsulation technology has also emerged as a strategy to protect flavor compounds during processing, ensuring a more robust sensory experience for consumers (Wang et al., 2022).

7.8.4.3 Sustainability and Waste Reduction Initiatives

Sustainability has become a critical focus in whey cheese production, prompting the development of various waste reduction initiatives. Strategies such as the valorization of whey into high-value products, including lactose and whey protein concentrates, have gained traction (López et al., 2024a). These initiatives not only enhance the sustainability of whey cheese production but also contribute to a circular economy by reusing waste materials. Moreover, the adoption of renewable energy sources and energy-efficient technologies in production facilities has been instrumental in reducing the carbon footprint associated with whey cheese manufacturing (González et al., 2023). According to Hernandez et al. (2023), there has been recent study that emphasizes the potential of life cycle assessment (LCA) as a tool for assessing the environmental effect of whey cheese production and directing producers toward more sustainable methods.

7.9 Hybrid Cheese

7.9.1 *Definition and Characteristics*

Hybrid cheeses represent an innovative category within the dairy industry, that combines attributes from two or more types of cheese, often incorporating different milk sources (e.g., cow, goat, sheep) or production techniques. This synthesis results in products exhibiting enhanced flavor complexity, where the unique flavor profiles of each milk type coalesce to yield distinctive taste experiences; for instance, the combination of cow and goat milk can produce a cheese that balances the creaminess inherent to cow's milk with the characteristic tanginess of goat's milk (Kumar & Sharma, 2023). Texturally, hybrid cheeses can range from soft and creamy to hard and crumbly, depending on the specific coagulation and aging processes utilized, thus offering a broad spectrum of culinary applications (Díaz et al., 2022). The nutritional profiles of hybrid cheeses are also notably diverse, often enhanced by the incorporation of different milk sources, which can lead to increased protein content and a wider array of vitamins and minerals (Fernández et al., 2023). Furthermore, the aging potential of hybrid cheeses is variable; some are optimized for immediate consumption, while others develop richer flavors and complex textures through prolonged aging (González et al., 2023). This versatility not only caters to the diverse preferences of consumers but also promotes the exploration of innovative production techniques that enhance flavor and texture, thereby contributing to the evolution of cheese-making practices in a sustainable manner (López et al., 2024a, b).

7.9.2 *Ingredients and Blending Techniques*

7.9.2.1 Combining Dairy and Plant-Based Ingredients

Hybrid cheeses often incorporate both dairy and plant-based ingredients to achieve a balance of traditional cheese characteristics with alternative nutritional or environmental benefits. Dairy ingredients, such as casein and milk fat, provide the core structure and richness, while plant-based ingredients like soy, almond, or coconut milk are included to meet consumer demand for plant-based alternatives and offer options for those with lactose intolerance or allergies (López et al., 2024a, b). The combination of these ingredients requires careful formulation, as plant proteins and fats exhibit different behaviors compared to dairy proteins, particularly in terms of emulsification and texture development (McSweeney et al., 2017). Successful hybrid cheese formulations utilize stabilizers, emulsifiers, and flavor-enhancing compounds to bridge the sensory and structural gaps between dairy and plant-based components, resulting in products that maintain the taste and mouthfeel of traditional cheese but with added nutritional benefits and lower environmental impact (Guinee, 2021).

7.9.2.2 Proportions and Processing Methods

The proportions of dairy to plant-based ingredients in hybrid cheeses vary based on the intended product, ranging from blends with a high proportion of dairy for traditional textures to nearly equal ratios for more distinct hybrid varieties (Guinee, 2021). Finding the optimal proportion is critical for achieving a balanced texture, flavor, and functional properties. Processing techniques such as homogenization and ultrafiltration are employed to ensure the even distribution of fats and proteins from both dairy and plant sources. These methods help achieve a consistent texture and prevent the separation of components, which is a common issue when combining ingredients with different physical properties (McSweeney et al., 2017). Enzymatic treatments and heat processing may also be used to enhance protein solubility and improve the meltability and stretchability of hybrid cheeses, particularly in varieties intended to mimic the functionality of traditional dairy *cheeses* (Fernández-García et al., 2020). These processing methods ensure that hybrid cheeses offer the sensory and functional attributes expected by consumers, while accommodating both dairy and plant-based ingredient.

7.9.3 Manufacturing Process

The process starts with selecting dairy milk (e.g., cow, goat, or sheep milk) and plant-based alternatives such as almond, soy, or oat milk. These ingredients are chosen based on the desired flavor, texture, and nutritional profile (McSweeney et al., 2017; López et al., 2024a, b). Both the dairy and plant-based milks are pasteurized separately to eliminate harmful microorganisms and extend the shelf life. This heat treatment also helps improve the stability of both ingredients before they are blended (Guinee, 2021). The dairy and plant-based milks are homogenized to ensure an even distribution of fat and protein molecules. Homogenization prevents the separation of these two components and creates a consistent texture (McSweeney et al., 2017). Lactic acid bacteria (LAB) starter cultures are introduced to the blended mixture to initiate fermentation. This step lowers the pH and contributes to the flavor and texture of the cheese. The starter culture selection may vary depending on the type of hybrid cheese being produced (Guinee, 2021). Rennet (animal or microbial) or other coagulants are added to induce curd formation. In hybrid cheeses, stabilizers may also be used to assist plant-based proteins in coagulating with dairy proteins. The mixture is then left to coagulate for a specific time to form curds (Fernández-García et al., 2020).

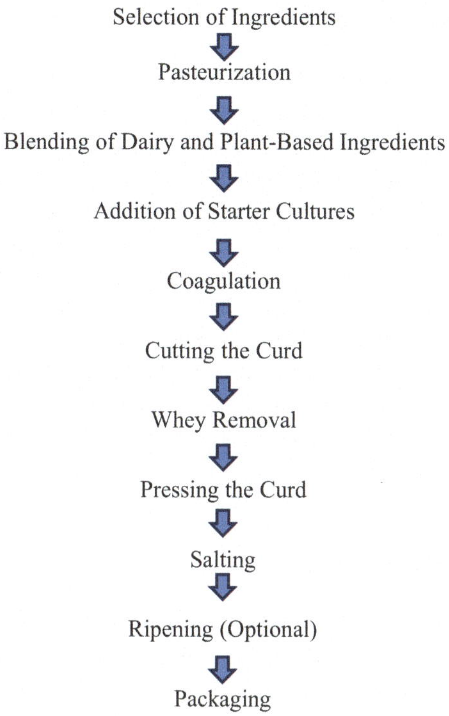

The whey is released by cutting the curd into smaller pieces once it has formed. Larger curds hold more moisture, while smaller curds produce drier cheeses (López et al., 2024a, b). The size of the curd pieces affects the moisture level of the finished product. To encourage the evacuation of whey, the curds are swirled and cooked gently. The amount of whey extracted influences the cheese's moisture content and texture. This stage of the production process for hybrid cheese may need to be adjusted to account for plant-based ingredients, which retain moisture in a different way than dairy curds (Guinee, 2021). To give the cheese its texture and shape, the curds are pressed into molds after the whey is removed. Consolidating the curds and removing extra whey are two benefits of pressing. The stress and Salt is added to the cheese either by dry salting or brining. This step enhances flavor, controls moisture, and acts as a preservative. Salt also influences the texture of the cheese by drawing out moisture from the curd (Guinee, 2021). Depending on the type of hybrid cheese, the product may undergo a ripening (aging) process. During ripening, microbial and enzymatic reactions take place, developing complex flavors and textures. Fresh hybrid cheeses may skip this step to retain fresh, mild flavors (Fernández-García et al., 2020). The final product is packaged using appropriate materials to ensure it remains fresh and protected from environmental factors. Packaging also plays a role in maintaining the hybrid cheese's moisture and preventing contamination (Guinee, 2021).

7.9.4 Recent Advancements

7.9.4.1 Improved Sensory Qualities

Recent advancements in hybrid cheese production have significantly enhanced sensory qualities, such as texture, flavor, and appearance, to closely resemble traditional dairy cheeses. Innovative processing techniques, including precision fermentation and the incorporation of novel plant-based ingredients, have improved emulsification and moisture retention. These advancements result in hybrid cheeses with superior meltability, stretchability, and overall mouthfeel, which are critical attributes for consumer acceptance (López et al., 2024a, b). Additionally, researchers have explored various blends of dairy and plant-based proteins to optimize texture and flavor profiles, ensuring that hybrid cheeses meet consumer expectations for quality while providing the versatility associated with both dairy and plant-based products (Fernández-García et al., 2020).

7.9.4.2 Nutritional Enhancements

In response to increasing health consciousness among consumers, there have been significant efforts to enhance the nutritional profile of hybrid cheeses. Manufacturers are now fortifying these products with plant-based proteins, dietary fiber, and essential vitamins such as B12 and calcium, thereby improving their overall nutritional value without compromising sensory characteristics (Fernández-García et al., 2020). These enhancements cater to the growing demand for lower-fat and allergen-free options, making hybrid cheeses more appealing to health-oriented consumers. By leveraging the nutritional benefits of both dairy and plant-based ingredients, hybrid cheeses not only provide functional attributes but also support a more balanced diet, contributing to their market viability (López et al., 2024a, b).

7.9.4.3 Market Acceptance and Consumer Trends

The market acceptance of hybrid cheeses has been increasingly favorable, supported by growing consumer trends toward sustainable and health-conscious food choices. Recent surveys indicate that approximately 40% of consumers are actively seeking out plant-based alternatives, with hybrid cheeses positioned as appealing options due to their combination of dairy and plant-based ingredients (Mintel., 2023). Studies have indicated that a majority of customers, specifically 63%, are prepared to pay a higher price for items that provide health benefits such reduced fat content and enhanced nutritional value (Nielsen, 2022). Furthermore, the market for plant-based cheese is expected to develop at a compound annual growth rate (CAGR) of 12.3% to reach $6.1 billion by 2026, reflecting the growing demand for lactose-free and allergy-friendly choices (MarketsandMarkets., 2022).

These trends suggest that consumers appreciate the versatility and flavor of hybrid cheeses while prioritizing sustainability, making them a significant segment within the broader cheese market. Manufacturers are thus focusing on product innovation and marketing strategies to meet this evolving consumer demand, which bodes well for the future of hybrid cheeses in the marketplace.

7.10 Dry Milk Cheeses

7.10.1 Definition and Characteristics

Dry milk cheese, often referred to as cheese powder or dehydrated cheese, is a concentrated form of cheese produced by removing moisture from traditional cheese products, resulting in a shelf-stable product with an extended shelf life. This type of cheese is typically made from high-quality milk, which undergoes pasteurization followed by coagulation and pressing to form curds. After its initial creation, the cheese is dried using techniques like spray drying or freeze-drying, which remove moisture while keeping the vital nutrients and flavors intact. The characteristics of dry milk cheese include a high-fat and protein content, concentrated flavor, and versatility, making it suitable for various applications in the food industry, such as snacks, seasonings, and ready-to-eat meals. Additionally, dry milk cheese can be reconstituted by adding water, allowing it to regain a texture similar to fresh cheese, making it an attractive option for both food manufacturers and consumers seeking convenience without sacrificing flavor. Its extended shelf life and ease of storage make dry milk cheese an economically viable product for various culinary applications.

7.10.2 Manufacturing Process

High-quality milk is the foundation of cheese production. Milk from cows, goats, or sheep is selected based on its compositional characteristics, including fat, protein, and lactose content.

Milk Selection
↓
Milk Concentration
↓
Drying
↓
Reconstitution
↓
Coagulation
↓
Curd Formation
↓
Cooking and Stirring
↓
Draining the Whey
↓
Salting
↓
Shaping
↓
Aging
↓
Packaging

The choice of milk significantly affects the sensory properties of the resulting cheese. To enhance the protein and fat concentration necessary for dry cheese production, milk undergoes concentration techniques such as ultrafiltration or evaporation. Ultrafiltration involves passing milk through a membrane that retains proteins and fat while allowing water and small solutes to pass, thereby increasing the solid content (Ganesan et al., 2020). The concentrated milk is then subjected to drying methods such as spray drying or drum drying, converting the liquid into dry milk powder. Spray drying involves atomizing the milk into fine droplets and rapidly evaporating moisture using hot air, while drum drying consists of spreading the milk on heated drums to form a dry sheet that is then scraped off.

The dried milk powder is reconstituted by adding a specific amount of water to restore its liquid form. This step is crucial for creating a suitable environment for curd formation, as the composition of the reconstituted milk should closely resemble that of fresh milk. The reconstituted milk is heated to an optimal temperature, typically between 30 °C and 35 °C. At this stage, starter cultures of lactic acid bacteria are introduced, which ferment lactose to produce lactic acid. This acidification process, combined with the addition of rennet, a natural enzyme, promotes coagulation and curd formation. Following coagulation, the curd forms a gel-like structure. The curds are then cut into small pieces, generally ranging from 1 to 2 cm, to facilitate whey drainage. The final cheese product's moisture content and texture are mostly determined by the size of the curd pieces (Fox et al., 2017a, b).

The cut curds are gently heated and stirred to promote further whey expulsion. Cooking temperatures typically range from 40 °C to 55 °C, depending on the desired texture of the cheese. This step is critical for achieving the appropriate moisture level and developing the desired consistency (Ganesan et al., 2020). After cooking, the whey is drained from the curds using cheesecloth or mechanical separators. Effective whey removal is essential for ensuring that the cheese achieves the correct texture and moisture content. The drained curds are mixed with salt to enhance flavor, inhibit microbial growth, and help control moisture retention. Salting methods can include direct salting, where salt is added directly to the curds, or brining, where the cheese is submerged in a saltwater solution. The salted curds are transferred into molds, where they are subjected to pressing. This process shapes the cheese and removes any residual whey, contributing to the cheese's density and firmness. The pressure applied during this step can vary based on the desired texture of the final product. The shaped cheese undergoes aging in a controlled environment, typically at temperatures between 10 °C and 15 °C and humidity levels of 80% to 90%. The aging process can last from several months to years, allowing biochemical reactions to occur that enhance flavor development and reduce moisture content. The aging time and conditions significantly influence the sensory attributes of dry cheese. Once the cheese has aged appropriately, it is cut into specified portions and packaged using materials designed to preserve freshness and extend shelf life. Techniques such as vacuum sealing or wrapping in moisture-proof film help minimize exposure to air and humidity, maintaining the cheese's quality.

7.10.3 Recent Advancements

7.10.3.1 Improved Drying Techniques

Advances in drying technology have significantly improved the efficiency and quality of dry cheese production. Novel drying methods such as low-temperature spray drying and microwave-assisted vacuum drying have been developed to preserve the bioactive compounds and sensory attributes of the cheese. These techniques minimize the loss of flavor volatiles and maintain a better texture, ensuring that the final product retains a closer resemblance to fresh cheese. Spray drying at lower temperatures also helps reduce the denaturation of proteins and improves the retention of essential nutrients (Ganesan et al., 2020).

7.10.3.2 Enhanced Reconstitution Methods

Reconstitution, the process of rehydrating dried cheese powders, has seen advancements through the use of emulsifying salts and optimized mixing processes. These improvements enhance the solubility and dispersion of dry cheese particles in water, resulting in a smoother texture and better mouthfeel. The addition of emulsifying agents, such as sodium phosphate and sodium citrate, has been shown to improve the stability of the reconstituted cheese matrix, leading to a more consistent product that resembles the original cheese in terms of flavor and texture (Fox et al., 2017a, b).

7.10.3.3 Innovations in Flavor and Texture Development

Innovations in biotechnology and enzymatic treatments have revolutionized the flavor and texture of dry cheese. Enzymatic modifications, such as proteolysis and lipolysis, allow for the controlled development of flavor compounds during the drying process, leading to more complex and desirable flavor profiles. Additionally, texturizing agents like hydrocolloids and modified starches are increasingly used to improve the texture of reconstituted cheese, mimicking the creamy, smooth properties of traditional cheese.

References

Appleby, P. N., Key, T. J., & Thorogood, M. (2016). Mortality in vegetarians and comparable non-vegetarians in the UK. *The American Journal of Clinical Nutrition, 104*(3), 753–757. https://doi.org/10.3945/ajcn.115.124086

Bava, L., Riva, G., & Sandrucci, A. (2018). Environmental sustainability of dairy production: Life cycle assessment of cow milk. *The Science of the Total Environment, 640-641*, 173–180. https://doi.org/10.1016/j.scitotenv.2018.05.021

Bergsma, J. (2017, March 1). *Vegan cheese analogue*. Patent WO2017150973A1, pp. 1–28.

Berkel, L., Moermond, J., & van der Goot, A. J. (2021). Improving the functionality of plant proteins through fermentation: Opportunities for dairy alternatives. *Food Research International, 141*, 110168. https://doi.org/10.1016/j.foodres.2021.110168

Borchers, A. T., et al. (2020). Nutritional considerations for plant-based diets: A review of vegan cheeses. *Nutrients, 12*(10), 2975. https://doi.org/10.3390/nu12102975

Braun, J., Rehfeldt, T., & Albrecht, A. (2016). Assessing the environmental impacts of food products: A comparison of two life cycle assessment methods. *Journal of Cleaner Production, 112*(1), 141–149. https://doi.org/10.1016/j.jclepro.2015.08.006

Brennan, M., et al. (2019). Nutritional evaluation of plant-based protein sources: Implications for vegan cheese production. *Nutrients, 11*(11), 2614. https://doi.org/10.3390/nu11112614

Chen, M., Lu, J., Liu, F., Nsor-Atindana, J., Xu, F., Goff, H. D., Ma, J., & Zhong, F. (2019). Study on the emulsifying stability and interfacial adsorption of pea proteins. *Food Hydrocolloids, 88*, 247–255.

Clark, M. A., Domingo, N. G., & Colgan, K. (2019). Global diets link environmental sustainability and human health. *Nature, 570*, 30–32. https://doi.org/10.1038/d41586-019-01942-0

Dannenberg, F., et al. (2021). Improvement of meltability and texture in plant-based cheese alternatives: A comparison of emulsification techniques. *Journal of Dairy Science, 104*(9), 9403–9415. https://doi.org/10.3168/jds.2021-04214
Díaz, M., Álvarez, A., & Martínez, J. (2022). Textural properties of hybrid cheeses: A comparative analysis. *Journal of Dairy Science, 105*(5), 3456–3465.
Fernández, M., Carreño, J., & García, E. (2023). Nutritional evaluation of hybrid cheeses: A comprehensive study. *Food Chemistry, 404*, 134112.
Fernández-García, E., Carreño, J., & Díaz, M. (2020). *Cheese technology and innovation: Advances in dairy processing*. Springer.
Finnegan, W. M., & McCarthy, P. (2018). The environmental impact of cheese: A life cycle assessment. *Journal of Dairy Science, 101*(4), 1–12. https://doi.org/10.3168/jds.2017-14246
Finnegan, W., Yan, M., Holden, N.M., et al. (2018). A review of environmental life cycle assessment studies examining cheese production. *Int J Life Cycle Assess 23*, 1773–1787. https://doi.org/10.1007/s11367-017-1407-7
Fox, P. F., & McSweeney, P. L. (2017). Cheese: An overview. In P. L. H. McSweeney (Ed.), *Cheese* (pp. 5–21). Academic Press.
Fox, P. F., McSweeney, P. L. H., & Cogan, T. M. (2017a). *Cheese: Chemistry, physics and microbiology* (Vol. 1). Academic Press.
Fox, P. F., McSweeney, P. L. H., Cogan, T. M., & Guinee, T. P. (2017b). *Fundamentals of cheese science*. Springer.
Ganesan, P., Muthuraman, T., & Kalidass, S. (2020). Impact of drying technology on the quality of dry cheese: A review. *Food Science and Technology International, 26*(3), 201–209.
González, J., Reyes, J. A., & Castro, J. (2023). Aging potential in hybrid cheeses: Impacts on flavor and texture. *International Dairy Journal, 143*, 104310.
Grasso, N., Roos, Y. H., Crowley, S. V., Arendt, E. K., & O'Mahony, J. A. (2021). Composition and physicochemical properties of commercial plant-based block-style products as alternatives to cheese. *Future Foods, 4*, 100048.
Grossmann, L., & McClements, D. J. (2021). The science of plant-based foods: Approaches to create nutritious and sustainable plant-based cheese analogs. *Trends in Food Science & Technology, 118*, 207–229. https://doi.org/10.1016/j.tifs.2021.09.033
Guinee, T. P. (2021). *Cheese: Chemistry, physics and microbiology*. Academic Press.
Hayek, M. N., Hristov, A. N., & Borkowski, J. (2021). Carbon sequestration potential of grazing land. *Agronomy for Sustainable Development, 41*(1), 1–12. https://doi.org/10.1007/s13593-020-00650-5
Hernandez, R., Torres, E., & Olivares, J. (2023). Life cycle assessment of whey cheese production: A case study in sustainability. *Food Research International, 165*, 112390.
Hernandez-Ledesma, B., Hsieh, C., & Weng, S. (2018). Health benefits of whey protein. *Journal of Dairy Science, 101*(5), 3306–3323.
IMARC Group. (2024). Cheese market: Global industry trends, share, size, growth, opportunity and forecast 2023–2028. IMARC Group. https://www.imarcgroup.com/cheese-market
Ismail, B. P., Senaratne-Lenagala, L., Stube, A., & Brackenridge, A. (2020a). Protein demand: Review of plant and animal proteins used in alternative protein product development and production. *Animal Frontiers, 10*, 53–63. https://doi.org/10.1093/af/vfaa040
Ismail, B. P., Senaratne-Lenagala, L., Stube, A., & Brackenridge, A. (2020b). Protein demand: Review of plant and animal proteins used in alternative protein product development and production. *Animal Frontiers, 10*, 53–63.
Khatiwada, B., Regmi, P., & Thapa, S. (2022). Fermentation of whey: A bioprocess for sustainable cheese production. *Applied Biochemistry and Biotechnology, 194*(7), 2268–2282.
Kovačević, J., Bechtold, T., & Pham, T. (2024). Plant-based proteins and their modification and processing for vegan cheese production. *Macromolecules, 4*(1), 23–41.
Kumar, P., & Sharma, R. (2023). Flavor profiles of hybrid cheeses: Understanding consumer preferences. *Cheese Science International, 11*(2), 99–108.

Kumar, P., Singh, S., & Agarwal, R. (2023). Enhancing flavor profiles in whey cheese through enzymatic hydrolysis. *International Dairy Journal, 132*, 105530.
López, J. A., Fernández, M. E., & Ruiz, R. (2024a). Innovations in cheese manufacturing: The rise of hybrid cheeses. *Journal of Food Engineering, 327*, 112432.
López, J. A., Fernández, M. E., & Ruiz, R. (2024b). Valorization of whey: Transforming waste into high-value products. *Bioresource Technology Reports, 22*, 100984.
MarketsandMarkets. (2022). *Plant-based cheese market by source, type, distribution channel, and region – Global forecast to 2026*. MarketsandMarkets.
McSweeney, P. L. H., Ottogalli, G., & Fox, P. F. (2017). *Cheese: Chemistry, physics, and microbiology* (Vol. 1). Springer.
Mintel. (2023). *Plant-based alternatives: Trends and insights*. Mintel Group Ltd.
Nielsen. (2022). *The future of food: Key trends in plant-based products*. Nielsen Holdings plc.
O'Brien, M., & Kelly, A. (2020). Recent innovations in cheese production. *International Journal of Dairy Technology, 73*(1), 1–15.
Parente, E., Cogan, T. M., & Powell, I. B. (2017). Starter cultures: General aspects. In *Cheese* (pp. 201–226). Academic Press.
Patel, A., & Vyas, A. (2023). Textural enhancement in whey cheese: The role of milk fat sources and emulsifiers. *Journal of Dairy Science, 106*(4), 2675–2689.
Pua, A., Tang, V. C. Y., Goh, R. M. V., Sun, J., Lassabliere, B., & Liu, S. Q. (2022). Ingredients, processing, and fermentation: Addressing the organoleptic boundaries of plant-based dairy analogues. *Foods, 11*(875). https://doi.org/10.3390/foods11060875
Salter, A. M. (2017). The role of livestock in the global food system: A new perspective. *Agricultural Systems, 157*, 204–212. https://doi.org/10.1016/j.agsy.2017.06.005
Walther, B., Guggisberg, D., Badertscher, R., Egger, L., Portmann, R., Dubois, S., Haldimann, M., Kopf-Bolanz, K., Rhyn, P., Zoller, O., et al. (2022). Comparison of nutritional composition between plant-based drinks and cow's milk. *Frontiers in Nutrition, 9*, 2645. https://doi.org/10.3389/fnut.2022.002645
Wang, X., He, Z., Zeng, M., Qin, F., Adhikari, B., & Chen, J. (2017). Effects of the size and content of protein aggregates on the rheological and structural properties of soy protein isolate emulsion gels induced by CaSO4. *Food Chemistry, 221*, 130–138.
Wang, Y., Li, X., & Zhang, J. (2022). Microencapsulation of flavor compounds in whey cheese: An innovative approach for flavor preservation. *Food Chemistry, 367*, 130709.
Willett, W., Rockström, J., & Loken, B. (2019). Food in the Anthropocene: The EAT–Lancet Commission on healthy diets from sustainable food systems. *The Lancet, 393*(10170), 447–492. https://doi.org/10.1016/S0140-6736(18)31788-4
Zhang, Y., Chen, Y., & Li, Y. (2023). Ultrafiltration and nanofiltration techniques for whey protein recovery: A review. *Journal of Food Engineering, 325*, 111295.

Chapter 8
Functional Properties and Rheology

Reshab Majumder and Saptadip Miatur

8.1 Introduction

The study of functional properties and rheology is very important in the field of food science and culinary arts since it provides information on the mechanical and physical features of different food products (Madadlou et al., 2005). Cheese is a prime illustration of this, with its diverse range of flavors, textures, and sensory experiences appealing to people all over the world. Cheese's rheological behavior and functional characteristics, which precisely define its texture, performance, and culinary diversity, are fundamental to its appeal. This chapter sets out to explore the complex interplay between cheese's functional qualities and rheology, exploring the scientific concepts and real-world applications that guide the product's development, quality, and sensory qualities.

Few foods are as fascinating as well as appealing in the world of culinary as cheese. Cheese is a food that people across the globe admire because of its wide variety of tastes, textures, and aromas. However, there is more to this delicious food than just fascinating science and creative cooking. Cheese's functional qualities and rheology, which are fundamental to its sensory experience, production methods, and culinary applications, are the focus of this complex interaction. This chapter delves into the fundamentals of cheese, revealing the rheological principles that control its behavior and the mysteries surrounding its useful qualities. It provides a more profound comprehension of this beloved dairy delicacy.

Cheese has a number of practical uses that include medical, nutritional, antibacterial, and other health advantages. Because of these qualities, cheese is not only a

R. Majumder (✉)
Jharkhand Milk Federation, Ranchi, Jharkhand, India

S. Miatur
Department of Food Process Engineering, IIT-KGP, Kharagpur, WB, India

S. Mansha Rafiq (ed.), *Technological Advances and Trends in Cheese Making*,
https://doi.org/10.1007/978-3-031-94476-5_8

tasty addition to food but also an important part of a balanced diet and a possible source of health advantages (López-Expósito et al., 2012). Cheese's ability to provide vital nutrients is one of its main therapeutic qualities (Saleem et al., 2024). High-quality proteins, calcium, phosphorus, and vitamins including riboflavin, vitamin B12, and vitamin A can all be found in abundance in cheese. These nutrients are necessary to sustain bone health, encourage muscle growth and repair, and support a number of bodily metabolic functions in addition to preserving general health. Moreover, cheese has bioactive substances that may be beneficial to health. For instance, it has been demonstrated that certain peptides produced during the breakdown of cheese proteins include antibacterial, antihypertensive, and antioxidant properties. These bioactive peptides may help maintain gut health and immunological function while also lowering the risk of chronic illnesses including hypertension and cardiovascular disease (Plante et al., 2021). Additionally, cheese has antibacterial qualities that assist prevent the formation of harmful bacteria and enhance food safety. The growth of pathogenic bacteria like Salmonella, Listeria, and E. coli can be inhibited by the lactic acid bacteria and other good microorganisms found in cheese. Furthermore, the high salt content and low pH of cheese provide an unfavorable environment for the growth of microorganisms (Giraldo et al., 2017).

Conjugated linoleic acid (CLA), another type of fatty acid found in cheese, has been linked to a number of health advantages, including anti-inflammatory, anticancer, and weight-management properties. Research indicates that CLA may help improve body composition by encouraging fat loss and maintaining lean muscle mass, as well as reducing inflammation and the risk of several cancers (Prandini et al., 2011). Cheese has various useful benefits in addition to its therapeutic, dietary, and antibacterial qualities. For instance, cheese's calcium and phosphorus content strengthens tooth enamel and promotes tooth remineralization, both of which are beneficial to dental health (Kashket & DePaola, 2002). Cheese is a gratifying and substantial snack or meal component since its protein and fat content can also aid in promoting satiety and lowering appetite.

The study of matter flow and deformation, or rheology, provides important insights into the structural dynamics of cheese and clarifies how it behaves in various processing stages and situations. Cheese has a complicated rheological profile that is impacted by temperature, moisture distribution, fat level, and protein structure, among other things. Comprehending the rheological behavior of cheese is crucial for simplifying production procedures, guaranteeing product uniformity, and boosting sensory attributes. Cream cheese's smooth spreadability, mozzarella's elastic stretchiness, and aged Gouda's firmness are just a few examples of how cheese's rheological characteristics affect its texture, mouthfeel, and versatility in the kitchen (Lucey et al., 2003). Researchers and manufacturers can uncover new opportunities for product innovation, quality enhancement, and global marketability by exploring the complexities of cheese's rheology. Let's take a closer look at the fundamentals of cheese's rheology and functional qualities as we go through its fascinating universe. We'll also discover the complex network of creativity and science that underlies cheese's culinary variety and sensory appeal. We learn to appreciate cheese as more than just a food via travel and exploration; it's a timeless representation of human ingenuity, creativity, and cultural history.

8.2 Relationship Between Functional Property and Rheology

Foods, especially cheeses, have a complicated interplay between their rheology and practical characteristics that profoundly affects the texture, flavor, and overall quality of the food. The term "functional properties" refers to a broad range of attributes, such as emulsification properties, protein structure, moisture content, melting behavior, and fat content (Guinee, 2016). In contrast, rheology is the study of matter's flow and deformation. As it relates to cheese and other meals, it describes how the foods react to different temperatures, pressures, and shear stresses. A great illustration of the close relationship between rheology and functionality is cheese (Cooke et al., 2013). Cheese's capacity to melt is one of its primary attributes, which affects the texture and mouthfeel of the cheese when eaten. For uses like pizza or grilled cheese sandwiches, cheeses with a higher moisture content and lower melting temperatures typically melt with a creamier texture. The viscosity and elasticity of these cheeses are examples of their rheological characteristics, which greatly influence how meltable they are. Higher elasticity cheeses, for example, have a tendency to create strings when melted, which adds to the distinctive stretchiness of melted cheese (Fox et al., 2017).

The functional characteristics and rheology of cheeses are significantly influenced by their composition, which includes their protein and fat content. Casein is one type of protein that is essential to the structure and texture of cheese. The firmness, elasticity, and meltability of cheese are influenced by the configuration of casein micelles in the matrix. These characteristics can be drastically changed by changes in the protein composition, such as the addition of additional proteins like whey protein or the presence of different casein types (Ren et al., 2013). Furthermore, cheese's fat content adds to its mouthfeel, richness, and lubricating qualities. Cheeses with higher fat contents have a tendency to melt more easily and have a smoother texture.

Cheese's moisture content is another significant component that affects its rheology and functional characteristics. Higher moisture content cheeses are typically softer and easier to spread, whilst lower moisture content cheeses are typically harder and stiffer. Cheese's rheological behavior is also influenced by moisture, which changes viscosity and flow characteristics. Cheeses with a higher moisture content, for instance, might flow more readily under shear stress, whereas cheeses with a lower moisture content might behave more elastically (Lee et al., 2004). Cheese's maturing process is essential to the development of its rheology and functional qualities. The cheese matrix undergoes a number of biochemical and physical alterations as it ages, such as moisture loss, lipolysis, and protein breakdown. The texture, flavor, and rheological characteristics of the cheese can all be significantly impacted by these modifications. For example, because of the breakdown of proteins and lipids during aging, aged cheeses frequently have a harder texture and more intense flavor. Fresh and aged cheeses can have different rheological behaviors, with aged cheeses showing more elasticity and resistance to deformation (Karami et al., 2009).

Additionally, cheese's rheology and functionality are influenced by its pH level. The charge and solubility of proteins in the cheese matrix are influenced by pH, which consequently affects the texture and meltability of the cheese. Higher pH cheeses typically have a softer feel and, because of their enhanced solubility in proteins, may melt more readily. On the other hand, because a lower pH encourages protein aggregation and cross-linking, which results in a denser cheese matrix, cheeses with a lower pH may be tougher and less likely to melt (Ong et al., 2020). Another crucial component of cheese functionality is its ability to emulsify, especially in processed cheese products (Hladká et al., 2014). Cheese's ability to melt smoothly and evenly without separating is enhanced by the addition of emulsifiers such phosphates or sodium citrate. By stabilizing the cheese matrix, these emulsifiers stop the fat and water phases from separating when the cheese is heated or melts. Cheeses that have been emulsified may behave differently rheologically than cheeses that have not; the one with emulsification will be more viscous and resistant to flow.

Cheese's rheology and functional qualities can also be influenced by the processing parameters employed in its manufacture. Cheese's structure, texture, meltability, and rheological behavior can all be impacted by variables like temperature, pressure, and shear rate. For instance, the length of time and temperature at which cheese is heated during manufacturing might affect how pliable and melted it is. Similar to this, the pressure used when pressing and shaping cheese can have an impact on its elasticity, stiffness, and density (Joshi et al., 2004). There is a complex and dynamic interaction between the rheology and functional characteristics of food, especially cheeses. The aging process, composition, pH level, emulsification characteristics, and processing circumstances are all important factors that affect cheese's texture, meltability, and rheological behavior. Comprehending these correlations is crucial for manufacturing superior cheeses possessing desired functional attributes and rheology. Furthermore, as this field of study advances, new understandings of the intricate interactions between cheese's composition, structure, and rheology are being revealed. This will advance cheese quality and drive new developments in cheese production methods.

8.3 Melting Property

Particularly in processed foods, the melting characteristics of food ingredients play a crucial role in defining their functional and sensory qualities (Atik & Huppertz, 2023). One of the most important factors affecting the mouthfeel, texture, and release of flavors in food products is melting. Cheese is unique among food ingredients because of its unique makeup and extensive use in cooking techniques. This chapter explores cheese's melting characteristics, looking at the underlying processes, variables influencing meltability, and effects on food's rheology and functional qualities.

8.3.1 Understanding Food Material Melting

When a food item is heated to a liquid or semi-liquid state, it is referred to as melting. In food science, this procedure is crucial since it affects texture, appearance, and flavor profiles. The types and amounts of lipids, proteins, and other elements of food products, in particular, determine their melting behavior. For example, the denaturation of proteins and the melting point of lipids have a big influence on how food melts. Cheese's diverse composition—which includes fat, water, proteins (mostly casein), and minerals—makes its melting behavior difficult. The way these ingredients interact with one another at high temperatures determines how cheese melts, which is why food scientists study it in great detail (Schenkel et al., 2013).

8.3.2 Cheese Melting Mechanisms

When exposed to heat, the cheese matrix softens and eventually liquefies, a process known as melting. The principal mechanisms consist of:

Melting of Fat: The milk fat in cheese melts at temperatures between 30 and 40 degrees Celsius. Fat melting aids in the softening of cheese, improving its spreadability and fluidity (Rowney et al., 2003).

Melting of protein: Casein, the primary protein found in cheese, experiences denaturation and partial solubilization when heated. The disruption of the protein matrix caused by this process facilitates the passage of fat and increases the meltability of the cheese (Lee et al., 2015).

Moisture Content: The water in cheese helps to transport heat, which helps to melt it. It also has an impact on the final texture of the melted cheese and the degree of protein interactions (Hennelly et al., 2005). Together, these processes control how cheese melts and can be used in a variety of culinary applications.

8.4 Factors Influencing Meltability of Cheese

Cheese's melting characteristics are influenced by a number of elements, such as:

Fat content: Higher fat contents often improve meltability because they lubricate the cheese matrix and lessen its rigidity. For example, Gouda and other high-fat cheeses melt more easily than low-fat cheeses.

Moisture content: Better meltability is correlated with higher moisture content. Proteins' glass transition temperature is lowered by moisture, which facilitates melting by acting as a plasticizer. Because of this, cheeses with a high moisture content, like mozzarella, have exceptional melting properties.

Protein Composition: Meltability is influenced by the type and structure of casein. Cheeses that have a higher percentage of α-casein, for instance, have a tendency to melt more evenly than those that are dominated by β-casein.

pH and acidity: Cheese's pH and acidity have an impact on the interactions and solubility of proteins. Cheeses that have a pH that is near to the isoelectric point of casein (about 4.6) have more protein-protein interactions, which results in less meltability. On the other hand, cheeses like cheddar that have a higher pH level melt more readily.

Ageing and ripening: Cheese's biochemical makeup, particularly the distribution of fat and the breakdown of protein, is impacted by aging. Because they have less cross-linked protein networks and a higher moisture content, younger cheeses typically melt more readily.

Comprehending these variables is essential for adjusting the characteristics of cheese to attain intended melting patterns in culinary uses.

8.5 Applications and Functional Properties

Cheese's functional properties in a variety of food applications are greatly influenced by its melting characteristics. Melting is a crucial factor in processed food composition because it influences texture, mouthfeel, and flavor release.

Pizza & Pasta: The required stretchiness and gooeyness of cheese in pizza and pasta recipes are dependent on melting behavior. Because of its superior elasticity and meltability, mozzarella is the flavored cheese.

Cheese Dips and Sauces: Cheeses with strong melting points, such as Cheddar, work best for velvety, smooth textures. The melting behavior guarantees a clump-free, uniform slurry.

Baking: The melting of cheese alters the texture and distributes taste in baked items. Cheeses with a smooth melt, such Gruyère and Swiss, are frequently used in quiches and gratins.

Cheese's functionality in various applications is improved by its capacity to anticipate and regulate its melting qualities, which guarantees consistent quality and customer satisfaction.

8.6 The Properties of Rheology in Melted Cheese

The study of melted cheese's flow and deformation behavior under various circumstances is known as rheology. Important rheological characteristics that are impacted by the structure and content of the cheese are viscosity, elasticity, and plasticity.

Viscosity: The flow properties of melted cheese are dictated by its viscosity. Cheeses that are heavy in fat and moisture have a lower viscosity, which makes them more spreadable and fluid.

Elasticity: Melted cheese's elastic qualities are crucial for applications where stretchability is needed, such pizza toppings. The stretchiness of the cheese is influenced by the protein network's capacity to reorganize.

Plasticity: The term "plasticity" describes a cheese's capacity to permanently deform without cracking. This characteristic is crucial for molding and shaping melted cheese in a variety of culinary uses.

Understanding and improving cheese performance under various mechanical and thermal circumstances is made possible by the analysis of these rheological characteristics (Ray et al., 2016).

8.7 Factors Impacting Functionality and Quality of Cheese

Cheese, an old dairy product with many applications, is valued for its numerous flavors, textures, and use in cooking. Cheese functions and quality are influenced by various factors such as the type of milk used and methods of production employed. The chapter discusses the major influences on cheese functionality and quality from a wide perspective providing the summary which is useful to academicians, manufacturers of cheese, and enthusiasts in the dairy business. Some of these aspects that affect the functionality as well as quality of cheese are discussed below.

8.7.1 Milk Composition and Quality

8.7.1.1 Milk Source

The quality of milk is essential for the manufacturing of cheese. The origin of milk, whether it from cows, goats, sheep, or buffalo, has a substantial impact on the qualities of the end product. The fat and protein contents of different types of milk have distinct characteristics that directly impact the texture, flavor, and nutritional properties of the resulting cheese (Fox et al., 2017). Goat milk has elevated quantities of medium-chain fatty acids, which give goat cheeses a unique acidic taste (Park et al., 2007).

8.7.1.2 Seasonal and Dietary Variations

Milk composition is influenced by seasonal variations and the food of breastfeeding animals. According to Dewhurst et al. (2006), cows who eat fresh pasture have greater amounts of conjugated linoleic acid and omega-3 fatty acids in their milk compared to cows that eat preserved forages. This variance may impact the nutritional composition and taste scharacteristics of the cheese. In addition, the microbiological composition of milk may be influenced by seasonal changes, which can in turn have an effect on the fermentation process and the formation of cheese flavor (Hayes et al., 2002).

8.7.2 Microbial Cultures

8.7.2.1 Starter Cultures

Starter cultures are necessary to begin the fermentation process. These cultures include distinct strains of lactic acid bacteria (LAB) that metabolize lactose into lactic acid, hence reducing the pH and causing the production of curds. The choice of starting cultures may have a substantial influence on the texture, taste, and safety of the cheese (Leroy & De Vuyst, 2004). Various strains of *Lactococcus lactis* may generate varied quantities of diacetyl, which contributes to the buttery taste seen in certain cheeses (Larpin-Laborde et al., 2010).

8.7.2.2 Adjunct Cultures

Adjunct cultures are used to augment the development of flavor and increase texture. These cultures, including of non-starter LAB, yeasts, and moulds, enhance the intricacy of cheese flavors and facilitate the ripening process (Gobbetti et al., 2002). *Penicillium roqueforti* is used in the manufacturing of blue cheese to cultivate the distinctive blue veins and intense flavor (Beresford & Williams, 2004).

8.7.3 Processing Techniques

8.7.3.1 Pasteurization

Pasteurization is a crucial process that guarantees the safety of milk by eradicating harmful germs. Nevertheless, it also impacts the indigenous microorganisms and catalysts that play a role in the maturation of cheese and the formation of its distinctive taste. Unpasteurized cheeses, known as raw milk cheeses, often exhibit more intricate and robust flavors as a result of the existence of indigenous microbial populations (Bachmann et al., 2011). On the other hand, cheeses made from pasteurized milk may need to incorporate more microbial cultures in order to produce the correct flavors.

8.7.3.2 Curd Formation and Cutting

The curd formation and cutting process plays a vital role in shaping the ultimate texture of cheese. Milk proteins, particularly casein, may be coagulated by using either rennet or acid. The selection of coagulant and the specific circumstances, including temperature and pH, have a substantial influence on the consistency and moisture level of the curd (Fox et al., 2000). Dividing the curd into smaller fragments facilitates the removal of whey, resulting in a cheese with a more solid consistency. Conversely, bigger curd pieces retain more moisture, which contributes to a softer texture.

8.7.3.3 Cooking and Stirring

Frequently, the curds are subjected to cooking and stirring in order to remove excess whey and enhance the texture of the cheese. The moisture content and stiffness of the cheese may be influenced by the temperature and time of cooking, as well as the degree of stirring. For instance, cheeses such as Parmesan are known for their hard and dry texture, which is achieved by high-temperature heating and rapid stirring (Lawrence et al., 1987).

8.7.3.4 Pressing

Applying pressure to the curds aids in removing more whey and forming the cheese. The cheese's moisture content and texture are influenced by the magnitude of pressure applied and the time of pressing. Increased pressure and extended pressing durations lead to the production of cheeses that have a lower moisture content and a more solid texture, such as Cheddar (McSweeney, 2004a).

8.7.4 Ripening and Aging

8.7.4.1 Environmental Conditions

The factors that determine the ripening process of cheese, such as temperature, humidity, and air circulation, play a crucial role in the formation of its flavor and texture. The ideal circumstances for ripening cheese differ based on the specific kind of cheese. High humidity is necessary for the development of surface molds on cheeses such as Brie and Camembert, however low humidity is preferred for hard cheeses like Parmesan to avoid undesired microbial growth (Fox et al., 2017).

8.7.4.2 Microbial Activity

Microbial activity is crucial in flavor formation throughout the ripening process. The enzymatic degradation of proteins and lipids by microorganisms results in the generation of diverse volatile chemicals that give rise to the distinct flavors seen in various types of cheeses. Proteolysis and lipolysis in blue cheeses result in the formation of intense, pungent flavors and smooth textures (McSweeney, 2004a, b).

8.7.5 Chemical and Biochemical Changes

8.7.5.1 Proteolysis

Proteolysis, the enzymatic degradation of proteins into smaller peptides and individual amino acids, is a fundamental biochemical process that occurs during the maturation of cheese. The primary driving force behind this process is the action of enzymes derived from milk, rennet, and microbes. The degree and distribution of protein breakdown fluctuate across various types of cheeses and have a substantial effect on their texture and taste. For instance, the process of significant proteolysis in soft cheeses such as Brie leads to the development of a smooth and creamy texture, as well as the emergence of robust and ammoniated flavors (Fox et al., 2000).

8.7.5.2 Lipolysis

Lipolysis, the enzymatic hydrolysis of triglycerides into free fatty acids and other metabolites, is a significant factor in the formation of cheese flavor. Lipases from both milk and bacteria act as catalysts in this process. The free fatty acids generated may undergo further reactions to create a range of volatile chemicals, which contribute to the intricate flavor profiles of cheeses such as blue and Camembert (McSweeney, 2004a, b).

8.7.5.3 Non-enzymatic Reactions

Non-enzymatic reactions, such as the Maillard reaction and caramelization, may significantly influence the flavor of cheese, particularly in aged and cooked varieties. The Maillard reaction, which occurs between amino acids and reducing sugars, generates a variety of flavor compounds that contribute to the nutty and caramel-like flavors found in aged cheeses such as Gouda and Parmesan (van Boekel, 2006).

8.7.6 External Factors

8.7.6.1 Packaging

The function of packaging in protecting the quality of cheese and prolonging its shelf life is of utmost importance. Appropriate packing materials and methods may effectively avoid contamination, loss of moisture, and oxidation. Vacuum packing is a frequently used method for preserving hard and semi-hard cheeses. Its purpose is to safeguard against spoiling microorganisms and retain the desired moisture level (Guinee, 2004).

8.7.6.2 Storage Conditions

Optimal storage conditions, including factors such as temperature and humidity, play a crucial role in preserving the quality of cheese. Inadequate storage may result in deterioration, alterations in taste, and modifications in texture. Storing cheese at elevated temperatures may speed up the growth of harmful microorganisms and the breakdown of fats, while low levels of moisture can lead to excessive drying and the development of cracks in the cheese (Fox et al., 2017).

8.8 Applications to Stretched, Vegan, Whey, Hybrid and Dry Milk Cheeses

Although cheese has been recognized as one of the most diversified and popular food items, its variations have increased manifold with development in the technology of foods and changing consumer preference. On the other hand, cheese is integral to humans since ancient times; in fact, it is among the oldest foods produced. Obviously, with changing technology and diets, it finds a place in almost every cuisine. The present paper proposes to discuss five broad categories of cheese: stretched cheese, vegan cheese, whey cheese, hybrid cheese, and dry milk cheese. All of these categories have varying features and uses that best adapt to different consumer needs and culinary practices in the most appropriate manner.

8.8.1 Stretched Cheeses

Stretched cheeses, otherwise referred to as pasta filata, include in their unique manufacturing process the heating and stretching of the curd to attain characteristic stringiness in texture (Fox et al., 2017). Mozzarella is arguably the best known of the stretched cheeses, widely used in Italian cookery-especially on pizza and in caprese salads.

8.8.1.1 Applications

Stretched cheese varieties are highly valued for their meltability and elasticity; this makes them especially ideal in recipes that require a kind of cheese to stretch and melt smooth. The most famous usage is probably mozzarella in pizza. It is vital to form a smooth and gooey layer when melted. Other varieties of stretched cheese, for instance, provolone and scamorza, are used in sandwiches and grilled dishes where the desired characteristic of the cheese would be that it should be melted and stringy (Kindstedt, 2004). But that is not all; the easy stretching of these cheeses also makes them very suitable for a gourmet type of cooking, mostly paired with tomato, basil, olive oil, and other fresh ingredients. For sure, such characteristics that ensure it melts without separating to oil and solid are crucial in the conventional and modern practices of culinary arts (Fox et al., 2017).

8.8.2 Vegan Cheeses

The growing interest in veganism and lactose intolerance is also fueling the innovation of vegan cheeses. These are dairy-free ingredients, made from plants like nuts, soy, and coconut oil, used to imitate traditional cheese (Henchion et al., 2017). They replicate the texture and taste of cheese but remain loyal to vegan dietary standards.

8.8.2.1 Applications

Vegan cheeses are used in many dishes as a direct replacement for traditional cheese. They can be sliced for sandwiches, shredded for pizzas, and melted into sauces to perform many similar functions that dairy cheese does. However, the main point of interest with vegan cheeses is the fact that they allow for options for those who are lactose intolerant or allergic to dairy and for those living a vegan lifestyle (Henchion et al., 2017). Finally, innovation in vegan cheese also underlines that in food technology itself, where such ingredients as nutritional yeast and fermented cashews are used to replicate an umami taste generally reminiscent of aged cheeses. In this way, vegan cheeses are no longer solely functional but increasingly sophisticated in flavor for gourmet applications (Shahbandeh, 2021).

8.8.3 Whey Cheeses

Whey cheeses are processed from the liquid by-product of cheese making, whey. The best-known example of whey cheese is ricotta, which is classically prepared from the whey left over from other cheese productions (Kosikowski & Mistry, 1997). Whey cheeses are an example of the industry's sustainability since they make use of what otherwise would be waste.

8.8.3.1 Applications

Whey cheeses, specifically ricotta, are very versatile in a lot of cooking and baking applications. In Italian cuisine, ricotta is used not only for savory dishes but also for desserts, as seen in the famous lasagna and cannoli respectively. Taste-wise, it has a creamy texture with a very mild flavor and thus finds much use in fillings and spreads. Another strong point for whey cheeses is their contribution to the reduction of food wastes, a trend that has been stressed and promoted in many aspects of food production today as sustainable and modern production (Kosikowski & Mistry, 1997). Beyond ricotta, other cheeses made from whey—including Norwegian brunost, or brown cheese—are distinct in flavor and application. Sweet and caramelized, brunost is mostly used topping bread, or in sauce for meats, mainly in Scandinavian cuisine (Olsen, 2009). Whey cheeses encompass a vast array of quite different applications, from the traditional to modern culinary arts.

8.8.4 Hybrid Cheeses

Hybrid cheeses are a new category that combines different milk types or dairy with non-dairy elements in the production of new cheese products. Consequently, such cheeses would result in different flavor profiles and textures that cannot otherwise be achieved from a single type of milk or ingredient (McSweeney, 2004a, b).

8.8.4.1 Applications

Hybrid cheeses are extremely popular in gourmet applications, with chefs creating unique combinations of milks—cow and goat milk, for example—to achieve cheeses with complex flavor and texture profiles. This form of cheese is often used in high-end dishes where their features can be best highlighted (McSweeney, 2004a, b). In addition, hybrid cheeses can also be created to suit other consumers who adhere to specific diets. For example, when combining dairy and plant-based elements, a lower lactose percentage can be attained, yet taking into account the creaminess of the cheese. Such cheese will be more easily picked up by lactose-intolerant consumers. Taste flexibility and the ability to adapt to different diets make hybrid cheese types rather in demand in the food industry these days (Farnworth, 2008).

8.8.5 Dry Milk Cheeses

Dry milk cheeses have their milk concentrated with most of the water removed. This type of cheese is usually aged, dense, and crumbly; the best example would be Parmesan (Kindstedt, 2004).

8.8.5.1 Applications

Dry milk cheeses are so valued for their intense flavor and long shelf-life that they are preferred for grating over pasta dishes, salads, and soups. Among them, one core of Italian cuisine is Parmesan because the sharp, umami variety improves a great many types of dishes (Kindstedt, 2004). Such cheeses find further application in baking because the low levels of moisture enable incorporation into doughs and batters without affecting their texture. The extended aging process results in complex flavors that work well when paired with wines and fruits on a cheese board (Fox et al., 2017).

8.9 Conclusion

The complex relationships that exist between the rheological and functional properties indicate that understanding these relations is important for the optimal use of cheese in various applications. Rheology, being the study of cheese deformation and flow under stress, affects functional properties like the texture of cheese, its melting, stretch, and general appeal. The various types of cheese, like vegan, hybrid, and other recent innovative varieties, indicate the importance of knowing the dynamic relationship between composition and processing in regard to rheological behavior for the further determination of overall functional properties that shape the quality and application potential of the cheese. With the consumer's preference continuing to shift toward greener, healthier options, further research into the rheological and functional properties of both traditional and alternative varieties will be key in meeting these demands while maintaining quality and optimized functionality. In other words, functionality and the rheological properties of cheese are closely related, and key compositional and processing factors impact performances. The food industry can enhance the quality of cheese, extend the application of the product, and meet the demands for traditional and alternative products of this growing segment through deeper study of such relationships.

References

Atik, D. S., & Huppertz, T. (2023). Melting of natural cheese: A review. *International Dairy Journal, 142*, 105648.

Bachmann, H. P., Puhan, Z., & Rüegg, M. (2011). Raw milk cheese. In P. L. H. McSweeney (Ed.), *Cheese: Chemistry, physics and microbiology* (pp. 417–439). Academic Press.

Beresford, T. P., & Williams, A. G. (2004). The microbiology of cheese ripening. In P. F. Fox, P. L. H. McSweeney, T. M. Cogan, & T. P. Guinee (Eds.), *Cheese: Chemistry, physics and microbiology* (Vol. 1, pp. 287–317). Academic Press.

Cooke, D., Khosrowshahi, A., & Mcsweeney, P. (2013). Effect of gum tragacanth on the rheological and functional properties of full-fat and half-fat Cheddar cheese. *Dairy Science & Technology, 93*(1), 45–62.

Dewhurst, R. J., Shingfield, K. J., Lee, M. R. F., & Scollan, N. D. (2006). Increasing the concentrations of beneficial fatty acids in milk produced by dairy cows in high-forage systems. *Animal Feed Science and Technology, 131*(3–4), 168–206.

Farnworth, E. R. (2008). *Handbook of fermented functional foods*. CRC Press.

Fox, P. F., Guinee, T. P., Cogan, T. M., & McSweeney, P. L. H. (2000). *Fundamentals of cheese science*. Springer.

Fox, P. F., Uniacke-Lowe, T., McSweeney, P. L. H., & O'Mahony, J. A. (2017). *Dairy chemistry and biochemistry*. Springer.

Giraldo, J., Althaus, R. L., Beltrán, M. C., & Molina, M. P. (2017). Antimicrobial activity in cheese whey as an indicator of antibiotic drug transfer from goat milk. *International Dairy Journal, 69*, 40-44.

Gobbetti, M., Corsetti, A., & Smacchi, E. (2002). The role of surface-active compounds on the behavior of lactic acid bacteria in dairy fermentations. *International Journal of Food Microbiology, 74*(1–2), 185–196.

Guinee, T. P. (2004). Salting and the role of salt in cheese. *International Journal of Dairy Technology, 57*(2–3), 99–109.

Guinee, T. P. (2016). Protein in cheese and cheese products: Structure-function relationships. Advanced dairy chemistry: volume 1B: proteins: applied aspects, 347–415.

Hennelly, P. J., Dunne, P. G., O'Sullivan, M., & O'Riordan, D. (2005). Increasing the moisture content of imitation cheese: effects on texture, rheology and microstructure. *European Food Research and Technology, 220*, 415–420.

Hladká, K., Randulová, Z., Tremlová, B., Pon꼍il, P., Mančík, P., Černíková, M., & Buňka, F. (2014). The effect of cheese maturity on selected properties of processed cheese without traditional emulsifying agents. LWT-Food Science and Technology, 55(2), 650–656.

Hayes, M. G., Ralyea, R. D., Murphy, S. C., Carey, N. R., Scarlett, J. M., & Boor, K. J. (2002). Identification and characterization of elevated microbial counts in bulk tank raw milk. *Journal of Dairy Science, 84*(1), 292–298.

Henchion, M., Hayes, M., Mullen, A. M., Fenelon, M., & Tiwari, B. (2017). Future protein supply and demand: Strategies and factors influencing a sustainable equilibrium. *Foods, 6*(7), 53.

Karami, M., Ehsani, M. R., Mousavi, S. M., Rezaei, K., & Safari, M. (2009). Microstructural properties of fat during the accelerated ripening of ultrafiltered-Feta cheese. *Food Chemistry, 113*(2), 424–434.

Joshi, N. S., Muthukumarappan, K., & Dave, R. I. (2004). Effect of calcium on microstructure and meltability of part skim Mozzarella cheese. *Journal of dairy science, 87*(7), 1975–1985.

Kashket, S., & DePaola, D. P. (2002). Cheese consumption and the development and progression of dental caries. *Nutrition reviews, 60*(4), 97–103.

Kindstedt, P. S. (2004). *American farmstead cheese: The complete guide to making and selling artisan cheeses*. Chelsea Green Publishing.

Kosikowski, F. V., & Mistry, V. V. (1997). *Cheese and fermented Milk foods* (Vol. 1). F.V. Kosikowski, L.L.C.

Larpin-Laborde, S., Imran, M., Gaillard, J. L., Spinnler, H. E., & Bonnarme, P. (2010). Diversity of monoterpene profiles of *Lactococcus lactis* and other bacterial species analyzed by headspace solid-phase micro extraction-gas chromatography-mass spectrometry. *Journal of Dairy Science, 93*(12), 4990–4996.

Lawrence, R. C., Creamer, L. K., & Gilles, J. (1987). Texture development during cheese ripening. *Journal of Dairy Science, 70*(8), 1748–1760.

Lee, S. K., Anema, S., & Klostermeyer, H. (2004). The influence of moisture content on the rheological properties of processed cheese spreads. *International journal of food science and technology, 39*(7), 763–771.

Lee, S. K., Klostermeyer, H., & Anema, S. G. (2015). Effect of fat and protein-in-water concentrations on the properties of model processed cheese. *International Dairy Journal, 50*, 15–23.

Leroy, F., & De Vuyst, L. (2004). Lactic acid bacteria as functional starter cultures for the food fermentation industry. *Trends in Food Science & Technology, 15*(2), 67–78.

López-Expósito, I., Amigo, L., & Recio, I. (2012). A mini-review on health and nutritional aspects of cheese with a focus on bioactive peptides. *Journal of Dairy Science, 92*(5): 419–438.
Lucey, J. A., Johnson, M. E., & Horne, D. S. (2003). Invited review: Perspectives on the basis of the rheology and texture properties of cheese. *Journal of dairy science, 86*(9), 2725–2743.
Madadlou, A., Khosroshahi, A., & Mousavi, M. E. (2005). Rheology, microstructure, and functionality of low-fat Iranian white cheese made with different concentrations of rennet. *Journal of Dairy Science, 88*(9), 3052–3062.
McSweeney, P. L. H. (2004a). Biochemistry of cheese ripening: Introduction and overview. In P. F. Fox, P. L. H. McSweeney, T. M. Cogan, & T. P. Guinee (Eds.), *Cheese: Chemistry, physics and microbiology* (Vol. 1, pp. 347–360). Academic Press.
McSweeney, P. L. H. (2004b). *Cheese problems solved*. Woodhead Publishing.
Ong, L., Pax, A. P., Ong, A., Vongsvivut, J., Tobin, M. J., Kentish, S. E., & Gras, S. L. (2020). The effect of pH on the fat and protein within cream cheese and their influence on textural and rheological properties. Food Chemistry, 332, 127327.
Olsen, E. (2009). The joy of brunost: The caramel cheese of Scandinavia. *The New York Times*.
Park, Y. W., Juárez, M., Ramos, M., & Haenlein, G. F. W. (2007). Physico-chemical characteristics of goat and sheep milk. *Small Ruminant Research, 68*(1–2), 88–113.
Plante, A. M., McCarthy, A. L., Lacey, S., & O'Halloran, F. (2021). Investigating The Bioactive Properties of Cheese-Fruit Combinations Following In Vitro Digestion Using an Elderly Model. *Current Research in Nutrition and Food Science Journal, 9*(2), 465–478.
Prandini, A., Sigolo, S., & Piva, G. (2011). A comparative study of fatty acid composition and CLA concentration in commercial cheeses. *Journal of Food Composition and Analysis, 24*(1), 55–61.
Ray, C. A., Gholamhosseinpour, A., Ipsen, R., & Hougaard, A. B. (2016). The effect of age on Cheddar cheese melting, rheology and structure, and on the stability of feed for cheese powder manufacture. *International Dairy Journal, 55*, 38–43.
Ren, D., Chen, B., Chen, Y., Miao, S., & Liu, J. (2013). The effects of κ-casein polymorphism on the texture and functional properties of mozzarella cheese. *International Dairy Journal, 31*(2), 65–69.
Rowney, M. K., Hickey, M. W., Roupas, P., & Everett, D. W. (2003). The effect of homogenization and milk fat fractions on the functionality of Mozzarella cheese. *Journal of Dairy Science, 86*(3), 712–718.
Saleem, G. N., Gu, R., Qu, H., Bahar Khaskheli, G., Rashid Rajput, I., Qasim, M., & Chen, X. (2024). Therapeutic potential of popular fermented dairy products and its benefits on human health. *Frontiers in Nutrition, 11*, 1328620.
Schenkel, P., Samudrala, R., & Hinrichs, J. (2013). Thermo-physical properties of semi-hard cheese made with different fat fractions: Influence of melting point and fat globule size. *International Dairy Journal, 30*(2), 79–87.
Shahbandeh, M. (2021). *Global vegan cheese market value 2022–2030*. Statista.
van Boekel, M. A. J. S. (2006). Formation of flavour compounds in dairy products. In T. H. Tamine (Ed.), *Dairy powders and concentrated products* (pp. 16–32). Wiley-Blackwell.

Chapter 9
Quality Evaluation of Cheese

Tridip Boruah, Morjina Parveen, David Daimary, and Gulzar Ahmad Nayik

9.1 Introduction

Cheese is widely produced in the world due to its high demand, which makes the cheese industry a highly growing industry. Because of its diversity in application, nutritional value, convenience, and good taste cheese is a very popular food. The diverseness is due to the increasing knowledge of the technology of cheese making and the biochemistry and microbiology of cheese ripening (Sameen et al., 2008). Throughout the production chain of cheese from raw material to final product the quality of the cheese must be controlled, which is a tough challenge for the cheese-making industry. The correct identification of sensory quality characteristics is important to improve the cheese-making process (Kraggerud et al., 2012). Consumers always demand a good quality product, for which it is important to assess the quality of the product before launching it in the market because the quality assessment can provide the pre-determined value and demand of the product in a market based on quality (Karoui et al., 2007). This helps individuals or companies to ensure an authentic identity and production in large quantities. Again, due to the risk factors like contamination, it is very important to study the quality assessment. To check the fat content, moisture levels, salt concentration, and pH chemical analysis of cheese is important. Different cheeses need different ageing times to become the best quality cheese for which ageing assessment is also necessary (Delgado et al., 2011). Along with the aging assessment self-life assessment of cheese is also

T. Boruah (✉) · M. Parveen · D. Daimary
P.G. Department of Botany, Madhab Choudhury College, Barpeta, Assam, India
e-mail: tridip_boruah@gauhati.ac.in

G. A. Nayik
Marwadi University Research Centre, Department of Microbiology, Marwadi University, Rajkot, Gujarat, India
e-mail: gulzarnaik@gmail.com

S. Mansha Rafiq (ed.), *Technological Advances and Trends in Cheese Making*,
https://doi.org/10.1007/978-3-031-94476-5_9

done to check the time of the storage of the cheese for how long a cheese can be stored and is safe to eat and also how long the quality of the cheese will be maintained. Sometimes, cheeses can also cause fatal deaths in human beings for which it is very important to evaluate the quality of the cheese and is getting more attention from the food safety organizations (Bhardwaj et al., 2015). Nowadays, making cheese is not a big task, people can make cheese at home and in shops on their own as it is a very easy and less time-consuming process. But only making it is not enough to make it available to eat, the quality of the cheese should also be good which is an important aspect of making cheese. The local shops where cheese is made and sold in the market, do not check the quality of the cheese with suitable measures whereas in other big cheese-making industries they take proper instrumental measures to check the quality of the cheese which is helpful to decide if it is right to make it available in the market for selling (Xiaoboa et al., 2010). Also, it is important to assess the quality because without checking the quality if the cheese is made in mass quantity it will cause a huge loss to the cheese-making industry if they are not healthy and safe for the consumers. But the local sellers do not give proper attention to it which can risk people's life. For instance, the "casu marzu" is made in Sardinia, Italy traditionally by the farmers, and is a cheese that is exposed to the environment by some types of cheese-flies which lay eggs in the cheese (Palatzidi et al., 2024). The eggs give rise to the maggots and eat the cheese and then excrete it which gives flavour to the cheese as they need. But it is dangerous to eat this cheese because the larvae of the maggots will remain in the cheese which is concerning for the health of people and this is why it is also illegal to sell this cheese in markets. For all these reasons the quality evaluation of cheese is very important which includes several instrumental techniques and gives importance to human health (Fig. 9.1). The techniques that are used to evaluate the quality of the cheeses, their applications, and the practical implementation of various cheese-based products were discussed in this chapter to provide a streamlined-based direction to aspiring budding researchers in this field.

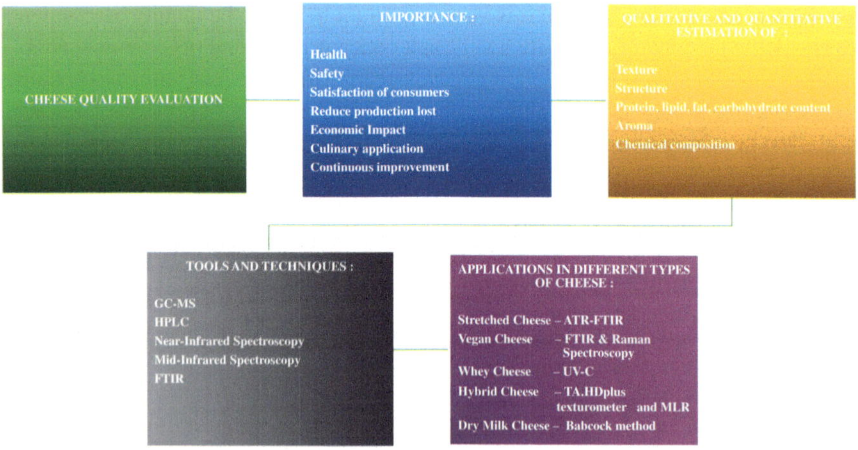

Fig. 9.1 Quality evaluation of cheese and its importance

9.2 Quality Evaluation of Cheese Using Novel Techniques

Quality analysis is an important topic that ensures the overall quality of cheese and its acceptability in the market. As milk is the precursor of cheese, it is also important to evaluate milk quality before processing it to make cheese. Milk constituents are also important, such are calcium, casein, fat, milk salt, and other enzymes. Milk salt plays a major role in cheese texture (Karoui et al., 2007). Different analytical strategies are used for the study of various characteristics of cheese (Table 9.1). Many such methods are discussed here.

9.2.1 Gas Chromatography-Mass Spectrometry (GC-MS)

There are various analytical techniques applied for the authentication of different volatile compounds. Among all other analytical methods gas chromatography is a preferred method for monitoring the biochemical composition of volatile compounds (Delgado et al., 2011). This analytical technique can be used for both quantitative and qualitative analysis of volatile compounds. Gas chromatography separates components of a sample by following the basic principle of chromatography. In GC-MS generally, the mobile phase is a gaseous substance, and the reaction between the stationary and mobile phase results in the separation of different constituents. A typical GC-MS consists of several analytical units, including a carrier gas tank, pressure regulator, GC column, detector or MS chamber, and a computer for data monitoring. GC-MS is extensively used in food industries for quality

Table 9.1 Tools and techniques for quality measurement of cheese

Sl. no.	Analytical methods	Qualitative/quantitative estimation of	References
1	GC-MS	Volatile compounds and chemical composition	Delgado et al. (2011), Ziino et al. (2005), and Cevoli et al. (2011)
2	HPLC	Organic acid, amino acid and other chemical compositions	Bhardwaj et al. (2015), Trihaas et al. (2005), and Yaman et al. (2023)
3	Near-infrared spectroscopy (NIR)	Texture, visual attributes such as the presence of holes, taste attributes, and other characteristics	Pellicer and Bravo (2011), Sørensen and Jepsen (1998), Xiaoboa et al. (2010), Reis et al. (2022), and Cattaneo et al. (2005)
4	Mid-infrared spectroscopy (MIR)	Compositional analysis of cheese, ripening stage, rheological analysis	Cattaneo et al. (2005), Woodcock et al. (2007), Martín-del-Campo et al. (2007), and Gori et al. (2012)
5	FTIR-ATR	Amino acids, organic acids, and fatty acids, reactions in cheese during ripening	Koca et al. (2007), Woodcock et al. (2007), and Gori et al. (2012)
6	Electronic nose	Aroma, compositional analysis curing ripening	Torri et al. (2010), Ali et al. (2020), and Tan and Xu (2020)

analysis, such as volatile composition analysis of cheese aroma can be monitored using this method. A recent study in cheese characterization conducted by Ziino et al. (2005) revealed the components of cheese using gas chromatography. Solid Phase Microextraction (SPME) interfaced with GC-MS was used as an analytical technique for the study of flavour compounds in cheese. This technology allows the isolation of different volatile compounds from liquid and solid materials, again it is a less time-consuming and simple technique. This method is used by Delgado et al. (2011) to analyze the formation of aroma during cheese maturation. In another study, Cevoli et al. (2011) compared GC-MS with electronic nose and found GC-MS less preferable, as this technique requires skilled personnel, as well as it, is time-consuming. However, a large number of studies related to GC-MS are there for the quality monitoring of cheese.

9.2.2 High Performance Liquid Chromatography (HPLC)

High performance liquid chromatography is one of the chromatographic techniques used for the separation of different compositions in a sample. The HPLC is considered to be an effective analytical method for the analysis of organic acids as compared to colorimetric and gas chromatographic methods. The analytical mode of HPLC is a complex process as the desired separation depends upon several variables such as mobile phase buffer, column temperature, and wavelength. However, it is one of the important analytical tools in different industries including the chemical industry, pharmaceutical, and many food industries (Bhardwaj et al., 2015). High performance liquid chromatography is widely applied in drug discovery and development as it can monitor different chemicals and their properties. A typical HPLC instrument consists of several parts such as a mobile phase, pump, solvent detector, and a computing unit to visualize the data. The basic mechanism of HPLC follows the typical principle of chromatography, in high performance liquid chromatography the sample solution is injected into the porous stationary phase material, and through the column the mobile phase or liquid phase is pumped at very high pressure. Trihaas et al. (2005) used HPLC to monitor the formation of soluble peptides during cheese ripening. Another study of organic acid analysis of cheese is done using high performance liquid chromatography as it is a reliable method for organic acid determination (Yaman et al., 2023).

9.2.3 Near-Infrared Spectroscopy (NIR)

Near-infrared spectroscopy (NIR) is one of the effective spectrometric analytical techniques used to study and determine organic substances in many food and pharmaceutical products. NIR spectroscopy was first used in medical and clinical research, now it is widely used to analyze dairy products like milk and its constituents such as protein, fat, moisture, and lactose (Pellicer & Bravo, 2011). This technique is effective in analyzing many products as it is less time-consuming and rapid, the major advantage of this spectrometric technique is that it is non-destructive and multiple parametric measurements can be obtained. However, NIR has the potential for the analysis of many other complex compounds in minimal sample preparation and at a low cost. In the quality assessment of cheese, NIR contributes to estimating the moisture, total solids, fat, proteins, and other cheese characteristics (Sørensen & Jepsen, 1998). In modern NIR spectroscopy, a large number of absorbance values are acquired for special spectral lengths. Then the information gathered is used to identify and predict the chemical constituents of the sample. A large number of spectral variable data is assessed for qualitative and quantitative data analysis (Xiaoboa et al., 2010). Analysis using NIR is obtained by probing the sample with electromagnetic radiation, and the wavelength usually ranges between 750 and 2500 nm. (Pasquini, 2018). However, specific suitable interacting methods are based on the type of sample because removing the noninformative variables helps better predict characteristics (Xiaoboa et al., 2010). There are several measuring modes of IR spectroscopy such as transflectance, diffuse reflection, transmission, and diffuse transmission.

NIR has been used by many for quality estimation of different types of cheese, a study on cellulose content in cheese has been conducted using NIR Spectroscopy. The cheese samples were collected and refrigerated before conducting the analysis. To conduct NIR spectroscopy the sample was brought to room temperature (20 °C), and using a force the sample was pressed to provide a homogeneous surface. This homogeneous surface is necessary for further analysis as a non-homogeneous surface could provide a wrong analytical figure. BUCHI NIRFlex N-500 FT-NIR spectrometer was used for further scanning of samples and chemometric software was used for analyzing the spectra (Reis et al., 2022). Similar strategies are applied to many studies related to the characterization of cheese (Cattaneo et al., 2005), again, calibration standards can be varied based on the matrix. More studies on the shelf-life of cheese, contamination, solubility, and hardness, have been done for the quality evaluation of cheese using NIR spectroscopy (Sørensen & Jepsen, 1998). The studies show that NIR spectroscopy can not only determine the quality of cheese but may also give results for certain other sensory properties.

9.2.4 Mid-infrared Spectroscopy (MIR)

Mid-infrared (MIR) spectroscopy is another type of spectrometric technique that is used for the analysis of different chemical functional groups, bonds, and other chemical properties. Fourier transform infrared (FTIR) spectroscopy represents MIR as FTIR uses the spectral range of MIR (400–4000 cm^{-1}). The analytical data is obtained by taking measurements of the absorbance of energy by transmitting electromagnetic radiation through the sample. The main application of MIR spectroscopy reflects on the identification of different chemical bonds and organic molecules. The standard spectral wavenumber used in MIR spectroscopy is 400–4000 cm^{-1} and 2500–50,000 nm wavelengths (Cattaneo et al., 2005). However, the figures can be particularly re-calibrated depending on the sample type. MIR spectroscopy is extensively used for quality control of different industrial products. The best advantages of this spectroscopic method are reliability, speed, and moderate instrument cost (Woodcock et al., 2007). However, MIR spectroscopy harbors a set of complex mechanisms to perform different studies. A recent survey in cheese ripening was done using Fourier transform infrared (FTIR) spectroscopy preferred MIR spectroscopy (Martín-del-Campo et al., 2007).

The FTIR harbors a light source, a beam splitter, and two types of mirrors, one movable and a fixed mirror together called an interferometer. The beam splitter splits the light into two separate beams and is directed towards the two mirrors. Each mirror reflects the beam again to the beam splitter recombining the two reflected beams. One of the beam's paths is fixed and another is constantly changing due to the movement of the mirror, resulting in the interfering of beams with each other. Further, the interference is processed by a detector, producing an interferogram. Using the interferogram and fast Fourier transform algorithm a fingerprint spectrum of sample or absorbance vs wavelength and other information can be obtained. Distinct advantages of FT-MIR over NIR have been found in studies that analyze cheese. However, the NIR has been widely used over MIR for ease of use and faster results (Woodcock et al., 2007). FTIR attenuated total reflectance (ATR) is performed in many studies to analyze the quality of cheese and other food items. In a study, the cheese slices are analyzed with FTIR (Magna IR-750 spectrometer) equipped with a deuterated triglycine sulfate (DTGS) detector and a ZnSe horizontal ATR accessory. This reflectance spectroscopy method has been used in several studies of cheese analysis (Martín-del-Campo et al., 2007). In another study using an FTS Excalibur 3500GX FTIR spectrometer with a MIRacle ATR accessory and a high-pressure clamp, and using infrared spectrum between 4000 and 700 cm^{-1} sample was analyzed. The instrument was equipped with a Permaglow MIR source, a KBr beam splitter, and 3 reflection diamond ATR crystals to increase the absorbance (Koca et al., 2007). Gori et al. (2012) used ATR-FTIR combined with other chemometric methods for the discrimination of grated cheese.

9.2.5 *Electronic Nose*

The quality of food products depends upon various characteristics, such valuable characteristics as the aroma and flavour of food items highly impact the quality of food items. Sensory attributes such as flavour and aroma are other important features of good quality food items. So, it is important to evaluate the aroma and distinct scent of food items like cheese for quality assessment. An electronic nose is a special instrument designed for analyzing simple to complex odours (Torri et al., 2010). The electronic nose is comprised of several electrochemical sensors which provide appropriate results instantly. Electronic nose is intensely used for quality estimation of agricultural and food products. Rapid sensing and analyzing, non-destructive, and simple as compared to other conventional sensing techniques, are some advantages of electronic nose. The sensing tool e-nose evaluates the sample with the help of gaseous contents related to the physiochemical properties of a sample. Again, the electronic nose is capable of recognizing volatile organic compounds released from natural or synthetic sources (Ali et al., 2020). Moreover, the use of electronic nose is not only limited to food analysis, it has a wide range of uses in medical, pharmaceutical, and cosmetics industries. The electronic nose has been used in several studies of cheese characteristics including the study of cheese classification, ripening, shelf-life detection, etc.

An electronic nose comprises three major components, a sample processing system, a detection unit, and a data processing system. The instrument harbors gaseous compound exchange and a selected coating substance for the transmittance of electrical current, a signal processing unit that helps in the analysis of data, and a computer unit that shows the final data. The data information concerning the pattern analysis is obtained from the gas odour of the sample. However, different hybrid instruments are there that consist of a single array of sensors and analyze integrated devices to help in the study of various samples. Metal oxide semiconductor (MOS) sensors are one of the most used technologies for electronic noses. MOS consists of metal-oxides or semiconductors such as zinc oxide, iron oxide, etc. as sensing material along with a heating element. Depending upon the sensing material there are two separate types of gas sensors, one is n-type which responds to reducing compounds and the other one is p-type which responds to oxidizing compounds. However, the MOS consumes a significant amount of energy as it requires temperatures between 150 °C and 400 °C. Conducting polymers (CP) sensors are another type of sensing tool for electronic nose after MOS. Similarly, the tool consists of conducting particles such as polythiophene, polyaniline, and polypyrrole. The exposure of the sample to the sensor causes a reaction, and the reaction results in changes in conductivity. Conducting polymers does not require external heat to process the analysis, which makes it one of the preferable techniques. Acoustic wave sensors contain a piezoelectric sub-unit such as ZnO, quartz crystal, that is coated with sensing substance, and two transducers (input & output). The tool generates acoustic waves, respective to the nature of the sample, which are further analyzed by the sensing tool (Tan & Xu, 2020).

9.3 Applications and Quality Evaluation of Various Cheese-Based Products

Several types of cheeses like, stretched, vegan, whey, hybrid, and dry-milk cheeses are made using different techniques for their increasing demand in the world. After making them their quality evaluation is also important. So, their processes, applications, and techniques of quality evaluation all are discussed here.

9.3.1 Stretched Cheese

9.3.1.1 Tools and Techniques Used in the Preparation of Stretched Cheese

The stretched curd cheese also called pasta filata cheese includes a wide range of cheeses in this group, for instance, Mozzarella, Provolone, Scamorza, Caciocavallo, etc. The origin places of this type of cheese are Italy, Greece, the Bulkans, and Turkey. The main characteristic feature of stretched or pasta filata cheeses is the unique heating and finishing process which occurs at the end of the process of making the cheese. Stretching is the main characteristic of these types of cheeses which has an evident and sophisticated impact on the cheeses. Among stretched cheeses pizza cheese is widely in demand and produced because of the increasing popularity of pizzas in the whole world over the last 20 years (Kindstedt et al., 2010). The process of making this cheese is explained below:

- Treatment of milk: This type of cheese is made with pasteurized milk which is heated at the temperature of 72 °C for 15 s. Higher temperature can also be used more than the normal temperature to increase the yield of the cheese. A standardized protein-to-fat (P/F) or casein-to-fat (C/F) ratio is important to produce a desired cheese of fat-in-dry matter content, which is an important aspect set down by the purchaser of the cheese. Casein protein can also be added to the cheese milk to increase the yield to 5.4 g/100 g in the form of low-heat, non-fat dry milk (NDM) or condensed skimmed milk. Addition of 3 g 100 m/L NDM to the milk increases the calcium level and decreases the wetness. Membrane filtration of milk increases the cheese solids and protein levels by 4 g 100/g (Kindstedt et al., 2010).
- Starter culture: Thermophiles for instance, *Lactobacillus delbrueckii* subsp. *bulgaricus*and *Streptococcus thermophilus* are the conventional starters for pizza cheese. To reduce the browning of pizza cheese *Lactobacillus helveticus*is used in the place of *Lactobacillus delbrueckii* subsp. *bulgaricus* and *Streptococcus thermophilus* because some of their strains can ferment the galactose when lactose is present in the milk and it decreases the assemblage of galactose in the ultimate cheese product. The pizza cheese can also be manufactured by mesophilic starters like *Lactococcus lactis* subsp. *lactis* and *Lactococcus lactis* subsp. *cremoris*, but they alter the flavour where we need the acetaldehyde flavour. The main motif of the starters is to produce more acid in the milk to convert the milk from unstretchable curd to stretcheable so, the thermophiles are used rather than the mesophiles. To attain the necessary combination of pH to produce lactic acid we can store the milk at the optimum temperature of ~30°Cfor making cheese (Kindstedt et al., 2010).
- Coagulant: Coagulation is an important step in producing the cheese as it is important to coagulate the milk to start the concentration which forms the chemical composition of the cheese. Different types of coagulants are used from different sources to coagulate the milk such as calf rennet, which is mainly composed of chymosin with a small percentage of bovine pepsin and is widely used all over

the world. In traditional cheese-making where starter culture plays an important role, the coagulant also plays a secondary but very important role in proteolysis, functional characteristics, and ageing (Kindstedt et al., 2010).

- Cooking and cheddaring: Cooking, draining, and cheddaring are important to cheese-making as they help to maintain the moisture and calcium content in the curd at stretching. This step is done in various steps, where the temperature of cooking and cheddaring is a very important factor. The optimum temperature where the thermophiles can produce lactic acid is 42–45 °C which is above the cooking or cheddaring temperature. When the temperature is reduced to 44 °C to 38 °C, the acid production becomes slower so, to overcome this additional 30 min is added to the total making time of the cheese. pH also plays an important role in the stretching of the cheese, if there is a high pH content the stretching shows effects on higher moisture content, calcium/protein ratio, and ageing which indicates more structured, fibrous, and chewy cheese (Kindstedt et al., 2010).
- Stretching and moulding: In the manufacturing of cheese, heating, and stretching have an important characteristic impact on the structure and chemical composition of the cheese. Single- or twin-screw mechanical mixtures are used for the process of stretching which contains hot water coupled with a steam injection system. The process of stretching is done in two stages: In the first stage, the curd enters the mixture and is warmed by the hot water present in it ranging the temperature from 50 °C to 55 °C which alters the curd into a plastic and workable condition. In the second stage, the plastic curd enters the series of auger which converts the curd into fibrous ribbon. Then the auger transported the fibrous ribbon to the moulding machine where by using different pressures the cheese is given different shapes according to the need. The molder also performs the cooling with the help of which the cheese will remain in its shape (Kindstedt et al., 2010).

9.3.1.2 Quality Evaluation Technique of Stretched Cheese

The quality and character of the stretched cheese or especially the pizza cheese is measured by high-resolution synchrotron transmission and ATR-FTIR microspectroscopy and this technique was used by Pax et al. (2019). Fourier transmission infrared (FTIR) spectroscopy is used to determine the protein, lipid, and starch content in cheeses. Different commercial samples of cheeses are taken and stored at 4 °C for 3 months until it is analyzed. After the first month, the manufactured cheeses can absorb water so, the storage period and temperature are chosen to reduce the variation in water absorbance. The standard lipid, protein, and moisture content of the cheeses is 22–23%, 23–25%, and 46–48% respectively which is assessed during the storage period. By using Bruker Vertex V80v spectrometer which is linked with a Hyperion 2000 FTIR microscope provided with a liquid nitrogen-cooled narrow-band mercury cadmium telluride (MCT) detector the S-FTIR microspectroscopic measurement is performed. For the measurement of transmission, the cheeses are cut from two different cheese blocks and the sections should be 8 μm thin and ~15 mm^3 in volume. The sections are then continuously frozen at the temperature of −80 °C for >2 h before cryo-sectioning at −25 °C. Then the samples are stored in CaF_2 for 24 h to completely dry them and using a 36× IR object the S-FTIR value is measured. The S-FTIR value is measured in the spectral range of 3800–700 cm^{-1} using 4 cm^{-1} spectral resolution. Again, to perform synchrotron ATR-FTIR the samples of cheeses are cut in the dimensions of 3 mm × 3 mm × 1 mm approximately and 20× objective, and a 250 μm diameter facet germanium (Ge) ATR crystal is used to obtain the spectral lines. For the investigations, the different spectral information of 400 and 3600 are collected. The obtained S-FTIR images show the lipid and protein content and their components in the cheese. The observed structure in the S-FTIR technique is matched with the standard structure of the proteins and lipids and if the structures are the same as the standards then it is said to be a good quality cheese (Pax et al., 2019).

9.3.2 *Vegan Cheese*

9.3.2.1 Tools and Techniques Used in the Preparation of Vegan Cheese

The concept of veganism is widely spread in the world and people are switching their diet from animal-based food products to plant-based food products. Like other food products, vegan cheese has also gained more attention in the last decade because of its nutritional value and has been eaten widely all over the world. The demand for this vegan cheese and its manufacturing has been increasing day by day. Vegan cheeses can be made from natural plant-derived materials like leaves, stems, flowers, tubers, fruit seeds, etc. It is found that plant-based lipids are healthier than the ingredients obtained from dairy products made from cow's milk. Moreover, plant-based cheeses are the alternative options for people who are vegetarian or cannot tolerate dairy products. Protein, starches, and vegetable oils are the main ingredients to make plant-based cheeses (Pham et al., 2024). The process of making vegan cheese is explained below:

The process of making plant-based cheeses or vegan cheeses are same as traditional cheese-making where the higher content of protein and fats derived from the plants are used for their physicochemical, mechanical, and sensorial properties. The making of vegan cheese in this way has similar properties as dairy cheese and to overcome the shortcomings of vegan cheeses they comprise an emulsifier, anti-foaming, and colouring agents, nutritional yeast, and supplements. The pH is also an important framework of cheese-making, which affects the protein-protein forces, and the optimum pH value ranges from 5.21 and 5.87. In the process of making of vegan cheeses waxy non-modified potato starch is used as a gelling agent because it helps to modify the texture and the nutritional value of the cheese. The required concentrations of starch, protein, lipids, and water are 10–24%, 0.5–8%, 15–35% and 35–74.5% respectively. The process of vegan cheese-making includes the gelatinization of starch and cooling the mixture to set in motion of sol-gel transition. The sol-gel transition helps to attain the viscoelasticity and textures of the cheese. It is important to maintain the protein dimension and gelling properties, where the gelation is increased by using Ca^{2+} salt which increases the aggregation of protein. Viscosity, emulsification, gelation, and reliability define the processing behaviour and sensory quality of cheese (Pham et al., 2024).

9.3.2.2 Quality Evaluation Technique of Vegan Cheese

With the production of the cheese, it is also important to evaluate the quality of the cheese. Khan et al. (2024) used the Fourier-Transform Infrared (FTIR) and Raman spectroscopy to evaluate the quality of plant-based or vegan cheeses. For FTIR, the fresh cheese of 28 days storage is taken and sliced about 0.5 g of each sample from the different parts of the cheese. After slicing they are pressed with high pressure to make contact between the crystals and the sample. 400–4000 cm^{-1} wave number and 4 cm^{-1} resolution are required to perform the FTIR. The absorption of the spectrum is measured by a single laser beam and to get a precise observation three spectra are measured. After the measurement, the functional groups of the samples are collated with the recent standards for the functional groups. Again, in Raman spectroscopy, the storage period of cheese samples 0 and 28 days are used to measure the spectra, and samples of the cheeses are cut into 50 μm thick. The Metrohm MIRA-M1 laser Raman spectrometer is used to measure the Raman spectra and the range of the spectrum is 300 to 3000 cm^{-1} with resolution power 1 cm^{-1} at 4.5 s. Eight points are chosen to get the spectrum and the average spectrum is taken to analyze the samples. To determine and understand the quality of the cheese the observed Raman spectra are compared to the standard Raman spectra of vegan cheese to check the texture and physicochemical characteristics of the cheeses (Khan et al., 2024).

9.3.3 Whey Cheese

9.3.3.1 Tools and Techniques Used in the Preparation of Whey Cheese

Whey is a protein and the main by-product of cheese-making industries, where it is also used to make cheeses as a sustainable way to not waste it. Some of the traditional whey cheeses made all over the world are Italy—Ricotta, Ricotta salata or Ricottone, and Ricotta fresca; Greece—Anthotyros, Myzithra, Manouri, xynomyzithra, and Urda (in Romania and Serbia as well as Israel); Turkey—Lor; Cyprus—Anari; Croatia and Serbia-Skuta; Norway—Gjetost and Brunost etc. Whey cheese is made through the denaturation process of whey protein, and it is done at a temperature of 88–92 °C. High pH and moisture content are the main properties of whey cheese which are important for the microbial growth in the whey protein

(Bintsis & Papademas, 2023). The process of making whey cheese from whey protein is explained here-.

At first, the whey is heated to the temperature of 88–92 °C while it is being stirred. The rate of heating and stirring is important in the making of whey cheese. At the temperature of around 80 °C the small particles of whey proteins are first seen. If the whey cheese is going to be consumed fresh then the final temperature should be lower and if dried whey cheese is made then a high temperature should be there. After selecting the required temperature the curd formed in the whey protein is left for 15–20 min on the surface to lessen the moisture, and after that it is taken out for molding and draining. If the yield is low then 30 mL/L of milk or cream can be added to the whey protein, and for high-quality cheeses, the milk percentage can be increased up to 10%. Salt and acidification agents are also important which help the whey to increase denaturation rate. 6 mL/L of citric acid is used as an acidification agent which helps decrease the pH of whey protein by about 5.2 (Bintsis & Papademas, 2023).

9.3.3.2 Quality Evaluation Technique of Whey Cheese

After manufacturing the required whey cheeses the evaluation of the quality of the cheese is also important. To enhance the shelf life and functional, and sensory characteristics several techniques have been used in whey cheeses over the last 10 years. Bintsis and Papademas (2023) explained this technique in their paper where some of the whey cheeses are stored under 100 MPa hyperbaric storage at different room temperatures to stop the growth of microbes in the cheese. This technique is more eco-friendly than normal refrigerated storage with a lower estimated carbon footprint of about 26-fold. In other research, it is found that applying UV-C light on the surface of cheese can delay the growth of mold during the storage period of the cheese which can help to maintain the shelf life of the cheese. The UV-C light can be used in the amounts of 1.627, 4.018, and 36.832 kJ/m^2 in the cheese surface, but extreme doses can also harm the cheese content so, we should be careful using UV-C light. Again, pulsed light and plasma treatment can also help to maintain the shelf life of Ricotta and heat treatment, homogenization, and hot packaging can maintain the shelf life of buffalo ricotta cheese, both of which are the types of whey cheese (Bintsis & Papademas, 2023).

9.3.4 Hybrid Cheese

9.3.4.1 Tools and Techniques Used in the Preparation of Hybrid Cheese

It is known that plant-based dairy products do not provide the nutritional value of animal-based dairy products. Sustainability is an important topic getting more attention nowadays for which the "Eat-Lancet" Commission declared to depend more on the plant-based diet than the animal-based diet for both health and sustainability. Like the other food products, plant-based cheeses are also getting more attention than animal-based cheeses. So, to reduce the use of animal-based components to make cheeses plant-based components are included with the animal-based component i.e. milk for making the so-called "hybrid cheeses" (Genet et al., 2023). The process of making hybrid cheese is described here:

Fermented side-stream ingredients are the components that are mixed with the dairy component to produce hybrid cheeses. The side-stream flours are mixed with the dairy pasteurized component i.e., milk in various percentages like 3.5%, 5%, and 7.5%. The fermentation step helps to maintain the range of pH and moisture content which are 4.79–5.60% and 45.86–61.29% respectively. The maintained pH and moisture content helps to increase the nutritional value of the cheese. The addition of side-stream flours can induce the coagulation of milk without using the rennet. The coagulated milk is then processed and made into the required hybrid cheeses (Palatzidi et al., 2024).

9.3.4.2 Quality Evaluation Technique of Hybrid Cheese

Quality evaluation is again important after making the cheese so, Garcia-Fontanals et al. (2023) explained this technique for the quality evaluation of the cheese. In this technique, the property of the texture of the cheese is measured by using TA.HDplus texturometer and the moisture content of the cheese are analyzed by Multiple Linear Regression (MLR) analysis techniques. The texture properties are determined by the spreadability test of the cheese by TA.HDplus texturometer which is an instrument coupled with a 5 kg load cell and Texture Technologies Corporation (TTC). One male and one female cone are present in the TTC spreadability fixture at an angle of 45°. At first, the sample is inserted into a female cone of the instrument and is pressed gently to remove the air present in it, after that to generate a level surface the extra sample is removed from it and then the readings are taken in graphical forms. The required temperature for this is 18.5 ± 0.5 °C, the samples are used after 48 h of cold storage. Again, the MLR technique is used to determine the moisture, colour, and texture of the cheese and to get the maximum firmness and spreadability of the cheese and are compared with the standard spreadability of the cheeses (Garcia-Fontanals et al., 2023).

9.3.5 Dry-Milk Cheese

9.3.5.1 Tools and Techniques Used in the Preparation of Dry-Milk Cheese

Dry milk is an alternative component to making cheeses other than using liquid milk for cost-effectiveness. The taste and texture of the cheese made from dry milk powder is almost the same as the liquid milk but sometimes it can be a failure due to the crumbly, grainy, or pasty texture of the dry milk. But if it is properly done from the start of the process then it can be very effective and also reduce the cost of using the dairy liquid milk. The dry milk powder is mixed with water according to the requirements, more powder can make the consistency of the sample more creamier. Then the reconstituted milk is heated in a pot, for soft cheese the temperature should be 30 °C and for hard cheese, the temperature should be 32 °C. After getting the desired temperature suitable starter cultures are added to the milk for acidification and kept at rest for about 1 h. Then rennet is diluted in cold water and added to the milk stirred gently and kept at rest for 30–60 min for the formation of curd in it. Coagulation of the milk will occur and solid curd will be formed in the milk then the curd is cut into small cubes and left it for 5–10 min to finely solidify the curd. The curd is then heated for 20–30 min by stirring it at a temperature around 105–125 °F which helps to reduce the whey from the curd and after cooling it the curd is drained properly. After draining the curd it is pressed for the solid structure for several hours and different types of flavours can also be added to the cheese for better taste enhancement in the cheese (Farkye et al., 2003).

9.3.5.2 Quality Evaluation Technique of Dry-Milk Cheese

There is a method called Babcock which was used by Marshall (1992) to determine the fat content in the cheese. In this method, at first, the sample is prepared by rehydrating the cheeses as the moisture content of dry milk cheese is less than other cheeses. A finely ground part of the cheese is weighed and distilled water is added to it to make it more functionable. The sample is taken in the Babcock bottle and 90–92% of sulphuric acid is added to the sample carefully as it can break down the protein and fatty components of the cheese which makes it easier to isolate the fat without much trouble. The heat produced by sulphuric acid can degrade the protein in the sample. Then the sample is centrifuged in the centrifuge machine for 5 min to separate the fat and for the availability of fat in the upper part of the sample. After the centrifugation water is added to the sample at 60 °C for the availability of the fat in the upper surface of the sample and then the bottle is heated at 55–60 °C and centrifuged again for 1 min. At last, the bottle is immersed at 55–60 °C for the settlement of the fat layer on the surface of the sample and the readings of fat content are taken and matched with the standard fat content of the dry-milk cheese (Marshall, 1992).

9.4 Conclusion

Quality evaluation is one of the most important aspects of the food industry including cheese manufacturing industries as it is important to monitor the product in different stages of processing. Alteration in the process of manufacturing can also bring changes to the final product, so analyzing the characteristics of the product at various stages is important to maintain consistency. Again, quality estimation provides adequate data for evaluating the product by comparing it with the industry standards. However, this data can also be used for future improvisation and benchmarking against competitors. Those data can be only obtained by using different spectroscopic and analytical methods that are used in cheese quality analysis. These evaluation techniques are used to evaluate the quality of cheese-based products by analyzing their texture, appearance, structure, chemical compositions, lipid, protein, fat, and moisture content, which are important to maintain the quality of the cheese. Advancements in these techniques have already brought adverse improvements in cheese manufacturing and further advances can be expected counting on future scopes.

References

Ali, M. M., Hashim, N., Aziz, S. A., & Lasekan, O. (2020). Principles and recent advances in electronic nose for quality inspection of agricultural and food products. *Trends in Food Science and Technology, 99*, 1–10. https://doi.org/10.1016/j.tifs.2020.02.028

Bhardwaj, S. K., Dwivedia, K., & Agarwal, D. D. (2015). A review: HPLC method development and validation. *International Journal of Analytical and Bioanalytical Chemistry, 5*(4), 76–81.

Bintsis, T., & Papademas, P. (2023). Sustainable approaches in whey cheese production: A review. *Journal of Dairy Science, 4*(2), 249–270. https://doi.org/10.3390/dairy4020018

Cattaneo, T.M.P., Giardina, C., Sinelli, N., Riva, M. and Giangiacomo, R. (2005). Application of FT-NIR and FT-IR spectroscopy to study the shelf-life of Crescenza cheese. *International Dairy Journal, 15*(6-9), pp.693–700.

Cevoli, C., Cerretani, L., Gori, A., Caboni, M. F., Toschi, T. G., & Fabbri, A. (2011). Classification of Pecorino cheeses using electronic nose combined with artificial neural network and comparison with GC–MS analysis of volatile compounds. *Food Chemistry, 129*(3), 1315–1319. https://doi.org/10.1016/j.foodchem.2011.05.126

Delgado, F. J., González-Crespo, J., Cava, R., & Ramírez, R. (2011). Formation of the aroma of a raw goat milk cheese during maturation analysed by SPME–GC–MS. *Food Chemistry, 129*(3), 1156–1163. https://doi.org/10.1016/j.foodchem.2011.05.096

Farkye, N. Y., & Drake, M. A. (2003). The effect of application of cold natural smoke on the ripening of cheddar cheese. *Journal of Dairy Science, 86*(6), 1910–1917.

Garcia-Fontanals, L., Llorente, R., Valderrama, J., Bravo, S., & Talens, C. (2023). Hybrid spreadable cheese analogues with faba bean and mealworm (Tenebrio molitor) flours: Optimisation using desirability-based mixture design. *Foods, 12*(7). https://doi.org/10.3390/foods12071522

Genet, B. M., Sedó Molina, G. E., Wätjen, A. P., Barone, G., Albersten, K., Ahrné, L. M., & Bang-Berthelsen, C. H. (2023). Hybrid cheeses—supplementation of cheese with plant-based ingredients for a tasty, nutritious and sustainable food transition. *Fermentation, 9*(7), 667.

Gori, A., Maggio, R. M., Cerretani, L., Nocetti, M., & Caboni, F. M. (2012). Discrimination of grated cheeses by Fourier transform infrared spectroscopy coupled with chemometric techniques. *International Dairy Journal, 23*(2), 115–120. https://doi.org/10.1016/j.idairyj.2011.11.005

Karoui, R., Pillonel, L., Schaller, E., Bosset, J. O., & De Baerdemaeker, J. (2007). Prediction of sensory attributes of European Emmental cheese using near-infrared spectroscopy: A feasibility study. *Food Chemistry, 101*(3), 112–1129. https://doi.org/10.1016/j.foodchem.2006.03.012

Khan, U. M., Sameen, A., Decker, E. A., Shabbir, M. A., Hussain, S., Latif, A., Abdi, G., & Adil, R. M. (2024). Implementation of plant extracts for cheddar-type cheese production in con-

junction with FTIR and Raman spectroscopy comparison. *Food Chemistry X, 22*. https://doi.org/10.1016/j.fochx.2024.101256
Kindstedt, P. S., Hillier, A. J., & Mayes, J. J. (2010). Technology, biochemistry and functionality of pasta filata/pizza cheese. In book: Technology of Cheesemaking. Ed (II) Barry A. Law, D and A. Y. Tamime, Blackwell Publishing Limited, West Sussex, UK, pp. 330–359
Koca, N., Rodriguez-Saona, L. E., Harper, W. J., & Alvarez, V. B. (2007). Application of Fourier transform infrared spectroscopy for monitoring short-chain free fatty acids in Swiss cheese. *Journal of Dairy Science, 90*(8), 3596–3603. https://doi.org/10.3168/jds.2007-0063
Kraggerud, H., Solem, S., & Abrahamsen, R. K. (2012). Quality scoring – A tool for sensory evaluation of cheese? *Food Quality and Preference, 26*, 221–230. https://doi.org/10.1016/j.foodqual.2012.04.006
Marshall, T. R. (1992). *Standard methods for the examination of dairy products*. American Public Health Association.
Martín-del-Campo, S. T., Picque, D., Cosío-Ramírez, R., & Corrieu, G. (2007). Middle infrared spectroscopy characterization of ripening stages of camembert-type cheeses. *International Dairy Journal, 17*(7), 835–845. https://doi.org/10.1016/j.idairyj.2006.10.003
Palatzidi, A., Nikoloudaki, O., Torreiro, M. G., Matteucci, C., Ferrentino, G., Scampicchio, M. M., Di Cagno, R., & Gobbetti, M. (2024). Novel formulations for developing fresh hybrid cheese analogues utilizing fungal-fermented brewery side-stream flours. *Current Research in Food Science, 9*. https://doi.org/10.1016/j.crfs.2024.100829
Pasquini, C. (2018). Near infrared spectroscopy: A mature analytical technique with new perspectives—A review. *Analytica Chimica Acta, 1026*, 8–36. https://doi.org/10.1016/j.aca.2018.04.004
Pax, A. P., Ong, L., Vongsvivut, J., Tobin, M. J., Kentish, S. E., & Gras, S. L. (2019). The characterisation of mozzarella cheese microstructure using high resolution synchrotron transmission and ATR-FTIR microspectroscopy. *Food Chemistry, 291*, 214–222. https://doi.org/10.1016/j.foodchem.2019.04.016
Pellicer, A., & Bravo, M. C. (2011). Near-infrared spectroscopy: A methodology-focused review. *Seminars in Fetal & Neonatal Medicine, 16*(1), 42–49. https://doi.org/10.1016/j.siny.2010.05.003
Pham, Q. H., Ho, T. M., Saris, P. J., Mäkelä-Salmi, N., Amiri, S., Zarei, M., & Yousefvand, A. (2024). Yogurts fortified with postbiotic powders derived from Lactobacillus acidophilus LA5: Physicochemical, rheological, antioxidant, and sensory properties. *LWT Food Science & Technology, 213*, 117043.
Reis, M.G., Agnew, M., Jacob, N. and Reis, M.M. (2022). Comparative evaluation of miniaturized and conventional NIR spectrophotometer for estimation of fatty acids in cheeses. Spectrochimica Acta Part A: Molecular and Biomolecular Spectroscopy, 279,121433.
Sameen, A., Anjum, F. M., Huma, N., & Nawaz, H. (2008). Quality evaluation of mozzarella cheese from different milk sources. *Pakistan Journal of Nutrition, 7*(6), 753–756. https://doi.org/10.3923/pjn.2008.753.756
Sørensen, L. K., & Jepsen, R. (1998). Assessment of sensory properties of cheese by near-infrared spectroscopy. *International Dairy Journal, 8*(10–11), 863–871. https://doi.org/10.1016/S0958-6946(98)00130-7
Tan, J., & Xu, J. (2020). Applications of electronic nose (e-nose) and electronic tongue (e-tongue) in food quality-related properties determination: A review. *Artificial Intelligence in Agriculture, 4*, 104–115. https://doi.org/10.1016/j.aiia.2020.06.003
Torri, L., Sinelli, N., & Limbo, S. (2010). Shelf life evaluation of fresh-cut pineapple by using an electronic nose. *Postharvest Biology and Technology, 56*(3), 239–245. https://doi.org/10.1016/j.postharvbio.2010.01.012
Trihaas, J., Vognsen, L., & Nielsen, P. V. (2005). Electronic nose: New tool in modelling the ripening of Danish blue cheese. *International Dairy Journal, 15*(6–9, 679), –691. https://doi.org/10.1016/j.idairyj.2004.07.023

Woodcock, T., Fagan, C. C., O'Donnell, C. P., & Downey, G. (2007). Application of near and mid-infrared spectroscopy to determine cheese quality and authenticity. *Food and Bioprocess Technology, 1*, 117–129.

Xiaoboa, Z., Jiewena, Z., Povey, M. J. W., Holmes, M., & Hanpin, M. (2010). Variables selection methods in near-infrared spectroscopy. *Analytica Chimica Acta, 667*(1–2), 14–32. https://doi.org/10.1016/j.aca.2010.03.048

Yaman, H., Aykas, D. P., & Rodriguez-Saona, L. E. (2023). Monitoring Turkish white cheese ripening by portable FT-IR spectroscopy. *Frontiers in Nutrition, 10*. https://doi.org/10.3389/fnut.2023.1107491

Ziino, M., Condurso, C., Romeo, V., Giuffrida, D., & Verzera, A. (2005). Characterization of "Provola dei Nebrodi", a typical Sicilian cheese, by volatiles analysis using SPME-GC/MS. *International Dairy Journal, 15*(6–9), 585–593. https://doi.org/10.1016/j.idairyj.2004.07.024

Chapter 10
Advanced Testing Techniques

Kuntal Roy, V. P. Aparnna, and Harsh Jindal

10.1 Introduction

Milk and milk derived ingredients are utilized to prepare a wide range of highly nutritious foods. This includes different types of cheeses, ice cream, yogurt, milk powder, whey protein products, butter, and other fermented dairy products. The dairy industry has seen a rise in milk consumption and the development of various product varieties, resulting in the implementation of advanced handling systems and processing techniques. Sensors are currently used in the dairy industry to measure process factors and make sure the end product fulfils quality standards (Vashishth et al., 2022).

Cheese is defined as a product made from curd obtained by coagulating casein in milk with rennet or similar enzymes, in the presence of lactic acid produced by added or adventitious microorganisms (Shenana & Patel, 2019). The process of cheesemaking can be traced back to 8000 BCE, coinciding with the domestication of sheep. Historically, cheese was prepared by allowing milk to curdle naturally in vessels formed from animal stomachs. The curdling of milk is caused by the presence of a milk coagulating agent known as rennet. Cheese-making techniques gradually disseminated over Europe, the Middle East, and Asia, undergoing refinement in monasteries during the Middle Ages (Dalby, 2009).

K. Roy (✉)
Faculty of Dairy Technology, Department of Dairy Technology, WBUAFS, Nadia, West Bengal, India
e-mail: kuntalwbuafs@ndri.res.in

V. P. Aparnna · H. Jindal
Dairy Technology Division, ICAR-National Dairy Research Institute, Karnal, Haryana, India
e-mail: vpaparnna33@ndri.res.in

S. Mansha Rafiq (ed.), *Technological Advances and Trends in Cheese Making*,
https://doi.org/10.1007/978-3-031-94476-5_10

Traditional stretched cheeses, novel vegan and hybrid kinds, and nutrient-rich whey and dry-milk cheeses are all contributing to the fast evolution of the worldwide cheese business (Lagrange et al., 2015). Worldwide cheese production in 2022–23 was 22.98 MMT, with 47% being produced in the EU (10.39 MMT), 29% from the US, and 5% from Russia (USDA, 2024). Asia has the lowest level of development when it comes to cheese production with only 2.2% of the world's total cheese output and 5% of the world's cheese consumption (Mikkelsen, 2000). Iran, India, China, Japan, and South Korea are the major cheese-producing countries in Asia accounting for more than 90% of the region's cheese production (Mikkelsen, 2000).

The composition of cheese, along with its physical and rheological properties, nutritional value, sensory attributes, microbial characteristics, and other features, are determined by the attributes of the raw milk, the methods used in cheesemaking, and the duration and conditions of the aging process. During its production and storage, cheese undergoes numerous chemical and biological changes. Consequently, it is imperative to assess and comprehend these attributes to produce a high-quality product. Time-consuming and chemical waste-generating, conventional "wet chemistry" methods for determining the chemical components of cheese necessitate the collection and transportation of samples to a laboratory (Lénárt et al., 2012). Despite their precision and accuracy, these methods are resource-intensive and burdensome.

The quality and safety of various cheese products can be ensured by employing sophisticated testing procedures that offer immediate and accurate information on the production process. The need for quick, non-destructive, and cost-effective techniques to measure the constituents of cheeses, such as fat, protein, and moisture, has led to their invention. The use of process analytical technologies (PAT) in At-line and In-line testing, has become essential in several industries. These methods provide notable benefits compared to traditional testing methods by facilitating continuous monitoring and prompt feedback (Azamfirei et al., 2023). At-line testing involves analyzing samples at or near the production line, allowing rapid adjustments without significant delays (Mali et al., 2016). Whereas, In-line testing integrates analytical instruments directly into the production process, facilitating constant monitoring and control (Simon et al., 2015). Both approaches are instrumental in maintaining the desired characteristics of various cheese types, such as texture, flavor, moisture content, and microbial safety.

This chapter explores the application of At-line and In-line testing methods utilized in the production and processing of various types of cheese. By examining case studies and technological advancements, we aim to demonstrate how these testing methodologies enhance product quality, ensure safety, and streamline the production process. The integration of these testing technologies supports compliance with regulatory standards and meets rising consumer expectations for high-quality, safe, and innovative cheese products.

Table 10.1 Comparison of At-line and In-line testing

Aspect	At-line testing	In-line testing
Location	Near the production line	Directly on the production line
Frequency	Intermittent, based on sampling	Continuous or near continuous
Disruption to process	May require pausing production to collect samples	No disruption, integrated into production
Complexity of tests	Can perform more complex, detailed tests	Limited to tests that can be automated and done quickly
Response time	Slower, due to sample transport and analysis time	Immediate, real-time feedback

10.2 At-line and In-line Testing

In-line and At-line testing are methods employed in manufacturing and quality control processes to guarantee that products adhere to the specified product standards. At-line testing is a term that denotes quality control tests that are conducted close to the production line, but not directly within the production process (Mali et al., 2016). Typically, these tests are conducted in a dedicated testing area adjacent to the production line or a neighboring laboratory. In contrast, In-line testing is a term that denotes quality control tests that are conducted directly on the production line without interrupting the manufacturing process (Simon et al., 2015). The integration of these evaluations into the production equipment facilitates real-time monitoring and control. A comparison between At-line and In-line testing is shown in Table 10.1.

Improved quality control and the guarantee of consistent product quality are among the advantages of In-line and At-line testing, which detect deviations in real-time. These techniques enhance efficiency by minimizing waste and delay, which enables rapid adjustments during production. Furthermore, they improve food safety by consistently monitoring critical control points. In-line and At-line testing also reduce the risk of product recalls and minimize the necessity for extensive end-product testing. Manufacturers can enhance their control over product quality, safety, and consistency by incorporating In-line and At-line testing into the cheese production process. This, in turn, results in increased consumer satisfaction and adherence to regulatory standards.

10.3 Process Analytical Technology

PAT is a framework that focuses on the design, analysis, and production processes by utilizing real-time measurements of essential quality and performance characteristics of raw materials, in-process components, and finished goods to guarantee the quality of the final product. This involves the utilization of measurement systems and sensors for the real-time oversight of processes. These instruments have progressed from single-variable measurement systems, like pH, temperature, pressure, and color, to more complex multivariate measurement systems. These sophisticated systems are capable of delivering insights regarding the biological, physical, and chemical characteristics of the materials undergoing processing (Panikuttira et al., 2018). Feedback from process analyzers or sensors can be utilized to enhance processes (Wood et al., 2016). The methodology employed is comprehensive, signifying that quality is woven into each phase of the procedure (FDA, 2004; Rathore et al., 2010).

Process Analytical Technology (PAT) can be utilized in three primary measurement modes:

- **At-line**: Samples are taken from the process line and analyzed nearby
- **On-line**: Samples are diverted from the process line for analysis and then returned to the stream after evaluation
- **In-line**: Samples are measured directly within the process stream

10.4 Application of At-line and In-line Testing in Cheese Production and Processing

In-line and At-line testing play a crucial role in ensuring uniformity and quality during the production of a range of dairy products, such as cheese. Every variety of cheese possesses distinct requirements and characteristics that must be carefully monitored and controlled during the production process. Multiple studies have examined the use of PAT testing methods in cheese preparations. A brief summary of these applications in stretched, vegan, whey, hybrid and dry milk cheese is described in Table 10.2.

10.5 Utilization of PAT Tools in Cheese Manufacture

Internal, post-production, laboratory-based quality assessment of completed commodities is frequently relied upon by the current food sector quality frameworks. This approach has the ability to necessitate more work and, in extreme circumstances, product retrieval. The procedure may be quickly adjusted to keep the product within defined limits by continuously monitoring the materials being processed (Hitzmann et al., 2015). PAT tools utilized for processing steps involved in cheese manufacture as shown in Table 10.3.

PAT tools can be utilized for assessing the milk quality, standardization of milk, monitoring coagulation kinetics, and product quality analysis. Some of the commercially available PAT tools for these testing methods have been shown in Table 10.4.

Table 10.2 Applications of In-line and At-line testing in stretched, vegan, whey, hybrid and dry-milk cheeses

Cheese variety	In-line testing applications	At-line testing applications
Stretched cheese	Monitoring moisture level during stretching to ensure proper texture and meltability Continuous measurement to control acidity, affecting flavor and stretchability Real-time temperature monitoring to maintain optimal conditions for stretching	Checking firmness and elasticity post-production to ensure consistency Ensuring no harmful bacteria are present, which can affect shelf life and safety
Vegan cheese	Monitoring the mixing process to ensure uniform distribution of ingredients Real-time measurement to control the flow properties, which affect the final texture	Verifying the presence of key nutrients like protein and fat or allergens lactose Regularly testing taste and mouthfeel to match dairy cheese equivalents
Whey cheeses	Monitoring the concentration of solids during whey processing to control yield and texture Ensuring the correct temperature is maintained during whey heating to optimize curd formation	Checking water content to ensure desired consistency and shelf life Verifying the smoothness and spreadability of the cheese
Hybrid cheese	Continuous monitoring to ensure even mixing of dairy and non-dairy components Real-time measurement of emulsion stability to prevent phase separation during production	Ensuring the blend meets nutritional targets for both dairy and non-dairy components Evaluating taste, texture, and appearance to ensure consumer acceptance
Dry-Milk cheese	Monitoring the rehydration process of dry milk to ensure proper curd formation Real-time tracking to achieve the desired dryness and texture	Particle size analysis to ensure uniformity in dry milk particles for consistent cheese production Post-production testing to confirm the cheese meets specifications for dry-milk products

10.6 Different At-line and In-line Analysis Methods for Cheese Quality Analysis

10.6.1 Spectroscopic Techniques for Quality Analysis of Cheese

Traditional analytical methods, including sensory analysis, chromatography, and physicochemical studies, are labour-intensive, expensive, and chemically dependent (Loudiyi et al., 2022). Hence, it is imperative to devise more efficient and economical techniques to precisely track variations in cheese quality during its production and preservation. Numerous non-destructive and non-invasive instrumental

Table 10.3 Process analytical technology (PAT) techniques application in cheese manufacture

PAT technique	Equipment	Wavelength range	Pros and cons	References
NIR light backscatter	CoAguLite by Reflectronics, Inc.	880 nm	Pros: Monitor milk coagulation and predict syneresis indices Analyse curd moisture, curd yield and whey fat loss Cons: Industrial-scale application is required	Fagan et al. (2008)
Fluorescence spectroscopy	UV photo-X fluorescence system model 53F by custom sensors and technology	Excitation 355, 281.3, 281.3 nm Emission 481.1, 340.6, 521.2 nm	Pros: Predict models with good fit developed Cons: A larger sample number analysis is necessary to confirm the sensitivity	Fagan et al. (2011)
NIR spectroscopy	Antaris MX FT-NIR process analyser from Thermo Scientific	1000–1850 nm	Pros: Predict the entire coagulation process	Lyndgaard et al. (2012)
NIR light backscatter	CoAguLab by Reflectronics Inc.	880 nm	Pros: Before the rheological response the coagulation-related information is obtained	Nicolau et al. (2015)

Van den Berg et al. (2013)
FT-MIR Fourier transforms mid-infrared, *IR* infrared, *NIR* near-infrared

approaches can be used to monitor food in real-time and while it is in the production line. Utilizing spectroscopic techniques can significantly mitigate challenges faced during cheese production (Karoui & De Baerdemaeker, 2007). Spectroscopic techniques encompass Fluorescence Spectroscopy, Mid-Infrared (MIR), Near-Infrared (NIR), and Nuclear Magnetic Resonance Spectroscopy (NMR). Typically, these methods are employed alongside chemometric tools to investigate the information contained in spectrum data (Loudiyi et al., 2022). Table 10.5 shows the application of fluorescence, MIR, NIR, and NMR spectroscopy in the monitoring of cheese during processing.

10.6.1.1 Real-Time Model Predictive Control of Milk Coagulation by NIR

The application of real-time modeling and spectroscopic monitoring during the milk coagulation phase in cheese production illustrates the effectiveness of model predictive control as a valuable tool for overseeing batch processes within the dairy

Table 10.4 PAT tools utilized for milk composition, standardization, and monitoring coagulation kinetics in cheese manufacture

Status	PAT tools	Manufacturer	Principle of working
Milk constituent analysis			
In-line	NIRQuest™ 512	Ocean optics, largo, FL, USA	NIR
In-line	MilkoStream™ FT	FOSS instruments	FT-IR
In-line	Optiquad™ M0450W	KROHNE Messtechnik GmbH Ludwig, Duisburg, Germany	Transmission, scattering, fluorescence and refraction of UV, VIS, NIR and IR
In-line	Milk-Inspector™	J&M Analytik AG, Willy-Messerschmitt, Essingen, Germany	Vis-NIR
On-line/ In-line	Matrix-F	Bruker instrument	FT-NIR
At-line	CombiFoss™ 7	FOSS instruments	Mid-infrared
Coagulation kinetics and prediction of coagulum cutting time			
In-line	FluorLite™ milk coagulation	Reflectronics, Inc.	Fibre optics—NIR diffuse reflectance and fluorescence
In-line	CoAguLite™	Reflectronics, Inc.	Light backscatter monitoring
In-line	CoAguLite™	Reflectronics, Inc.	Fibre-optic NIR diffuse reflectance
At-line	CoaguSens™	Rheolution instruments, Montreal, QC, Canada	Mechanical
At-line	ElastoSense™	Rheolution instruments	Mechanical
Standardization of Milk			
In-line	RelcoTM optimizer	Relco	Works in conjunction with an In-line milk analyser
In-line	APV CompoMasterTM	SPX Flow Technology, West Sussex, UK	Works in conjunction with an In-line milk analyser
In-line	Tetra PakTM	Tetra Pak, Pully, Switzerland	Mass flow measurements

sector (Roupas, 2008). Observing the coagulation process of milk is crucial for maintaining the quality of cheese. The moment at which the cheese curd is cut referred to as the endpoint, significantly influences both the moisture level and the overall yield of the final cheese product (Lyndgaard et al., 2012).

A study by Thygesen and van den Berg (2012) involved monitoring various cheese batches through real-time NIR reflection measurements during the coagulation of milk. Subsequently, the NIR data underwent condensation through the application of principal component analysis. The variations in component scores over time were subsequently employed to create kinetic models for the three processes of milk coagulation: kappa-casein proteolysis, micelle aggregation, and network formation. The NIR data showed a strong fit to the kinetic models, with a R^2 value greater than 0.99. By utilizing advanced models and a sophisticated algorithm, this cutting-edge technology offers instantaneous analysis of the milk coagulation

Table 10.5 Application of fluorescence, MIR, NIR, and NMR spectroscopy to monitor chemical modification of cheese during processing and for predicting their functional properties

Study	Fluorophore/ wavelength range/ frequency (MHz)	Chemometric tools	References
Application of fluorescence spectroscopy			
Differentiation of cheese as per the manufacturing process	Tryptophan and Vitamin A	PCA, FDA	Herbert et al. (2000)
Analysis of the changes during cheese ripening of tryptophan fluorescence spectra		CCA, PCA	Mazerolles et al. (2001)
Texture analysis		CCA, PCR, PCA	Lebecque et al. (2001)
Analysis of proteins during ripening	Tryptophan	CCA, CCSWA	Mazerolles et al. (2002)
Analysis of oxidative stability and maillard reaction	Excitation: 270 to 550 nm Emission: 310 to 590 nm	PARAFAC	Christensen et al. (2003)
Monitor meltability and structural changes in correlation with rheological parameters		FDA, CCA, PCA	Karoui et al. (2003)
Analysis of the changes at molecular and macroscopic levels		CCSWA	Kulmyrzaev et al. (2005a)
Differentiation of cheeses as per origin		FDA, PCA	Karoui et al. (2005b)
Investigate the reliability of dynamic testing rheology, MIR and SFS to evaluate the meltability and the viscoelastic behaviour of a semi-hard cheese	Offset: 80 nm	PCA, CCSWA	Boubellouta and Dufour (2012)
Monitor the changes during the heating of cheeses varying in salt contents and type	Offsets: from 20 to 80 nm with (Dk ¼ 120)	ICA	Loudiyi et al. (2017)
Investigate the potential to monitor gentle heating and cooling of cheeses varying in their salt contents and type			Loudiyi et al. (2018)
Application of MIR spectroscopy			
Analysis of changes of FTIR spectra during ripening in the methylene region and of vitamin A fluorescence spectra	3000–2780 cm^{-1}	CCA, PCA	Dufour et al. (2000)
Investigate the changes during ripening of the FTIR spectra in the amide I and II regions		CCA, PCA	Mazerolles et al. (2001)
Monitor the changes in proteins during ripening	3000–900 cm^{-1}	CCSWA, CCA	Mazerolles et al. (2002)
Geographic origin determination	4000–900 cm^{-1}	FDA, PCA	Karoui et al. (2004)

(continued)

Table 10.5 (continued)

Study	Fluorophore/ wavelength range/ frequency (MHz)	Chemometric tools	References
Analyse changes at molecular and macroscopic levels	3000–2800 cm^{-1} 1700–1500 cm^{-1} 1500–900 cm^{-1}	CCSWA	Kulmyrzaev et al. (2005b)
Differentiate cheeses based on origin		PCA, CCSWA, FDA	Karoui et al. (2005a)
Fat, protein, and moisture determination		PLSR	Rodriguez-Saona et al. (2006)
Analysis of fat, NaCl, NPN, WSN, and pH		PLS	Karoui et al. (2006a)
Investigate the ripening differences between the sampling zones and the manufacturing process			Karoui et al. (2006b)
Selected sensory attributes determination			Fagan et al. (2007a)
Predict meltability and instrumental texture attributes	4000–640 cm^{-1}	PLSR, PCA	Fagan et al. (2007b)
Flavour quality analysis of cheese		PCA, SIMCA	Subramanian et al. (2009)
Evaluate viscoelastic behaviour and meltability		PCA, CCSWA	Boubellouta and Dufour (2012)
Application of NIR spectroscopy			
Fat and moisture determination	800–1100 nm	PLSR	Mazerolles et al. (2000)
Process cheese (a quality control tool during production)	Fat: 2.13 and 2.33 mm Protein: 2.18 mm Moisture: 1.94 mm	LR	Adamopoulos et al. (2001)
Predict the total solids content	400–1700 nm (visible/NIR)	PCA, PLSR	da Costa Filho and Volery (2005)
Maturation determination	400–2498 nm	PCA, PLSR	Downey et al. (2005)
Evaluation of self-life	1200–4000 nm	PCA	Cattaneo et al. (2005)
Utilization of multivariate statistical methods to segregate cheeses based on origin		PCA, FDA, CCSWA	Karoui et al. (2005b)
Texture analysis (sensory and instrumental)	750–2498 nm	PLSR	Blazquez et al. (2006)
Selected free amino acids development during cheese ripening	780–2500 nm	PLSR	Skeie et al. (2006)
Chemical parameter determination	1000–2500 nm	PLSR	Karoui et al. (2006a)

(continued)

Table 10.5 (continued)

Study	Fluorophore/ wavelength range/ frequency (MHz)	Chemometric tools	References
Predict fat, vitamins, minerals, carotenoids, total antioxidant capacity, dry matter, pH and colour	400–2500 nm (visible/NIR)	MPLS	Lucas et al. (2008)
Identification of cheese depending on the milk used	1100–2000 nm	MPLS	Revilla et al. (2009)
Volatile compound determination		MPLS	González-Martín et al. (2014)
Predict mineral content	Transmittance mode: 850–1050 nm Reflectance mode: 866–2530 nm	PLSR	Manuelian et al. (2017)
Application of NMR spectroscopy			
Geographical discrimination and ripening of cheeses	500.13	O-PLS, PCA, SIMCA, PLS-DA	Consonni and Cagliani (2008)
Analysis of water distribution, salt uptake and moisture loss in cheese	43.85	NNLS regression	Altan et al. (2011)
Quality evaluation of mozzarella cheese	400	HCA, PCA, DA,	Mazzei and Piccolo (2012)
Classification of cheese based on maturation and cheese cultures	399.95	PLSR, PCA	Piras et al. (2013)

process in every batch, accurately determining the time it takes to transition between the various coagulation phases. This information can help in creating a system for determining cutting time automatically and setting alarms for batches that do not meet specifications. This can effectively reduce the occurrence of failed batches.

10.6.1.2 Predicting Curd and Whey Composition Using Visible-Near Infrared Spectroscopy Sensor

The phenomenon of syneresis is essential in cheese production, significantly influencing the quality of the end product. Inadequate moisture levels in curd can negatively influence the cheese ripening process (Fox et al., 2017) and affect its overall yield (Fagan et al., 2007c). Even a minor reduction in moisture content can greatly affect cheese yield and profitability. Although monitoring and controlling syneresis is essential, the approach remains largely reliant on empirical methods. Creating monitoring and control technology for syneresis is essential for enhancing the automation of the cheese-making process.

Taifi et al. (2005) investigated how ultrasonic velocity and attenuation can be utilized to detect the onset of syneresis in a milk gel. The observation indicated that the alteration in velocity at elevated frequencies appeared to signal syneresis, accompanied by a notable increase in attenuation. Everard et al. (2007) examined the syneresis process in a cheese vat under stirring conditions by monitoring color variations with a computer vision system. The method demonstrated its capability to distinguish between the effects of pH levels and stirring speed. By integrating known elements and adjusting them to different operational conditions, it is possible to predict with precision the moment syneresis takes place. However, a limitation was that the images were only taken from the surface of the curd and whey mixture, where the distribution of curd particles is significantly affected by the stirring speed.

Castillo et al. (2005) and Fagan et al. (2007c) investigated the potential of near-infrared light backscatter at 980 nm for monitoring syneresis. The team found that a sensor designed to detect the reflection of near-infrared light in a cheese vat, with a broad field of vision relative to the size of the curd particles, demonstrated potential for monitoring both milk coagulation and curd syneresis. Fagan et al. (2007c) effectively employed the LFV syneresis sensor technology to precisely predict the fat content in whey, curd yield, and moisture levels in curd. Nonetheless, the early predictions were limited by the small dataset used and the reduced ability to forecast accurately.

Fagan et al. (2008) expanded this research to explore the potential of the LFV sensor in predicting key parameters in cheese production. Equations were developed to predict the moisture levels in curd and the fat levels in whey at the end of syneresis. Furthermore, the moisture level of curd during syneresis was assessed in relation to the duration of processing. The LFV sensor technology is said to precisely gauge the speed and extent of syneresis and whey losses, leading to enhanced consistency and efficiency in cheese production, irrespective of moisture content. Although models have shown success in forecasting cheese-making indices through light backscatter at 980 nm, broadening the spectral range may improve the models further. This is due to the fact that other near-infrared regions and the visible spectrum can offer additional insights into the condition of the intermediate product in the cheese vat. Castillo et al. (2005) employed NIR transmission and side scatter spectra of whey to precisely determine the fat content within the whey. The findings revealed that the light extinction coefficients for whey fat peaked at around 830 nm, whereas they were minimized near 425 nm. The model utilizing a waveband ratio where the denominator falls within 375–475 nm and the numerator is in the range of 725–1025 nm demonstrated the highest R^2 value. Nonetheless, the infrared spectrum frequently contains regions where data overlaps, complicating the interpretation process.

Chemometrics employs mathematical and statistical techniques to effectively address the problem of overlapping peaks, enabling the extraction of valuable information from the spectra. Using plots, such as loading plots, can be a valuable tool for distinguishing between overlapping bands and assigning them correctly. This method allows for the precise identification of spectral regions that show the greatest promise for observing syneresis. This method of employing chemometrics to derive valuable quantitative insights from samples through NIR spectral data has been widely applied in both the pharmaceutical sector (2008) and the food production industry (Woodcock et al., 2008). Partial least squares regression and principal component analysis are two widely used chemometric techniques. PLS regression and PCA can be employed to streamline spectral data into a few linear combinations derived from the original dataset. The reduced group of variables, referred to as PLS loadings (L) or principal components (PC), offers an assessment of the nuanced variations in the spectra among various samples. Consequently, a considerable amount of the variation in the data can be accounted for by utilizing a more limited selection of variables. These variables can be employed to formulate equations that effectively forecast various characteristics.

Employing visible-near infrared spectroscopy alongside chemometrics to predict key cheese-making indicators during syneresis can enhance overall process control in cheese production and optimize the management of curd moisture levels and whey losses. Fagan et al. (2009) conducted a study examining the potential of visible-near infrared spectra, obtained through a light backscatter sensor, combined with chemometrics, to predict the moisture content of curd and the fat levels in whey within a cheese vat. The researchers found that employing the Jack-knifing technique to reduce the impact of irrelevant wavelengths in model development led to improved model precision. The models were crafted with meticulous attention to detail to demonstrate the capabilities of this technology. This method could be utilized to forecast important indicators of syneresis, including curd moisture and whey fat levels, in a stirred cheese vat. This innovation is essential for improving the overall process management in cheese manufacturing, providing numerous benefits for both consumers and producers.

10.6.1.3 Near-Infrared Spectroscopy Prediction of Cheese Composition

Vis-NIR spectroscopy is an advantageous, quick, and cost-effective technique for estimating the moisture, fat, and protein levels in cheese. Portable or handheld spectrometers eliminate the necessity of collecting, storing, transporting, and processing samples. Nevertheless, the estimation of additional chemical constituents such as salt, minerals, and water-soluble nitrogen is less precise and necessitates additional investigation. To ensure accurate prediction equations and avoid biased results, it is crucial to maintain a balanced variability in the cheese samples used for calibrating the system. This prevents the equations from being valid exclusively for a given cheese type or having a bias towards a particular cheese type.

While there may be notable variations across different spectrometers, these variances do not appear to be strongly influenced by the instrument's spectrum extension. Various indicators about moisture, fat, and protein can be found in different parts of the cheese spectrum. Therefore, forecasts made using specific portions of the visible, IR-A (short-wave infrared, 740–1400 nm), and IR-B (long wavelength infrared, 1100–2500 nm) regions can be nearly as precise as those made using a much broader spectrum. The use of Vis-NIR spectroscopy prediction of cheese composition is discussed in Table 10.6, where Bittante et al. (2022) explained the

Table 10.6 Application of Vis-NIR spectroscopy prediction of cheese composition

Component	Wavelength (nm)	Band assignment	References
Fat	1190–1220	Second overtone V(CH)	Blazquez et al. (2004)
Water	1201	Combinations first overtone Vs(OH), δ(OH)	Karoui et al. (2006b)
Fat	926–928	Third overtone V(CH)	Lucas et al. (2008)
Water	970–990	Second overtone V(OH)	Lucas et al. (2008)
Fat	1150	Second overtone Vs(CH_3)	Pi et al. (2009)
Fat	1650	Combination V(CH)	Pi et al. (2009)
Protein	1890	First overtone CH and CONH group	Pi et al. (2009)
Fat, protein	1898	Second overtone V(C=O)	Pi et al. (2009)
Fat, protein	1135–1200	Second overtone Vs(CH_2)	Lénárt et al. (2012)
Protein	1480–1550	First overtone Vs(NH)	Lénárt et al. (2012)
Fat, protein	1159	Second overtone Vas(CH2)	Lénárt et al. (2012)
Protein	2220–2222	Second overtone V(NCHO)	Lénárt et al. (2012)
Fat	2300–2380	Combinations V(CH) and δ(CH)	Ottavian et al. (2012)
Fat	1233–1333	Second overtone Vs(OH)	Lénárt et al. (2012)
Water	1900–1954	Combinations V(OH) and δ(OH)	Ottavian et al. (2012)
Protein	2030–2080	V(C=O), second overtone (NH_2)	Lénárt et al. (2012)
Fat	2070	Combination V(OH/HOH), second overtone δ(OH)	Lénárt et al. (2012)
Protein	1584	First overtone V(NH)	Lénárt et al. (2012)
Protein	1620–1740	First overtone Var(CH)	Lénárt et al. (2012)
Fat, protein	1729–1739	First overtone Vas(CH_2)	Lénárt et al. (2012)
Water	1450–1486	First overtone V(OH)	Botelho et al. (2013)
Fat	1687–1786	First overtone V(CH)	Botelho et al. (2013)
Protein	2148–2180	Second overtone, combinations δ(NH), V(CH/C=O)	Madalozzo et al. (2015)
Fat	1400–1409	Combinations V(CH_2) and δ(CH_2)	Madalozzo et al. (2015)

band assignment of Vis-NIR spectroscopy instrument as visible region (350–780 nm), third overtone region (900–1100 nm), second overtone region (1100–1400 nm), first overtone region (1400–1850 nm) and combination region (1850–2500 nm). The mathematical preprocessing techniques, chemometric models, and validation methodologies are interconnected and should be collectively assessed. Chemometric models that utilize Bayesian or machine learning techniques are more effective than partial least squares (PLS) regression in predicting features that are challenging to forecast. To prevent too optimistic outcomes and assess the accuracy of forecasts in an industry setting, it is crucial to have a truly independent validation data collection.

10.6.1.4 At-line Prediction of Mineral Composition Using Near-Infrared Technologies

Milk and dairy products serve as significant sources of vital macro- and micronutrients that are crucial for human health. The challenge in reliably forecasting mineral concentration is heightened, however, because fresh cheeses typically have a lower mineral composition than other types of cheese. In order to tackle this issue, cheese manufacturers may need to create on-site instruments to assess the mineral composition during the cheese production process. Implementing this will ensure adherence to labelling regulations and facilitate the inclusion of more comprehensive health assertions on their products.

Studies have utilized infrared spectroscopy to create a model for assessing the mineral composition of fresh cheeses in real-time. This technology provides fast, unbiased, and non-invasive analysis at a far reduced expense compared to conventional laboratory methods, which are also time-consuming. The developed spectroscopy models using near-infrared reflectance (NIR, 866–2530 nm) and transmittance (NIT, 850–1050 nm) are used to accurately estimate the primary mineral content of fresh cheeses. This was achieved through the use of partial least squares regression analysis. Subsequently, these models underwent cross-validation and external validation.

To verify this methodology, reference methodologies and NIR and NIT spectroscopy were implemented to analyze mozzarella and Stracchino cheese samples. The results showed that the most precise prediction model was for the sodium content of mozzarella cheese using NIT spectroscopy, with a coefficient of determination for external validation of 0.75. In the same vein, the accuracy of the phosphorus content prediction in Stracchino cheese was comparable when utilizing both NIR and NIT spectroscopy. The study identified a substantial limitation: the low mineral concentration in cheese restricts the efficacy of near-infrared technologies in accurately predicting mineral content, despite these findings. As a result, the investigation determined that none of the prediction models were sufficiently precise to supplant the existing reference analysis methods (Manuelian et al., 2017).

10.6.1.5 Inorganic Salt and Moisture Content Prediction Using Dielectric Spectroscopy

It is essential to evaluate the efficacy of various online instruments, as they influence the precision of measurements. The predictive capabilities of two distinct near-infrared spectroscopy (NIRS) instruments for the identification of main and minor nutritional characteristics in cheese were studied by Franzoi et al. (2021). To compare the data from the At-line instruments with the data from the reference methods, prediction models and a cluster approach were implemented. The findings indicated that the prediction performance was satisfactory across a variety of cheese categories. It was proposed that prediction models that encompass a broad spectrum of cheese varieties may prove advantageous for general process control, while a cluster approach may be advantageous for optimizing particular product attributes.

10.6.2 At-line Measurement of Lactose

Vegan cheese is created entirely from plant-based materials and is free of lactose. Without the use of any animal ingredients, vegan cheese aims to imitate the texture, melting point, and flavor of dairy cheese. Lactose should not be present in such cheese as it is only present in animal milk and thus considered an allergen when present in vegan products. For the purpose of At-line lactose biosensor allergy detection in cheese processing plants or loss monitoring, the dairy industry has developed a system that combines a sequential injection analyzer with a third-generation enzyme biosensor. In order to improve plant efficiency, maximize saleable goods, and decrease effluent discharge from dairy-processing plants, Glithero et al. (2013) created an automated At-line lactose biosensor for wastewater stream monitoring.

In the past, researchers in the dairy industry have tried to introduce lactose sensors by utilizing multi-enzyme cascades that employed β-galactosidase and galactose oxidase, occasionally in conjunction with glucose oxidase or peroxidase (Eshkenazi et al., 2000). These methods were able to estimate lactose within a range of 10–60 g/L. Experimental monitoring of lactose concentrations during real fermentations was made easier by combining a flow injection analyzer (FIA) with a lactose biosensor, as demonstrated by Ferreira et al. (2003). While Ferreira et al. (2004) found that their linear range (1–30 g/L) and response times (3 min) were better than stated values in the literature, they did mention that temperature affects needed to be compensated for, air bubbles needed to be avoided, and the oxygen electrode's membrane needed to be carefully observed.

Using third-generation amperometric biosensors, which eliminate the need for a co-substrate (first generation) or a mediator (second generation) (Stoica et al., 2006; Safina et al., 2010; Ludwig et al., 2010), allows for direct electron transfer (DET) from the active site of the enzyme to the electrode surface, making the process of lactose analysis faster and easier. Spectrographic graphite or screen-printed carbon electrodes (SPCE) were used to immobilize cellobiose dehydrogenase (CDH) from white rot fungus, *Trametes villosa* or *Phanerochaete sordida*, resulting in these biosensors of the third generation. The sensitivities and reaction times of CDH-modified electrodes are significantly greater and faster than those of enzyme cascade biosensors (40 s; 45–340 μM; Eshkenazi et al., 2000). The study was conducted by Safina et al. in 2010.

Cellobiose dehydrogenase is a monomeric enzyme that features external glycosylation and consists of two domains connected by a flexible polypeptide linker region (Zamocky et al., 2006). A recently discovered redox partner protein plays a role in the degradation of lignocellulose, as detailed by Ludwig et al. (2010) and Li et al. (2012). In the dehydrogenase domain containing flavin adenine dinucleotide (FAD), the sugar substrate is oxidized, leading to the formation of fully reduced flavin (FADH2) and a lactone. The reoxidation of the decreased DHCDH can occur via various pathways (Ludwig et al., 2010). In the absence of a soluble electron acceptor, electrons from the reduced DHCDH are sequentially transferred to the cytochrome domain (CYTCDH), which is equipped with heme b. This transfer occurs through an internal pathway within the DHCDH, moving from FADH2 to the heme b located on the surface of the CYTCDH. Following this, the electrons are transferred directly to the electrode surface, which acts as the final destination for the electrons. CDH is produced by both basidiomycete fungi (white rot, class I CDH) and ascomycete fungi (brown rot and plant diseases, class II and III CDH) (Harreither et al., 2011). Class I CDHs of the white rot type are noted for their acidic nature, thriving best within a pH range of 4 to 5. They exhibit significant catalytic efficiency in the degradation of cellodextrins and lactose, which are linked by 1,4-β-linkages. The turnover rates of these enzymes are impressive, and they exhibit a strong selectivity for monosaccharides and oligosaccharides linked by 1,4-α-linkages (Zamocky et al., 2006). Conversely, class II and III CDHs demonstrate broader ranges of substrates they can interact with, such as glucose, and maintain pH levels that are nearer to neutral (Harreither et al., 2012). Given that milk and other dairy products lack significant amounts of alternative sugar substrates and exhibit a high turnover rate with class I CDHs, utilizing a biosensor that incorporates class I CDHs is the optimal solution for lactose detection in the dairy industry.

The current constraints in lactose monitoring within dairy processing plants are twofold: firstly, the absence of an automated At-line analyzer capable of accurately measuring lactose due to the restricted capabilities of sensors; and secondly, the requirement for automatic dilution to adapt samples to an appropriate linear range and acceptable pH. The characteristics of CDH-modified electrodes for lactose detection have the potential to address these concerns (Glithero et al., 2013).

10.6.3 At-line/In-line Measurement of the Stretchability

Mozzarella, Caciocavallo, and Scamorza are varieties of cheese that undergo a distinctive process of heating and stretching to attain a unique texture. This technique gives the cheese a fibrous and flexible texture, which is perfect for melting and generating a stretchy outcome when heated. The optimal stretching properties of these cheeses are determined by various parameters, namely the pH level during coagulation (ranging from 5.2 to 5.5), the temperature at which stretching occurs (between 80 and 85 °C), and the correct technique of kneading. PAT approaches can effectively control the temperature and pH, while the kneading of the cheese curd can be accurately assessed utilizing At-line and In-line testing.

In a study conducted by Young (2012), the texture of bread dough was examined in real-time throughout the continuous production process. This same approach can also be used to assess the stretchability of cheese. In both cases, the machinery measures the force applied to the mixer blades, which is proportional to the curd's changing resistance throughout kneading. Graphs displaying the torque data are continuously generated throughout the necessary mixing duration. Keep in mind that the data and curves you get will be specific to the tools you used and won't be exactly the same when you use other tools and techniques. Nevertheless, as torque is typically measured over a period of time, there might be certain resemblances in the forms of the curves.

Some examples of laboratory-scale equipment are the Brabender® Farinograph®, Newport Scientific (now Perten) DoughLab, and Chopin Mixolab. Contemporary mixers come with this feature already integrated, allowing users to observe the peaks and energy levels over a specific time frame on the display. These measurements can be correlated with the quality of the final product. Stretch cheese possesses a significant attribute known as extensibility. The Brabender® Extensograph® and Chopin Alveograph are two instruments that can be used to measure extensibility and elasticity.

One technique entails vertically elongating a slice of cheese between two fixtures and documenting the load/extension curve while it is being stretched. Another technique for evaluating extensibility entails applying pressurized air uniformly from underneath to create a bubble, as demonstrated by the Alveograph. This device measures the pressure within the bubble and represents it over time in a curve known as an alveogram. Inflation persists until the bubble eventually collapses. During inflation, a pressure sensor measures the pressure, and the displacement of the piston is utilized to calculate the air volume. The pressure, volume, and time are directly measured to find the cheese's rheological properties (Young, 2009).

10.7 Conclusion

The advancements in the manufacturing and quality control processes of cheese production, such as At-line and In-line testing methodologies, have proven to be crucial. These methodologies offer significant benefits in terms of efficiency, safety, and product quality. By implementing these testing methods, manufacturers can effectively monitor and control their processes in real-time. This enables them to quickly identify and address any deviations, resulting in reduced waste, downtime, and the potential for product recalls. The combination of At-line and In-line testing, along with the use of advanced spectroscopic and sensor technologies, marks a major advancement in the field of cheese manufacturing. These methods result in increased consumer satisfaction, adherence to regulatory standards, and enhanced efficiency in the production process. In the ever-changing dairy industry, ongoing research and development in these areas are expected to result in even more advanced tools and techniques, ultimately improving the quality and safety of cheese products on a global scale.

References

Adamopoulos, K. G., Goula, A. M., & Petropakis, H. J. (2001). Quality control during processing of feta cheese—NIR application. *Journal of Food Composition and Analysis, 14*(4), 431–440.

Altan, A., Oztop, M. H., McCarthy, K. L., & McCarthy, M. J. (2011). Monitoring changes in feta cheese during brining by magnetic resonance imaging and NMR relaxometry. *Journal of Food Engineering, 107*(2), 200–207.

Azamfirei, V., Psarommatis, F., & Lagrosen, Y. (2023). Application of automation for in-line quality inspection, a zero-defect manufacturing approach. *Journal of Manufacturing Systems, 67*, 1–22.

Bittante, G., Patel, N., Cecchinato, A., & Berzaghi, P. (2022). Invited review: A comprehensive review of visible and near-infrared spectroscopy for predicting the chemical composition of cheese. *Journal of Dairy Science, 105*(3), 1817–1836.

Blazquez, C., Downey, G., O'Donnell, C., O'Callaghan, D., & Howard, V. (2004). Prediction of moisture, fat and inorganic salts in processed cheese by near infrared reflectance spectroscopy and multivariate data analysis. *Journal of Near Infrared Spectroscopy, 12*(3), 149–157.

Blazquez, C., Downey, G., O'Callaghan, D., Howard, V., Delahunty, C., Sheehan, E., et al. (2006). Modelling of sensory and instrumental texture parameters in processed cheese by near infrared reflectance spectroscopy. *Journal of Dairy Research, 73*(1), 58–69.

Botelho, B. G., Mendes, B. A., & Sena, M. M. (2013). Development and analytical validation of robust near-infrared multivariate calibration models for the quality inspection control of mozzarella cheese. *Food Analytical Methods, 6*, 881–891.

Boubellouta, T., & Dufour, É. (2012). Cheese-matrix characteristics during heating and cheese melting temperature prediction by synchronous fluorescence and mid-infrared spectroscopies. *Food and Bioprocess Technology, 5*, 273–284.

Castillo, M., Payne, F. A., Lopez, M. B., Ferrandini, E., & Laencina, J. (2005). Optical sensor technology for measuring whey fat concentration in cheese making. *Journal of Food Engineering, 71*(4), 354–360.

Cattaneo, T. M. P., Giardina, C., Sinelli, N., Riva, M., & Giangiacomo, R. (2005). Application of FT-NIR and FT-IR spectroscopy to study the shelf-life of Crescenza cheese. *International Dairy Journal, 15*(6–9), 693–700.

Christensen, J., Povlsen, V. T., & Sørensen, J. (2003). Application of fluorescence spectroscopy and chemometrics in the evaluation of processed cheese during storage. *Journal of Dairy Science, 86*(4), 1101–1107.

Consonni, R., & Cagliani, L. R. (2008). Ripening and geographical characterization of Parmigiano Reggiano cheese by 1H NMR spectroscopy. *Talanta, 76*(1), 200–205.

da Costa Filho, P. A., & Volery, P. (2005). Broad-based versus specific NIRS calibration: Determination of total solids in fresh cheese. *Analytica Chimica Acta, 544*(1–2), 82–88.

Dalby, A. (2009). *Cheese: A global history*. Reaktion Books.

Downey, G., Sheehan, E., Delahunty, C., O'Callaghan, D., Guinee, T., & Howard, V. (2005). Prediction of maturity and sensory attributes of Cheddar cheese using near-infrared spectroscopy. *International Dairy Journal, 15*(6–9), 701–709.

Dufour, E., Mazerolles, G., Devaux, M. F., Duboz, G., Duployer, M. H., & Riou, N. M. (2000). Phase transition of triglycerides during semi-hard cheese ripening. *International Dairy Journal, 10*(1–2), 81–93.

Eshkenazi, I., Maltz, E., Zion, B., & Rishpon, J. (2000). A three-cascaded-enzymes biosensor to determine lactose concentration in raw milk. *Journal of Dairy Science, 83*(9), 1939–1945.

Everard, C. D., O'callaghan, D. J., Fagan, C. C., O'Donnell, C. P., Castillo, M., & Payne, F. A. (2007). Computer vision and color measurement techniques for inline monitoring of cheese curd syneresis. *Journal of Dairy Science, 90*(7), 3162–3170.

Fagan, C. C., Everard, C., O'Donnell, C. P., Downey, G., Sheehan, E. M., Delahunty, C. M., & O'Callaghan, D. J. (2007a). Evaluating mid-infrared spectroscopy as a new technique for predicting sensory texture attributes of processed cheese. *Journal of Dairy Science, 90*(3), 1122–1132.

Fagan, C. C., Everard, C., O'Donnell, C. P., Downey, G., Sheehan, E. M., Delahunty, C. M., et al. (2007b). Prediction of processed cheese instrumental texture and meltability by mid-infrared spectroscopy coupled with chemometric tools. *Journal of Food Engineering, 80*(4), 1068–1077.

Fagan, C. C., Leedy, M., Castillo, M., Payne, F. A., O'Donnell, C. P., & O'Callaghan, D. J. (2007c). Development of a light scatter sensor technology for on-line monitoring of milk coagulation and whey separation. *Journal of Food Engineering, 83*(1), 61–67.

Fagan, C. C., Castillo, M., O'Donnell, C. P., O'Callaghan, D. J., & Payne, F. A. (2008). On-line prediction of cheese making indices using backscatter of near infrared light. *International Dairy Journal, 18*(2), 120–128.

Fagan, C. C., Castillo, M., O'Callaghan, D. J., Payne, F. A., & O'Donnell, C. P. (2009). Visible-near infrared spectroscopy sensor for predicting curd and whey composition during cheese processing. *Sensing and Instrumentation for Food Quality and Safety, 3*, 62–69.

Fagan, C. C., Ferreira, T. G., Payne, F. A., O'Donnell, C. P., O'Callaghan, D. J., & Castillo, M. (2011). Preliminary evaluation of endogenous milk fluorophores as tracer molecules for curd syneresis. *Journal of Dairy Science, 94*(11), 5350–5358.

FDA. (2004). *Guidance for industry, PAT-A framework for innovative pharmaceutical development, manufacturing and quality assurance*. http://www.fda.gov/cder/guidance/published.html

Ferreira, L. S., Souza, M. B., Jr., Trierweiler, J. O., Hitzmann, B., & Folly, R. O. M. (2003). Analysis of experimental biosensor/FIA lactose measurements. *Brazilian Journal of Chemical Engineering, 20*, 07–13.

Ferreira, L. D. S., Trierweiler, J. O., De Souza, M. B., Jr., & Folly, R. O. M. (2004). A lactose FIA-biosensor system for monitoring and process control. *Brazilian Journal of Chemical Engineering, 21*, 307–315.

Fox, P. F., Guinee, T. P., Cogan, T. M., & McSweeney, P. L. (2017). *Fundamentals of cheese science*. Springer.

Franzoi, M., Ghetti, M., De Lorenzi, C., & De Marchi, M. (2021). Effectiveness of two different At-line instruments for the assessment of cheese composition, major minerals and fatty acids content. *International Dairy Journal, 123*, 105184.

Glithero, N., Clark, C., Gorton, L., Schuhmann, W., & Pasco, N. (2013). At-line measurement of lactose in dairy-processing plants. *Analytical and Bioanalytical Chemistry, 405*, 3791–3799.

González-Martín, I., Hernández-Hierro, J. M., González-Pérez, C., Revilla, I., Vivar-Quintana, A., & Ortega, I. L. (2014). Potential of near infrared spectroscopy for the analysis of volatile components in cheeses. *LWT - Food Science and Technology, 55*(2), 666–673.

Harreither, W., Sygmund, C., Augustin, M., Narciso, M., Rabinovich, M. L., Gorton, L., et al. (2011). Catalytic properties and classification of cellobiose dehydrogenases from ascomycetes. *Applied and Environmental Microbiology, 77*(5), 1804–1815.

Harreither, W., Nicholls, P., Sygmund, C., Gorton, L., & Ludwig, R. (2012). Investigation of the pH-dependent electron transfer mechanism of ascomycetous class II cellobiose dehydrogenases on electrodes. *Langmuir, 28*(16), 6714–6723.

Herbert, S., Riou, N. M., Devaux, M. F., Riaublanc, A., Bouchet, B., Gallant, D. J., & Dufour, É. (2000). Monitoring the identity and the structure of soft cheeses by fluorescence spectroscopy. *Le Lait, 80*(6), 621–634.

Hitzmann, B., Hauselmann, R., Niemoeller, A., Sangi, D., Traenkle, J., & Glassey, J. (2015). Process analytical technologies in food industry-challenges and benefits: A status report and recommendations. *Biotechnology Journal, 10*, 1095–1100.

Karoui, R., & De Baerdemaeker, J. (2007). A review of the analytical methods coupled with chemometric tools for the determination of the quality and identity of dairy products. *Food Chemistry, 102*(3), 621–640.

Karoui, R., Laguet, A., & Dufour, É. (2003). Fluorescence spectroscopy: A tool for the investigation of cheese melting-correlation with rheological characteristics. *Le Lait, 83*(3), 251–264.

Karoui, R., Dufour, É., Pillonel, L., Picque, D., Cattenoz, T., & Bosset, J. O. (2004). Determining the geographic origin of Emmental cheeses produced during winter and summer using a technique based on the concatenation of MIR and fluorescence spectroscopic data. *European Food Research and Technology, 219*, 184–189.

Karoui, R., Dufour, É., Pillonel, L., Schaller, E., Picque, D., Cattenoz, T., & Bosset, J. O. (2005a). The potential of combined infrared and fluorescence spectroscopies as a method of determination of the geographic origin of Emmental cheeses. *International Dairy Journal, 15*(3), 287–298.

Karoui, R., Martin, B., & Dufour, É. (2005b). Potentiality of front-face fluorescence spectroscopy to determine the geographic origin of milks from the Haute-Loire department (France). *Le Lait, 85*(3), 223–236.

Karoui, R., Mouazen, A. M., Dufour, É., Pillonel, L., Picque, D., Bosset, J. O., & De Baerdemaeker, J. (2006a). Mid-infrared spectrometry: A tool for the determination of chemical parameters in Emmental cheeses produced during winter. *Le Lait, 86*(1), 83–97.

Karoui, R., Mouazen, A. M., Ramon, H., Schoonheydt, R., & De Baerdemaeker, J. (2006b). Feasibility study of discriminating the manufacturing process and sampling zone in ripened soft cheeses using attenuated total reflectance MIR and fiber optic diffuse reflectance VIS–NIR spectroscopy. *Food Research International, 39*(5), 588–597.

Kulmyrzaev, A., Dufour, E., Noël, Y., Hanafi, M., Karoui, R., Qannari, E. M., & Mazerolles, G. (2005a). Investigation at the molecular level of soft cheese quality and ripening by infrared and fluorescence spectroscopies and chemometrics—Relationships with rheology properties. *International Dairy Journal, 15*(6–9), 669–678.

Kulmyrzaev, A. A., Levieux, D., & Dufour, É. (2005b). Front-face fluorescence spectroscopy allows the characterization of mild heat treatments applied to milk. Relations with the denaturation of milk proteins. *Journal of Agricultural and Food Chemistry, 53*(3), 502–507.

Lagrange, V., Whitsett, D., & Burris, C. (2015). Global market for dairy proteins. *Journal of Food Science, 80*(S1), A16–A22.

Lebecque, A., Laguet, A., Devaux, M. F., & Dufour, É. (2001). Delineation of the texture of Salers cheese by sensory analysis and physical methods. *Le Lait, 81*(5), 609–624.

Lénárt, J., Szigedi, T., Dernovics, M., & Fodor, M. (2012). Application of FT-NIR spectroscopy on the determination of the fat and protein contents of lyophilized cheeses. *Acta Alimentaria, 41*(3), 351–362.

Li, X., Beeson, W. T., Phillips, C. M., Marletta, M. A., & Cate, J. H. (2012). Structural basis for substrate targeting and catalysis by fungal polysaccharide monooxygenases. *Structure, 20*(6), 1051–1061.

Loudiyi, M., Karoui, R., Rutledge, D. N., Lavigne, R., Montel, M. C., & Aït-Kaddour, A. (2017). Contribution of fluorescence spectroscopy and independent components analysis to the evaluation of NaCl and KCl effects on molecular-structure and fat melting temperatures of Cantal-type cheese. *International Dairy Journal, 73*, 116–127.

Loudiyi, M., Karoui, R., Rutledge, D. N., Montel, M. C., Rifa, E., & Aït-Kaddour, A. (2018). Fluorescence spectroscopy coupled with independent components analysis to monitor molecular changes during heating and cooling of Cantal-type cheeses with different NaCl and KCl contents. *Journal of the Science of Food and Agriculture, 98*(3), 963–975.

Loudiyi, M., Temiz, H. T., Sahar, A., Haseeb Ahmad, M., Boukria, O., Hassoun, A., & Aït-Kaddour, A. (2022). Spectroscopic techniques for monitoring changes in the quality of milk and other dairy products during processing and storage. *Critical Reviews in Food Science and Nutrition, 62*(11), 3063–3087.

Lucas, A., Andueza, D., Rock, E., & Martin, B. (2008). Prediction of dry matter, fat, pH, vitamins, minerals, carotenoids, total antioxidant capacity, and color in fresh and freeze-dried cheeses by visible-near-infrared reflectance spectroscopy. *Journal of Agricultural and Food Chemistry, 56*(16), 6801–6808.

Ludwig, R., Harreither, W., Tasca, F., & Gorton, L. (2010). Cellobiose dehydrogenase: A versatile catalyst for electrochemical applications. *ChemPhysChem, 11*(13), 2674–2697.

Lyndgaard, C. B., Engelsen, S. B., & van den Berg, F. W. (2012). Real-time modeling of milk coagulation using in-line near infrared spectroscopy. *Journal of Food Engineering, 108*(2), 345–352.

Madalozzo, E. S., Sauer, E., & Nagata, N. (2015). Determination of fat, protein and moisture in ricotta cheese by near infrared spectroscopy and multivariate calibration. *Journal of Food Science and Technology, 52*, 1649–1655.

Mali, A. S., Jagtap, M., Karekar, P., & Maruška, A. (2016). A brief review on process analytical technology (PAT). *International Journal of Current Pharmaceutical Research, 8*(1), 10–15.

Manuelian, C. L., Currò, S., Visentin, G., Penasa, M., Cassandro, M., Dellea, C., Bernardi, M., & De Marchi, M. (2017). Technical note: At-line prediction of mineral composition of fresh cheeses using near-infrared technologies. *Journal of Dairy Science, 100*(8), 6084–6089.

Mikkelsen, P. (2020). World cheese production. In: New World Cheese Market Report 2000–2020.

Mazerolles, G., Duboz, G., & Hugot, S. (2000). Détermination des taux d'humidité et de matière grasse de fromages de type pâte pressée par spectroscopie proche infrarouge en mode transmission. *Le Lait, 80*(3), 371–379.

Mazerolles, G., Devaux, M. F., Duboz, G., Duployer, M. H., Riou, N. M., & Dufour, É. (2001). Infrared and fluorescence spectroscopy for monitoring protein structure and interaction changes during cheese ripening. *Le Lait, 81*(4), 509–527.

Mazerolles, G., Devaux, M. F., Dufour, E., Qannari, E. M., & Courcoux, P. (2002). Chemometric methods for the coupling of spectroscopic techniques and for the extraction of the relevant information contained in the spectral data tables. *Chemometrics and Intelligent Laboratory Systems, 63*(1), 57–68.

Mazzei, P., & Piccolo, A. (2012). 1H HRMAS-NMR metabolomic to assess quality and traceability of mozzarella cheese from Campania buffalo milk. *Food Chemistry, 132*(3), 1620–1627.

Nicolau, N., Buffa, M., O'Callaghan, D. J., Guamis, B., & Castillo, M. (2015). Estimation of clotting and cutting times in sheep cheese manufacture using NIR light backscatter. *Dairy Science & Technology, 95*, 495–507.

Ottavian, M., Facco, P., Barolo, M., Berzaghi, P., Segato, S., Novelli, E., & Balzan, S. (2012). Near-infrared spectroscopy to assist authentication and labeling of Asiago d'allevo cheese. *Journal of Food Engineering, 113*(2), 289–298.

Panikuttira, B., O'Shea, N., Tobin, J. T., Tiwari, B. K., & O'Donnell, C. P. (2018). Process analytical technology for cheese manufacture. *International Journal of Food Science & Technology, 53*(8), 1803–1815.

Pi, F., Shinzawa, H., Ozaki, Y., & Han, D. (2009). Non-destructive determination of components in processed cheese slice wrapped with a polyethylene film using near-infrared spectroscopy and chemometrics. *International Dairy Journal, 19*(10), 624–629.

Piras, C., Marincola, F. C., Savorani, F., Engelsen, S. B., Cosentino, S., Viale, S., & Pisano, M. B. (2013). A NMR metabolomics study of the ripening process of the Fiore Sardo cheese produced with autochthonous adjunct cultures. *Food Chemistry, 141*(3), 2137–2147.

Rathore, A. S., Bhambure, R., & Ghare, V. (2010). Process analytical technology (PAT) for biopharmaceutical products. *Analytical and Bioanalytical Chemistry, 398*, 137–154.

Revilla, I., González-Martín, I., Hernández-Hierro, J. M., Vivar-Quintana, A., González-Pérez, C., & Lurueña-Martínez, M. A. (2009). Texture evaluation in cheeses by NIRS technology employing a fibre-optic probe. *Journal of Food Engineering, 92*(1), 24–28.

Rodriguez-Saona, L. E., Koca, N. U. R. C. A. N., Harper, W. J., & Alvarez, V. B. (2006). Rapid determination of Swiss cheese composition by Fourier transform infrared/attenuated total reflectance spectroscopy. *Journal of Dairy Science, 89*(5), 1407–1412.

Roupas, P. (2008). Predictive modelling of dairy manufacturing processes. *International Dairy Journal, 18*(7), 741–753.

Safina, G., Ludwig, R., & Gorton, L. (2010). A simple and sensitive method for lactose detection based on direct electron transfer between immobilised cellobiose dehydrogenase and screen-printed carbon electrodes. *Electrochimica Acta, 55*(26), 7690–7695.

Shenana, M. E., & Patel, A. R. (2019). Starter cultures: Classification, traditional production technology and potential role in the cheese manufacturing industry. In *Microbiology for food and health* (pp. 51–92). Apple Academic Press.

Simon, L. L., Pataki, H., Marosi, G., Meemken, F., Hungerbühler, K., Baiker, A., et al. (2015). Assessment of recent process analytical technology (PAT) trends: A multiauthor review. *Organic Process Research & Development, 19*(1), 3–62.

Skeie, S., Feten, G., Almøy, T., Østlie, H., & Isaksson, T. (2006). The use of near infrared spectroscopy to predict selected free amino acids during cheese ripening. *International Dairy Journal, 16*(3), 236–242.

Stoica, L., Ludwig, R., Haltrich, D., & Gorton, L. (2006). Third-generation biosensor for lactose based on newly discovered cellobiose dehydrogenase. *Analytical Chemistry, 78*(2), 393–398.

Subramanian, A., Harper, W. J., & Rodriguez-Saona, L. E. (2009). Cheddar cheese classification based on flavor quality using a novel extraction method and Fourier transform infrared spectroscopy. *Journal of Dairy Science, 92*(1), 87–94.

Taifi, N., Bakkali, F., Faiz, B., Moudden, A., Maze, G., & Décultot, D. (2005). Characterization of the syneresis and the firmness of the milk gel using an ultrasonic technique. *Measurement Science and Technology, 17*(2), 281–287.

Thygesen, J. H., & van den Berg, F. W. (2012). Subspace methods for dynamic model estimation in PAT applications. *Journal of Chemometrics, 26*(8–9), 435–441.

USDA. (2024). *Cheese | USDA Foreign Agricultural Service*. Retrieved from https://fas.usda.gov/data/production/commodity/0240000. Accessed on 21 June 2024.

Van den Berg, F., Lyndgaard, C. B., Sørensen, K. M., & Engelsen, S. B. (2013). Process analytical technology in the food industry. *Trends in Food Science and Technology, 31*(1), 27–35.

Vashishth, R., Pandey, A. K., Kaur, P., & Semwal, A. D. (2022). Smart technologies in food manufacturing. In *Smart and sustainable food technologies* (pp. 125–155). Singapore.

Wood, C., Alwati, A., & Halsey, S. (2016). Near infrared spectroscopy as a multivariate process analytical tool for predicting pharmaceutical co-crystal concentration. *Journal of Pharmaceutical and Biomedical Analysis, 129*, 172–181.

Woodcock, T., Downey, G., & O'Donnell, C. P. (2008). Better quality food and beverages: The role of near infrared spectroscopy. *Journal of Near Infrared Spectroscopy, 16*(1), 1–29.

Young, L. S. (2009). The ICC handbook of cereals, flour, dough and products: Methods, tests and applications. *Quality Assurance and Safety of Crops & Foods, 1*(1), 74–74.

Young, L. S. (2012). Applications of texture analysis to dough and bread. In *Breadmaking* (pp. 562–579). Woodhead Publishing.

Zamocky, M., Ludwig, R., Peterbauer, C., Hallberg, B. M., Divne, C., Nicholls, P., & Haltrich, D. (2006). Cellobiose dehydrogenase-a flavocytochrome from wood-degrading, phytopathogenic and saprotropic fungi. *Current Protein and Peptide Science, 7*(3), 255–280.

Chapter 11
Advances in Packaging and Storage

Syed Insha Rafiq, Bazilla Gayas, Deepak Joshi, Ayush Pandey, and Syed Mansha Rafiq

11.1 Introduction

Cheese is a well-known fermented milk-based products produced worldwide. There are around 4000 varieties of cheeses with different textures and flavors (Akarca et al., 2015; Li et al., 2018). Cheese packaging is one of the critical facets of cheese production and reason being the long-distance between the manufacturer and the consumer. The environment inside the cheese package plays an important role in maintaining its quality. The packaging used for cheese should have good moisture and oxygen barrier properties. Using optimal packaging will help to decrease the oxygen transfer and moisture loss thereby protecting against microorganisms, and improve the overall sensory quality of cheese. Cheese stored without packaging mainly get spoiled by bacteria, moulds and yeasts resulting in development of off flavours thereby decreasing the quality of cheese. Thus, packaging plays an important role in avoiding the spoilage by microorganisms. The most common packaging used for cheese include polyethylene, polyamide, polypropylene and polystyrene. These packaging materials can cause migration of the materials into the cheese and are non-biodegradable hence there has been the development of new packaging materials based on biopolymers and natural lipids. These packaging material-based films and coatings act as semi permeable barriers to oxygen, carbon dioxide and water vapour thereby reducing water losses and guarantee food quality (Garcia &

S. I. Rafiq (✉)
Department of Food Engineering, Bhaigurdas College of Engineering and Technology, Sangrur, Punjab, India

B. Gayas
Department of Food Technology, Chandigarh university, Chandigarh, Punjab, India

D. Joshi · A. Pandey · S. Mansha Rafiq
Department of Food Science and Technology, National Institute of Food Technology Entrepreneurship and Management, Sonipat, Haryana, India

S. Mansha Rafiq (ed.), *Technological Advances and Trends in Cheese Making*,
https://doi.org/10.1007/978-3-031-94476-5_11

Barrett, 2002; Fajardo et al., 2010). Moreover, these coatings biodegradable and can be consumed with the cheese avoiding waste generation and act as carriers of antimicrobial agents thereby enhancing shelf life (Cerqueira et al., 2010; Rodriguez-Turienzo et al., 2012; Robertson, 2013; Wang et al., 2007). Various techniques like vacuum packaging and modified atmosphere packaging (MAP) have been adopted to improve the shelf life of different cheese varieties. Nowadays antimicrobial, smart and intelligent packaging systems are used for cheese packaging.

11.2 Biodegradable Packaging of Cheese

Biodegradable packaging based on polymeric packaging materials such as cellulose, starch, chitosan, alginate, pullulan, polylactic acid etc. are most commonly used as packaging materials. To have extended shelf life and increase food safety, food packaging industries have developed and adopted new antimicrobial materials. The existing biopolymers other than chitosan, do not present antimicrobial activity. To obtain the desired antimicrobial properties various antimicrobial agents are added to the polymeric matrix to develop antimicrobial biodegradable materials. These antimicrobial biodegradable materials reduce microbial proliferation thereby controlling food spoilage and enhance the shelf life. These also help provide a better food quality (Otto et al., 2021).

11.2.1 Edible Films and Coatings

As evidenced by recent advancements in edible coatings and films for cheese, the use of edible materials in food packaging has expanded quickly, and the cheese industry is undoubtedly one of the industries where these materials have a strong chance of being applied. In addition to being edible, edible coatings and films can limit weight loss and stop microbiological spoiling by controlling the pace at which oxygen and carbon dioxide are exchanged and acting as a vehicle for antimicrobial substances. However, bio-nanocomposite films have recently received more attention due to the fact that synthetic films' non-biodegradability has caused serious environmental concerns, as well as consumers' desire for safe food products and longer shelf lives (Alirezalu et al., 2021; Hermawan et al., 2019). According to Ramos et al. (2012a), edible films and coatings have various advantages over traditional coatings, including improved spreading, diffusivity, and solubility. They can also serve as carriers of antimicrobial agents. By serving as a semipermeable barrier to carbon dioxide, oxygen, and water vapor and preventing water loss, edible films and coatings made of biodegradable substances such as proteins, lipids, polysaccharides, and waxes can preserve the quality of cheese (Chong et al., 2019; Jafarzadeh et al., 2018). Additionally, cheese can be protected from undesirable microbial development and lipid degradation by using edible films and coatings as a carrier for antioxidant and antibacterial compounds (Rodriguez-Turienzo et al.,

2012). Other beneficial features may be offered by edible films or coatings. Cheese is eaten with these coatings, which have nutritional, biodegradable, and organoleptic qualities. According to Costa et al. (2018), they can add flavoring and coloring to cheese or have nutritious qualities on their own. Additionally, they can serve as carriers of active chemicals that have antibacterial and antioxidant qualities, and by regulating the pace at which these molecules diffuse into cheese, they can enhance cheese maintenance. These coatings' antibacterial and antioxidant qualities are not the only ways they extend the shelf life of cheese products but also some barrier qualities that protect against environmental causes that deteriorate Additionally, Zhong et al. (2014) assessed the impact of dipping, enrobing, spraying, and electrostatically spraying mozzarella cheese with an edible coating made of chitosan, soy protein isolate, and sodium alginate. The four coating techniques did not substantially differ in their ability to preserve cheese, despite the fact that sodium alginate-coated cheese was claimed to have superior physicochemical qualities during storage.

In addition to extending the shelf life, edible films and coatings can preserve the flavor, color, and nutritional content of cheese while also enhancing food safety and quality. Furthermore, they offer additional beneficial characteristics (Robertson, 2013; Rojas-Graü et al., 2009): they can be eaten with cheese, preventing waste by eliminating the need to dispose of leftover packaging (even if not used) because they decompose naturally (Wang et al., 2007). Cheese's organoleptic and nutritional qualities can be improved by the film or coating's composition, which may be advantageous in and of itself, or by its ability to add ingredients like flavoring, coloring, or sweeteners that are consumed with the coating. They can also serve as a carrier of active ingredients, such as those with antioxidant and antimicrobial qualities, which will help preserve the cheese while also controlling the rate at which those substances diffuse into it. For instance, crucial in certain nations where natamycin movement is restricted. Additionally, these active ingredients can be added to edible films and coatings by:

- microencapsulation, which allows for effective retention and controlled release into the cheese (Cui et al., 2016; Embuena et al., 2016);
- multilayers, which allow for the use of various materials and, consequently, an improvement of barrier properties; and
- application in small cheese portions (e.g., cheese slices), which functions as individual packaging for each portion, providing an additional benefit over non-edible packaging (Balaguer et al., 2014).

11.2.2 Impact on the Characteristics of Cheese

The physical properties of the coating materials including barrier, mechanical, and optical properties of the coatings are influenced by the materials employed and their primary attributes (such as solubility, density, viscosity, and surface tension). The type of cheese (i.e., its moisture content, whether it is matured or fresh) and the storage circumstances (i.e., temperature and relative humidity) should be taken into consideration when choosing film and coating materials. Additionally, the coating or film composition, the eventual usage (coating or film), and the requirements of the cheese manufacturer should all be taken into account when choosing application techniques (such as spraying, immersion, or wrapping) (Martins et al., 2012; Soradech et al., 2012). The methods used to create and apply edible coatings and films vary. While films are dried separately to create a stand-alone substance that is then utilized to cover the food, edible coatings, which come in liquid form, are applied directly to the food surface where a thin layer forms after drying. The most common materials used to create edible films and coatings for use on cheese are proteins (Ramos et al., 2012b; Rodriguez-Turienzo et al., 2012), polysaccharides (Bourbon et al., 2011; Martins et al., 2012; Pereira et al., 2010), and lipids (Cerqueira et al., 2009; Fadini et al., 2013). The primary benefit of polysaccharides and proteins is their water solubility; for lipids and waxes, on the other hand, an organic solvent or hot-melting techniques need be applied. The solubility of the antimicrobial agent and any potential interactions with the coating that can impact how well it works when applied to cheese products are also crucial considerations if an antimicrobial agent is added to the coating composition.

11.2.3 Types of Biopolymers Used in Cheese

11.2.3.1 Polysaccharides

A number of polysaccharides are being used as biomaterials for edible packaging including chitosan, alginate etc. Chitosan is very popular for cheese packaging due to its antimicrobial properties. Chitosan has excellent coating and film forming capacity and has been studied in different types of cheese for enhancing the shelf life. This has been used as a carrier of anti-microbial agents. In a study by Coma et al. (2002) Emmental cheese was inoculated with *Listeria innocua* on a chitosan based coating at 1% (v/v) and was found to have a positive antimicrobial effect. In another study by Duan et al. (2007) chitosan coating with 60% of lysozyme was applied to sliced mozzarella cheese. After 14 days of storage a reduction in *Escherichia coli*, *Pseudomonas fluorescens* and *L. monocytogenes* was observed as compared to control samples stored at 10 °C. Fajardo et al. (2010) tested different concentrations of natamycin in a chitosan-based coating of semi-hard cheese and observed a decrease in moulds/yeasts at 0.5 mg/mL of natamycin, after 27 days of storage at 4 °C, as compared to the control. Sodium alginate is another polysaccharide with excellent water permeability and mechanical properties (Wang et al., 2007) and has been found to increase the shelf life of various cheese varieties. In one study by Mastromatteo et al., 2014 Mozarella cheese packed in alginate-based coatings with modified atmosphere packaging (MAP) was found to have extended shelf life up to 160 days as compared to control (53 days).

Stretched cheeses, such as mozzarella, benefit greatly from biodegradable packaging due to their high moisture content and need for freshness. Biodegradable films made from materials like polylactic acid (PLA) and starch-based composites are used to package mozzarella and other stretched cheeses, providing a moisture barrier and extending shelf life without contributing to plastic waste. Vegan cheeses, often made from nuts, soy, or other plant-based ingredients, align well with biodegradable packaging solutions. Packaging materials like PLA, chitosan, and alginate not only preserve the cheese but also match the eco-friendly appeal of vegan products. These biopolymers offer excellent barrier properties and can be designed to be compostable or biodegradable, ensuring that the packaging aligns with the sustainable values of vegan consumers. Whey cheeses, such as ricotta, utilize byproducts of cheese production and are often packaged in biodegradable materials to emphasize sustainability. Gelatine-based films, for instance, have been studied for their ability to preserve the quality of whey cheeses by providing a good moisture barrier and maintaining texture. Such films also decompose naturally, reducing environmental impact. Hybrid cheeses, which combine dairy and plant-based ingredients, benefit from innovative biodegradable packaging that reflects their hybrid nature. Packaging made from biopolymer blends can enhance shelf life and reduce the environmental footprint. These material often include additives like natural fibers or antimicrobial agents to improve performance and food safety. Cheeses made from dry milk, often found in regions with limited access to fresh milk, are typically packaged in materials that can extend shelf life while maintaining product integrity. Biodegradable options like PLA or starch-based films are particularly useful here, as they provide necessary protection and align with sustainability goals by reducing reliance on conventional plastics.

11.2.3.2 Lipids

Lipids are commonly used as edible coatings in food products and in recent years, their use has increasingly involved combining them with other materials to enhance the hydrophobic properties of the coatings. Depending on their chemical structure, lipids can be categorized into resins, acetoglycerides, natural waxes, fatty acids, and various types of vegetable oils (Galus & Kadzińska, 2015). These lipids may be liquid or solid at room temperature, and this property plays a key role in their application, either directly on the cheese surface or in combination with other materials to form a more complex coating. Waxes, for instance, have seen limited application in cheese packaging, although beeswax has been successfully utilized. Yilmaz and Dagdemir (2012) demonstrated the effectiveness of beeswax in coating Kashar cheese with both single and double layers applied by immersion. After storing the cheese for 4 months at 4 °C and 85% relative humidity, the beeswax coating proved to be an effective protective barrier, reducing mold contamination to 1.89 log (CFU/g) by the 120th day, compared to the control, which showed a mold count of 4.60 log (CFU/g). Sensory evaluation using a 9-point scale for texture, taste, and odor revealed that the coated cheese had improved qualities (7.62 for texture, 7.61

for taste, and 7.62 for odor), outperforming vacuum-packed Kashar cheese, which scored 7.23, 7.34, and 7.46, respectively. Additionally, a decrease in moisture loss was observed for the coated cheese, with dry matter values of 61.36%, compared to 67.85% for uncoated cheese.

Ünalan et al. (2013) explored the use of a zein-wax composite film, combined with lysozyme, catechin, and gallic acid as bioactive compounds, for coating Kashar cheese. Several wax types were tested, including carnauba, candelilla, and beeswax, all in combination with zein. The zein-carnauba wax films were particularly effective in reducing Listeria monocytogenes in cheese samples. Moreover, films containing lysozyme, catechin, and gallic acid exhibited the strongest antimicrobial effects, with a reduction in L. monocytogenes to 5.59 log (CFU/g) compared to the control (8.13 log CFU/g), and showed enhanced antioxidant properties, preventing lipid oxidation during storage at 4 °C for 8 weeks.

Emulsion-based edible coatings, formed by combining lipids with polysaccharide or protein solutions and homogenizing them using mechanical shear, also offer an interesting strategy for creating effective coatings. For example, Ramos et al. (2012a, b) used a mixture of sunflower oil and whey protein isolate, while Cerqueira et al. (2010) employed galactomannan with corn oil. These emulsified coatings were applied to semi-hard cheese by dipping. The results showed a significant reduction in oxygen consumption (1.30 mL/kg/h) and a decrease in CO_2 production (1.5 mL/kg/h) at 20 °C, compared to uncoated cheese. Furthermore, weight loss was reduced by 3.8% and 3.1% for cheese stored at 4 °C and 20 °C, respectively. Emulsified coatings are advantageous due to their ease of application and the hydrophobic properties they impart, which help control water loss and improve the shelf life of cheeses.

11.2.3.3 Proteins

Whey proteins are commonly used in edible coatings. These proteins, derived from cheese production, are available as whey protein concentrate (WPC) or whey protein isolate (WPI), depending on the protein content (20–80% and >90%, respectively). Whey proteins are effective film-forming agents and have the added benefit of offering low barrier properties to oxygen, volatile aromas, and lipids (Ramos et al., 2012a, b). Ramos et al. (2012a, b) investigated WPI-based coatings incorporating various antimicrobial compounds to extend the shelf life of semi-hard cheeses. Their findings showed that the coatings reduced moisture loss by 10% and minimized changes in hardness and color during storage compared to uncoated cheese. The WPI-based coatings acted as barriers to water vapor, reducing moisture loss and maintaining cheese texture. Other WPI-based coatings, incorporating chitooligosaccharides (COS) and lactic acid, were tested and showed enhanced antimicrobial effects against bacteria, though their action against yeasts and molds was less pronounced compared to a commercial coating (polyvinyl acetate with natamycin). Another study demonstrated that WPI-based coatings with 3% mint essential oil effectively reduced the growth of microorganisms in Lor cheese stored at 4 °C for 15 days. This coating led to a complete reduction of Staphylococcus aureus, Listeria monocytogenes, and yeast-mold growth, with a notable decrease in E. coli (2.01 log CFU/g), compared to the control. The bactericidal and antimicrobial effects were even more pronounced when 4% mint essential oil was used, as it completely eradicated microbial growth throughout the 15-day storage period (Kavas & Kavas, 2014).

11.3 Antimicrobials for Use in Coatings

Incorporating antimicrobial agents into coatings is a well-established method for reducing the growth of spoilage microorganisms such as bacteria, molds, and yeasts in cheese. While antimicrobials can be added directly to the cheese during production or applied to the surface of the cheese in their free form, their effectiveness can be compromised due to partial inactivation or rapid diffusion into the food matrix (Reps et al., 2002). One solution to this issue is the incorporation of antimicrobial agents into the coating matrix, which prevents their migration and ensures sustained concentrations on the cheese surface for longer periods, thus improving the efficiency of the coating (Fajardo et al., 2010).

Common antimicrobial agents used in edible coatings include nisin, natamycin, lysozyme, organic acids, and essential oils. Kavas et al. (2015) studied the effect of thyme and clove essential oils (1.5% v/v) incorporated into WPI-based films applied to Kashar cheese. They found that thyme essential oil exhibited stronger antimicrobial activity compared to clove essential oil. Additionally, incorporating nisin into chitosan-cellulose films has been shown to provide antibacterial effects, especially against Listeria monocytogenes (Divsalar et al., 2018).

Due to their edible nature, coatings and films offer significant advantages over non-edible alternatives, providing similar benefits such as reduced weight loss, controlled respiration rates, and enhanced protection against spoilage. Over the past decade, the development of edible coatings and films for cheese has transitioned from research to commercial applications. With the continued use of emerging technologies like nanotechnology, the market for edible coatings is expected to grow further, offering sustainable solutions for preserving cheese quality and safety.

11.4 Modified Atmosphere Packaging (MAP) with Bio-composites

Modified atmosphere packaging (MAP) is a promising technique for extending the shelf life of dairy products, including cheese. However, the success of MAP depends on various factors, such as the type of cheese, the use of starter cultures, microbial contamination, and storage conditions (Pantaleao et al., 2007; Papaioannou et al.,

2007). MAP works by replacing the air surrounding the food product with a gas mixture that typically includes CO_2, O_2, and N_2. MAP has been particularly effective for soft cheeses, due to their delicate textures and the cushioning effect of the gases. Lucera et al. (2014) studied the impact of an active coating made from sodium alginate (2%) and potassium sorbate (1%) in combination with MAP on the shelf life of low-moisture Mozzarella. Their findings showed that the combination of MAP and active coatings could extend the shelf life of the cheese to 40 days at 8 °C and 160 days at 4 °C. In a separate study, Gammariello et al. (2011) tested the use of a bio-nanocomposite coating containing silver-montmorillonite nanoparticles (NPs) along with MAP for preserving Fiordi latte cheese. Their results demonstrated that the nanocomposite coating extended the shelf life of the cheese by more than 5 days, highlighting the potential of combining MAP with innovative coatings to enhance the preservation of cheese quality.

11.5 Cognitive Packaging System

11.5.1 Active Packaging

Active packaging incorporates active constituents like enzymes, organic acids, salts, fatty acids, antibiotics, nano-particles of metals, antioxidants, essential oils from plants, volatile compounds from fungicides, bacteriocins and plants into food packaging or employs functionally active polymers. It is a packaging system that actively modifies its environment to improve the quality of food products (Prasad & Kochhar, 2014; Mei et al., 2020; Jafarzadeh et al., 2020). Active packaging techniques include release absorbers, systems aimed at preserving and enhancing the safety and quality of food products. Absorbing systems effectively remove undesirable compounds, including oxygen, from cheeses. Antioxidants, antimicrobials, carbon dioxide and preservatives are systematically introduced or released into packaged cheese or its headspace through films, corks, labels or sachets (Bodbodak & Rafiee, 2016). Multiple studies indicate that active packaging enhances the shelf life by decreasing weight and water content, maintaining the sensory attributes and inhibiting microbial proliferation, of the cheese during storage. The use of antimicrobial nanoparticles in packaging materials can enhance the shelf life y prolonging the lag phase and decreasing the rate of spoilage by growth of microbes. The anti-microbial nano-packaging technology is developed through the integration of anti-microbial fillers (nano size) into packaging films, the formulation of packaging materials from bio/polymers with intrinsic antimicrobial properties, and the application of coatings containing antimicrobial agents to the packaging films, facilitating the gradual release of these agents onto cheese surfaces. ZnO nanoparticles are regarded as effective materials owing to their photocatalytic and antimicrobial characteristics. Youssef et al. (2016) researched the anti-microbial properties of bio-nanocomposite films made from ZnO nanoparticles, chitosan and carboxymethyl cellulose. The application of nanoparticles resulted in an increase in the diameter of the inhibition zone. The films demonstrated enhanced inhibitory effects against the tested pathogenic fungi and bacteria, indicating a variable spectrum of activity. The most significant effect was noted against Bacillus cereus, while Staphylococcus aureus displayed an inhibition zone measuring between 5 and 15 mm. Additionally, these packaging films extended the shelf life of white soft cheese. Ligaj et al. (2020) synthesized polylactide films containing zero-valent iron through the casting method. Their findings demonstrate that films with 3% (w/w) zero-valent iron effectively inhibited the proliferation of both Gram-negative bacteria (E. coli) and positive bacteria (S. epidermidis and B. subtilis). Additionally, the polylactide film with pure poly(lactic) acid and iron oxide showed no antimicrobial activity; however, a minor growth inhibition was noted in the sample containing 1% zero-valent iron. Following goat cheese was encapsulated in these films, yielding promising results. El-Sayed et al. (2020) demonstrated that the nanocomposite solutions and Roselle calyx extract (RE) containing zinc oxide, guar gum and chitosan displayed the most significant inhibitory effect against A. terries and L. Escherichia coli in a 3% RE and ZnO formulation. Bio-nanocomposite films containing ZnO and RE were subsequently applied to Ras cheese. The formulated coatings significantly prolonged the shelf life of cheese by around 90 days through the inhibition of spoilage and bacterial growth. Wang et al. (2019) formulated an edible film that integrates carvacrol (CA) and whey protein isolate nanofibers (WPNFs) for the preservation of Cheddar cheese. The findings indicated that the edible films displayed antimicrobial properties against L monocytogenes, E coli, S enteritidis and S aureus. The antioxidant activity was further enhanced. WPNFs coatings improved the properties of fresh Cheddar cheese during storage period. Amjadi et al. (2019) created active packaging using chitosan nanofibers combined with zinc oxide nanoparticles. The bio-nanocomposite film's efficacy was assessed using cheese as the food model. Inoculated bacteria growth was decreased. The cheese samples were encapsulated in nanocomposite film throughout the storage period to control fluctuations in color metrics and pH levels. Youssef et al. (2019) asserted that applying a coating of Ras cheese with a film made of 2% TiO2-NPs and polyvinyl alcohol/chitosan can inhibit mold growth on the cheese's surface. Jafarzadeh et al. (2019) employed ZnO nanoparticles within semolina films in a distinct investigation. Antimicrobial activity was observed against E coli, A niger, C albicans and S aureus. The antimicrobial activity increased with the addition of ZnO-NPs. The antibacterial efficacy was lesser against Gram-negative bacteria than against Gram-positive bacteria. The films demonstrated favorable outcomes for the packaging of Mozzarella cheese. Organic antibacterial agents exhibit limited applicability in active packaging due to insufficient thermal stability. In contrast, inorganic metal nanoparticles are ideal for active packaging due to their remarkable high surface reactivity, significant surface-to-volume ratio and thermal stability. However, it is essential to address the safety concerns and migration characteristics of metal nanoparticles in food packaging applications. Recent studies (Jafarzadeh & Jafari, 2020; Jafarzadeh et al., 2021) have examined the safety, migration and toxicological aspects of metal nanoparticles in food products.

11.5.2 Innovative Nanocomposites in Active Cheese Packaging

Nanomaterials, exhibit distinct properties than larger systems. Modern packaging technologies has improved the bioactivities of system, including antioxidant and antimicrobial properties, through the incorporation of hydrophilic or lipophilic materials that release during storage periods (Rehman et al., 2020; Bahrami et al., 2020). The term "edible nano-coatings" denotes an innovative approach in food technology, employing biodegradable synthetic or natural polymers, along with polymers which are non-toxic, for the purpose of food preservation. Recent advancements in liquid and lipid based nano-systems have positioned them as viable edible coating systems at ambient temperature.

11.5.2.1 Nanoparticles

Nano particles in case of edible food packaging, refers to a variety of compounds, including metals (Thirumurugan et al., 2013; Dairi et al., 2019), C-induced materials (Biswas et al., 2019), nanolayers (Ramesh et al., 2015; Roy et al., 2019), mixed metal oxides (Kooti et al., 2015), metal oxides (Swaroop & Shukla, 2019; Espitia et al., 2016), all defined by dimensions under 100 nm, in addition to nano-coatings. Nanomaterials, due to their biological and physicochemical properties, can impart various characteristics. Nanocomposites are of two types based on the nature of inorganic or organic interactions: (a) covalent and (a) non-covalent interactions, which are formed via self-assembly methods or sol–gel method. The properties of nanocomposites are dictated by the attributes of the nanoparticles (inorganic phase) and polymeric materials (organic phase) (Salehabadi et al., 2014). Prominent solid-state inorganic nanoparticles (NPs) employed in coating techniques encompass (Zumeta et al., 2010), titanium dioxide (TiO2), gold (Au) (Amendola et al., 2014; Thirumurugan et al., 2013), silver (Ag) (Wu et al., 2018), oxides of copper (CuO and Cu2O) (Reyes et al., 2018; Almasi et al., 2018;), ferrous-ferric oxide (Fe3O4) (Adami & Reverchon, 2012; Zhou et al., 2013; Sanaeifar et al., 2017) and magnesium oxide (Ropp, 2013). As per the research of Al-Nabulsi et al. (2020) ZnO nanoparticles incorporated in chitosan inhibits the growth on E coli O157:H7 on white brined cheese. The nanocomposites (ZnO and chitosan) markedly diminished the counts of E. coli O157.H7–2.5 log CFU/g at 10 °C.

11.5.2.2 Nanoemulsions

Nanoemulsions function as a vehicle for active compounds. Nano emulsions are thermodynamically viable but kinetically favprable because of their physicochemical properties (Jamali et al., 2020). These inedible coatings have attracted considerable attention due to their natural composition, effectiveness, and non-toxicity; but this nano-system remains under-commercialized. The interaction profile (polymer + nano-emulsions + foods) can produce an edible coating. Nano-emulsions are used as a coloring and flavoring agents, antimicrobial, and antioxidants for applications in dairy products (such as cheese), meats and fruits freshly cut (Aswathanarayan & Vittal, 2019). Nano-emulsion-derived oregano essential oil (OEO) was employed as an antimicrobial agent on less-fat sliced cheese for shelf life extenxion. The resistance to antimicrobial effectiveness and water vapor against inoculated S aureus were incrementally enhanced. The anti-fungal efficacy of nano-emulsions with essential oils can be affected by many factors. The characteristics of nano-emulsions can differ depending on the types of oleaginous liquids, aqueous liquids, and emulsifiers, which can subsequently affect antifungal effectiveness. Essential oils are aromatic compounds derived from plants, noted for their antifungal characteristics. The antifungal effectiveness of essential oils against Mucor, Pichia, Aspergillus, Eurotium, Debaryomyces, Penicillium etc. is recorded. The storage temperature, the immersion duration of cheese slices in nano-emulsions and volume of essential oil may influence the antifungal effectiveness of nano-emulsions loaded with essential oil. The antifungal efficacy of essential oils can be influenced by changes in climate, extraction techniques, agricultural practices, concentrations of bioactive compounds, chemical composition, fungal genera and inoculated spores. A study documented the antifungal efficacy of nanoemulsions containing oregano essential oil (OEO) and its utilization in Minas Padr˜ao cheese (Bedoya-Serna et al., 2018). The findings revealed the in vitro antifungal effectiveness of OEO nanoemulsions against the proliferation of Fusarium sp., Cladosporium sp., and Penicillium sp., on slices of Minas Padrão cheese. The researchers determined that the Penicillium species demonstrated superior resistance to the antifungal properties of OEO relative to other fungi.

11.5.2.3 Nanofibers

Electrospinning resulted in the formation of a nanofibers. In this high electric field is utilized to convert polymeric solutions into nanofibers (Rostamabadi et al., 2020). Unique functionalities are imparted because of large surface area. Nanofibers can be employed in food coating technologies to create active packaging with antimicrobial characteristics (Khoshnoudi-Nia et al., 2020). Nanofibers may be infused with plant extracts and bioactive compounds, like anti-microbials or anti-oxidants in edible coatings. The active compounds in nanofibers may be liberated during the storage of cheese. The nanofibers can be arranged in two configurations: (a) random and (b) aligned (Nikmaram et al., 2017; Bharathi et al., 2020). The aligned configuration exhibits superior chemical and physical properties compared to the other configuration. In edible coatings the aligned configuration is beneficial, as the nanofibers and cells are oriented uniformly. This configuration leads to the direct production of extracellular matrix in oriented form.

Nanofibers in packing of cheese can encapsulate bioactive materials, facilitating the development of nanocomposites for intelligent packaging and edible coatings solutions. CNPs (Chitosan nanoparticles) integrated into gelatin nanofibers served as antimicrobial packaging to address Staphylococcus and Listeria contamination in cheese, alongside oil of moringa (Lin et al., 2019). This packaging admixture demonstrated notable antibacterial efficacy against S. aureus and L. monocytogenes. The activity was most effective between 4 °C and 25 °C, without affecting the sensory quality of the cheese. Tatlisu et al. (2019) demonstrated the efficacy of nanofiber mats loaded with thymol (an antimold coating). This may inhibit Kashar cheese proliferation by A parasiticus. The antimold efficacy of thymol was assessed through 56% encapsulation, and the in-situ activities exhibited significant inhibition of mold proliferation on cheese. Electrospun nanofibers infused with phytochemicals were utilized to improve antibacterial efficacy on cheese slices (Goksen et al., 2020). This research focused on integrating cineole-rich extracts from bay and rosemary into active edible and antimicrobial films. Active nanofibers were synthesized using electrospinning techniques, exhibiting anti-microbial efficacy with sustained release from the fibers, as indicated by a significant reduction of approximately 2 logarithmic units of S. aureus and L. monocytogenes. The proliferation of aerobic mesophilic bacteria was restricted in cheese slices enveloped in films of zein nanofiber infused with essential oils. Research on nanofibers incorporated with nisin-loaded poly-γ-glutamic acid/chitosan (NGC) nanoparticles has been conducted to evaluate their efficacy as anti-listerial agents for cheese (Cui et al., 2017). The findings indicate that NGC nanofibers may serve as a viable active packaging solution for cheese preservation.

11.5.3 Diagnostic Methods: Paper-Based and Smartphone

Although significant research has been conducted on intelligent packaging and sensors, there has been a lack of comprehensive reviews focusing specifically on paper-based and smartphone diagnostics. Paper-based sensors provide a cost-effective solution by employing low-cost base materials and enabling communication with smartphones, thus presenting a promising opportunity for the advancement of intelligent packaging applications within the food industry.

11.6 Conclusion

Innovative nanocomposites in active cheese packaging are being explored to enhance the preservation of cheese. These nanocomposites, often incorporating materials like nanofibers, nanoparticles, and biopolymers, offer benefits such as improved barrier properties, antimicrobial effects, and extended shelf life. They can actively interact with the cheese, providing protection against spoilage and pathogens, while also maintaining the cheese's quality and safety. There is a lack of comprehensive reviews on paper-based sensors and smartphone diagnostics in intelligent packaging. These sensors offer a cost-effective solution for the food industry, using low-cost materials and enabling smartphone communication. The future of edible coatings in the cheese industry will depend on factors like regulatory compliance, cost-effectiveness, and production scalability and manufacturers must follow good manufacturing practices to ensure quality and safety.

References

Alirezalu, K., Pirouzi, S., Yaghoubi, M., Karimi-Dehkordi, M., Jafarzadeh, S., & Mousavi Khaneghah, A. (2021). Packaging of beef fillet with active chitosan film incorporated with ε-polylysine: An assessment of quality indices and shelf life. *Meat Science, 176*, 108475. https://doi.org/10.1016/j.meatsci.2021.108475

Almasi, H., Jafarzadeh, P., & Mehryar, L. (2018). Fabrication of novel nanohybrids by impregnation of CuO nanoparticles into bacterial cellulose and chitosan nanofibers: Characterization, antimicrobial and release properties. *Carbohydrate Polymers, 186*, 273–281. https://doi.org/10.1016/j.carbpol.2018.01.067

Balaguer, M. P., Fajardo, P., Gartner, H., Gomez-Estaca, J., Gavara, R., Almenar, E., & Hernandez-Munoz, P. (2014). Functional properties and antifungal activity of films based on gliadins containing cinnamaldehyde and natamycin. *International Journal of Food Microbiology, 173*, 62–71. https://doi.org/10.1016/j.ijfoodmicro.2013.12.013

Bourbon, A. I., Pinheiro, A. C., Cerqueira, M. A., Rocha, C. M. R., Avides, M. C., Quintas, M. A. C., & Vicente, A. A. (2011). Physico-chemical characterization of chitosan-based edible films incorporating bioactive compounds of different molecular weight. *Journal of Food Engineering, 106*(2), 111–118. https://doi.org/10.1016/j.jfoodeng.2011.03.024

Cerqueira, M. A., Lima, Á. M., Souza, B. W. S., Teixeira, J. A., Moreira, R. A., & Vicente, A. A. (2009). Functional polysaccharides as edible coatings for cheese. *Journal of Agricultural and Food Chemistry, 57*(4), 1456–1462. https://doi.org/10.1021/jf802726d

Cerqueira, M. A., Sousa-Gallagher, M. J., Macedo, I., Rodriguez-Aguilera, R., Souza, B. W. S., Teixeira, J. A., & Vicente, A. A. (2010). Use of galactomannan edible coating application and storage temperature for prolonging shelf-life of "regional" cheese. *Journal of Food Engineering, 97*(1), 87–94. https://doi.org/10.1016/j.jfoodeng.2009.09.019

Costa, M. J., Maciel, L. C., Teixeira, J. A., Vicente, A. A., & Cerqueira, M. A. (2018). Use of edible films and coatings in cheese preservation: Opportunities and challenges. *Food Research International, 107*, 84–92. https://doi.org/10.1016/j.foodres.2018.02.013

Cui, H. Y., Wu, J., Li, C. Z., & Lin, L. (2016). Anti-listeria effects of chitosan-coated nisin-silica liposome on Cheddar cheese. *Journal of Dairy Science, 99*(11), 8598–8606. https://doi.org/10.3168/jds.2016-11658

Fadini, A. L., Rocha, F. S., Alvim, I. D., Sadahira, M. S., Queiroz, M. B., Alves, R. M. V., & Silva, L. B. (2013). Mechanical properties and water vapour permeability of hydrolysed collagen–cocoa butter edible films plasticised with sucrose. *Food Hydrocolloids, 30*(2), 625–631. https://doi.org/10.1016/j.foodhyd.2012.08.011

Fajardo, P., Martins, J. T., Fuciños, C., Pastrana, L., Teixeira, J. A., & Vicente, A. A. (2010). Evaluation of a chitosan-based edible film as carrier of natamycin to improve the storability of Saloio cheese. *Journal of Food Engineering, 101*(4), 349–356. https://doi.org/10.1016/j.jfoodeng.2010.06.029

Galus, S., & Kadzińska, J. (2015). Food applications of emulsion-based edible films and coatings. *Trends in Food Science & Technology, 45*(2), 273–283. https://doi.org/10.1016/j.tifs.2015.07.011

Garcia, E., & Barrett, D. M. (2002). Preservative treatments for fresh-cut fruits and vegetables. In *Fresh-cut fruits and vegetables: Science, technology, and market* (pp. 267–303) https://www.scopus.com/inward/record.uri?eid=2-s2.0-85056729442&partnerID=40&md5=99157af39d5e98322968f4b50eaddb1a

Jafarzadeh, S., Jafari, S. M., Salehabadi, A., Nafchi, A. M., Uthaya Kumar, U. S., & Khalil, H. P. S. A. (2020). Biodegradable green packaging with antimicrobial functions based on the bioactive compounds from tropical plants and their by-products. *Trends in Food Science & Technology, 100*, 262–277. https://doi.org/10.1016/j.tifs.2020.04.017

Jafarzadeh, S., Mohammadi Nafchi, A., Salehabadi, A., Oladzad-abbasabadi, N., & Jafari, S. M. (2021). Application of bio-nanocomposite films and edible coatings for extending the shelf life of fresh fruits and vegetables. *Advances in Colloid and Interface Science, 291*, 102405. https://doi.org/10.1016/j.cis.2021.102405

Kavas, G., & Kavas, N. (2014). The effects of mint (Mentha spicata) essential oil fortified edible films on the physical, chemical and microbiological characteristics of lor cheese. *Journal of Food, Agriculture and Environment, 12*(3–4), 40–45. https://www.scopus.com/inward/record.uri?eid=2-s2.0-84908101169&partnerID=40&md5=bddc10837e7b1c319f81b265e8831935

Lucera, A., Mastromatteo, M., Conte, A., Zambrini, A. V., Faccia, M., & Del Nobile, M. A. (2014). Effect of active coating on microbiological and sensory properties of fresh mozzarella cheese. *Food Packaging and Shelf Life, 1*(1), 25–29. https://doi.org/10.1016/j.fpsl.2013.10.002

Martins, J. T., Cerqueira, M. A., & Vicente, A. A. (2012). Influence of α-tocopherol on physicochemical properties of chitosan-based films. *Food Hydrocolloids, 27*(1), 220–227. https://doi.org/10.1016/j.foodhyd.2011.06.011

Mastromatteo, M., Conte, A., Faccia, M., Del Nobile, M. A., & Zambrini, A. V. (2014). Combined effect of active coating and modified atmosphere packaging on prolonging the shelf life of low-moisture mozzarella cheese. *Journal of Dairy Science, 97*(1), 36–45. https://doi.org/10.3168/jds.2013-6999

Pereira, R. N., Souza, B. W. S., Cerqueira, M. A., Teixeira, J. A., & Vicente, A. A. (2010). Effects of electric fields on protein unfolding and aggregation: Influence on edible films formation. *Biomacromolecules, 11*(11), 2912–2918. https://doi.org/10.1021/bm100681a

Ramos, Ó. L., Pereira, J. O., Silva, S. I., Fernandes, J. C., Franco, M. I., Lopes-da-Silva, J. A., Pintado, M. E., & Malcata, F. X. (2012a). Evaluation of antimicrobial edible coatings from a whey protein isolate base to improve the shelf life of cheese. *Journal of Dairy Science, 95*(11), 6282–6292. https://doi.org/10.3168/jds.2012-5478

Ramos, Ó. L., Santos, A. C., Leão, M. V., Pereira, J. O., Silva, S. I., Fernandes, J. C., Franco, M. I., Pintado, M. E., & Malcata, F. X. (2012b). Antimicrobial activity of edible coatings prepared from whey protein isolate and formulated with various antimicrobial agents. *International Dairy Journal, 25*(2), 132–141. https://doi.org/10.1016/j.idairyj.2012.02.008

Reps, A., Drychowski, L. J., Tomasik, J., & Winiewska, K. (2002). Natamycin in ripening cheeses. *Pakistan Journal of Nutrition, 1*(5), 243–247. https://doi.org/10.3923/pjn.2002.243.247

Rodriguez-Turienzo, L., Cobos, A., & Diaz, O. (2012). Effects of edible coatings based on ultrasound-treated whey proteins in quality attributes of frozen Atlantic salmon (Salmo salar). *Innovative Food Science & Emerging Technologies, 14*, 92–98. https://doi.org/10.1016/j.ifset.2011.12.003

Rojas-Graü, M. A., Soliva-Fortuny, R., & Martín-Belloso, O. (2009). Edible coatings to incorporate active ingredients to fresh-cut fruits: A review. *Trends in Food Science and Technology, 20*(10), 438–447. https://doi.org/10.1016/j.tifs.2009.05.002

Soradech, S., Nunthanid, J., Limmatvapirat, S., & Luangtana-Anan, M. (2012). An approach for the enhancement of the mechanical properties and film coating efficiency of shellac by the formation of composite films based on shellac and gelatin. *Journal of Food Engineering, 108*(1), 94–102. https://doi.org/10.1016/j.jfoodeng.2011.07.019

Ünalan, İ. U., Arcan, I., Korel, F., & Yemenicioğlu, A. (2013). Application of active zein-based films with controlled release properties to control Listeria monocytogenes growth and lipid oxidation in fresh Kashar cheese. *Innovative Food Science & Emerging Technologies, 20*, 208–214. https://doi.org/10.1016/j.ifset.2013.08.004

Zhong, Y., Cavender, G., & Zhao, Y. (2014). Investigation of different coating application methods on the performance of edible coatings on mozzarella cheese. *LWT, 56*(1), 1–8. https://doi.org/10.1016/j.lwt.2013.11.006

Zumeta, I., Díaz, D., & Santiago, P. (2010). Synthesis of TiO2 nanoparticles with narrow size distribution and their evaluation in the photocatalytic oxidative degradation of Bis(4-nitrophenyl) phosphate. *The Journal of Physical Chemistry C, 114*(26), 11381–11389. https://doi.org/10.1021/jp1021399

Chapter 12
Low-Fat Cheese

Pavel Rout and Shreya Saha

12.1 Introduction

The CODEX Commission on International Trade has mandated that for cheese to be labelled as reduced fat, it must have a maximum fat reduction of 50% from a specified reference variety. Consequently, calling a cheese "zero-fat Cheddar" would not meet the international trade labelling criteria. Nevertheless, individual countries have the autonomy to set their own internal criteria. In the United States, low-fat cheese must have 6 g or less of fat per 100 g, and reduced-fat cheese must have at least 25% less fat than the traditional fat content of that type of cheese. Fat free cheese is cheese that has less than 0.5 g of fat per 100 g of cheese. "Low-fat cheese" in this sense refers to cheese with at least a 25% reduction in fat compared to the reference type, which also includes fat-free cheeses.

Most of the significant challenges associated with low-fat cheeses occur when the fat reduction reaches 50% or higher. Cheese competitions often categorize all low-fat variants together, leading to an unfair competition. Low-fat soft cheeses immediately compete with low-fat semi-soft or semi-hard cheeses. Creating low-fat versions of mild-tasting, high-moisture cheeses such as low-fat cream cheese or mozzarella is easier than making full-flavored, low-fat cheddar or gouda cheese.

P. Rout (✉)
Dairy Chemistry Division, ICAR-NDRI, Karnal, Haryana, India

S. Saha
Dairy Microbiology Division, ICAR-NDRI, Karnal, Haryana, India

S. Mansha Rafiq (ed.), *Technological Advances and Trends in Cheese Making*,
https://doi.org/10.1007/978-3-031-94476-5_12

12.2 Types of Low-Fat Cheese

The realm of low-fat cheese manufacturing caters to four distinct applications, each demanding specific cheese attributes. On the other hand, most consumers, both retail and institutional, expect a single cheese to have all the traits they want in all of these situations. This way of thinking has made it harder for low-fat cheeses to become popular. The following are these four applications:

12.2.1 Table Cheese

Table cheeses are consumed in their natural state without further heating or alteration. They retain their original identity and are savored for their flavor and textural qualities, such as firmness, chewiness, and mouthfeel. Creating a low-fat version of this cheese type while maintaining the desired flavor, body, and mouthfeel attributes throughout the ripening process poses significant challenges.

12.2.2 Condiment Cheese

Condiment cheeses are intended for heating or incorporation into other foods while retaining their essence as cheese. For instance, Mozzarella on pizza or cheese in burgers, lasagna, or chili. Emphasis is placed on flow and chew characteristics, where flavor importance may vary depending on the dish's overall taste profile. These cheeses often involve added ingredients to enhance functionality, yield, and typically have shorter shelf-life requirements. Cost-effectiveness is crucial due to their use in dishes where pricing and competition with alternative ingredients matter.

12.2.3 Ingredient Cheese

Ingredient cheeses lose their visual identity when utilized in other food products. They serve as flavor contributors in items like salad dressings or as components in dough for crackers or bread. Here, flavor attributes hold primary importance, alongside the cheese's ability to integrate smoothly into dough or paste for ease of use in food production.

12.2.4 *Process Cheese*

Low-fat or skim-milk cheeses are used to produce higher-fat process cheeses. These emphasize machinability and physical-chemical attributes over flavor. Key parameters include intact casein levels and residual CCP (calcium caseinate phosphate). Manipulating flavor in processed cheese typically involves adding flavourful ingredients rather than relying on developing flavor in the base low-fat cheese.

12.3 Methods of Preparation

To improve the texture of low-fat cheese, increasing its moisture level is a key method to reduce its hardness. Amelia et al. (2013) discovered that full-fat Cheddar cheese had an average moisture level of 35.65%, while Cheddar cheese with 50% reduced fat and 75% reduced fat had moisture contents of 49.23% and 52.52%, respectively. Amelia et al. (2013) proposed that variations in fat dispersion and water pocket sizes in the cheese structure explained the differences in texture between experimental low-fat Cheddar (with 58.07% moisture) and commercial 75% reduced-fat Cheddar.

Several techniques are available to enhance cheese moisture, including reducing cooking temperatures, accelerating acidification, curd washing, salting, and using fat mimetics or replacers. However, increasing moisture content might have negative consequences. Increasing moisture levels while keeping the salt content constant reduces the salt-in-moisture (S/M) ratio in cheese, which, therefore impacts the activities of starters and enzymes. Consequently, this leads to a notable decrease in protein breakdown in low-fat Cheddar cheese, which hinders the process of taste formation (Laloy et al., 1996; Mistry & Kasperson, 1998). Thus, any changes made to enhance texture should also consider the possible effects on cheese flavor.

12.3.1 *Standardization of Milk for Low-Fat Cheese Manufacture*

To get the desired fat level in the finished cheese, it is necessary to standardize the milk to meet specific casein-to-fat (C/F) or protein-to-fat (P/F) ratios as suggested by Banks (2004) and Mistry and Kasperson (1998). The recommended C/F ratios for low-fat Cheddar cheese with 15.1% or 20.2% fat are 2.050 and 1.285, respectively. This indicates that the cheese milk should have around 1.5% to 2% fat. The fat content of the cheese milk is controlled at around 0.5% to enable the production of low-fat Cheddar. Casein can be used in milk for producing low-fat cheese using various forms including non-fat dry milk (NFDM), condensed skim milk (CS), or milk protein concentrates (MPC) produced using membrane filtering. Anderson et al. (1993) standardized milk with CS to produce low-fat Cheddar, whereas McGregor and White (1990) and St-Gelais et al. (1998) utilized UF retentate for the same purpose. Augmenting whole milk with protein to enhance uniformity might lead to solubility and spread issues due to the substantial quantities of protein required. Standardizing with NFDM increases the remaining lactose content in the cheese, perhaps leading to unwanted fermentation. Ghandi and Farkye conducted a research to produce low-fat cheese by combining whole milk with UF skim milk retentate. Short-chain fatty acids crucial for flavor are lost during the process of removing fat from cheese milk to enhance its consistency.

12.3.2 Acidification

Preacidifying milk is the first step in cheesemaking to control cheese moisture and texture before adding rennet. The hardness of cheese, as determined by Texture Profile Analysis (TPA), is influenced by the kind of acid utilized in the acidification process. Farkye et al. (1995) found that Queso Blanco-type cheese manufactured with lactic acid had lower TPA hardness, chewiness, and gumminess than cheese created with acetic acid. Acidification decreases the pH of milk, transforming calcium from forms that cannot dissolve into forms that can dissolve. Subsequently, the soluble calcium is extracted from the curd particles present in the whey, contributing to the decrease in the cheese's total calcium concentration (Lucey & Fox, 1993). The decrease in calcium levels results in softer cheese, which is beneficial for low-fat cheese types that often have a stiff, rubbery consistency.

Preacidifying milk for Mozzarella cheese manufacturing has been shown to reduce post-melt chewiness (Metzger et al., 2001). The extraction of casein from casein micelles and the release of calcium are influenced by pH and temperature (Dalgleish & Law, 1988). The highest casein dissociation, around 30%, was seen at pH 5.8 and 4 °C. Preacidifying milk at 4 °C may enhance calcium solubilization, perhaps resulting in increased cheese suppleness. Preacidifying to a pH of 5.8 is not practical for making low-fat cheese when aiming for a final pH range of 5.2–5.4 for Cheddar. This restriction occurs due to insufficient time for the formation and activity of the starting culture in the cheese.

Calcium is needed for rennet to work, and it also gives cheese its structure and ability to hold its shape (Lucey & Fox, 1993). Calcium is a key part of cross-linking proteins in cheese, which changes the structure of the cheese directly (Lawrence et al., 1987). Mozzarella cheese gets harder and less meltable when it has more calcium in it (Lucey & Fox, 1993; Metzger et al., 2001). By adding calcium chelators like sodium citrate or sodium EDTA to cheese milk with 0.6% fat at concentrations of 0.02% and 0.2%, the amount of calcium in low-fat cheese was lowered. This made the calcium levels 477.5 mg/kg and 502.5 mg/kg, which is lower than the 622.5 mg/kg found in normal low-fat cheese. As a result, the TPA hardness scores went down as the fruit got riper (Poveda et al., 2013).

12.3.3 Curd Washing

Curd washing, which involves adding water to curds, significantly impacts the moisture and calcium levels within cheese (Buch Kristensen, 1999). The absorption of moisture by curds is influenced by the temperature of the washing water; lower temperatures lead to increased moisture uptake. In the production of low-fat cheese, rinsing curds with cold water is a common practice aimed at elevating moisture content and softening the cheese. Augmenting moisture through this method presents yield advantages in cheese manufacturing.

By washing the curd, soluble calcium and lactic acid are removed. This stops the formation of calcium lactate crystals during the hardening phase. It also helps get rid of any leftover lactose in the curd and cheese, which stops fermentation that isn't wanted (Banks, 2004; Mistry, 2001; Rodrı́guez, 1998). It's important to note, though, that curd washing gets rid of flavor compounds that dissolve in water. This makes cheese taste bland, with hints of fatty and brothy off-flavors (Johnson & Chen, 1995).

12.3.4 Altering Cooking Temperature

The normal cooking temperature for making full-fat Cheddar cheese is around 39 °C. For making cheeses with less fat, on the other hand, temps lower than 39 °C are best. In 1989, Banks et al. suggested cooking 25% and 50% reduced-fat Cheddar cheese at 37 °C and 35 °C, respectively, to increase the amount of liquid. Kosikowski and Mistry (1997) said that the best cooking temperature for making low-fat Cheddar cheese was between 34 °C and 36 °C. A lower cooking temperature has a direct effect on starting activity, yield, and the process of proteolysis. It also makes the cheese creamier.

12.3.5 Salting

In the production of low-fat cheese, the decrease in fat content is commonly compensated by raising the moisture levels, leading to a decrease in salt concentration within the moisture phase of the cheese (Mistry, 2001). The salt concentration, especially the Salt-in-Moisture (S/M) ratio in low-fat Cheddar cheese, significantly impacts flavor formation. Research indicates that a 1.8% salt content in low-fat Cheddar cheese enhances its distinctive Cheddar flavor while simultaneously diminishing bitterness. Augmenting the S/M ratio reduces the emergence of bitterness and unwanted flavors (Banks et al., 1993). Conversely, a decrease in S/M ratio fosters excessive bacterial growth, amplifying proteolysis during cheese aging (Kelly et al., 1996). This could potentially contribute to inadequate flavor development and flavor irregularities in low-fat Cheddar cheese. Hence, it is crucial to manage this parameter effectively to ensure the production of cheese with desirable characteristics (Rodríguez, 1998).

Johnson and Chen (1995) reported devised a method to create high-quality low-fat Cheddar by adjusting various production stages to optimize moisture retention. The curd formation process involves allowing firm curd development for 50 minutes before cutting. The scald temperature is decreased to 37 °C, and the curd-to-whey mixture is stirred at this temperature for just 5 minutes. Draining of the curd occurs at an elevated pH of 6.45, followed by salting at a high pH of 5.9. This method yields a low-fat cheese containing approximately 14.5–15% fat (classified as 50% reduced-fat cheese), with a moisture content of 48–49%, lactic acid levels at 1.9–2.1%, and a pH range of 5.0–5.1 within 10 days. Texture development occurs over 2 months at temperatures between 7 °C and 8 °C, leading to the development of the characteristic mature Cheddar flavor.

Additionally, an alternative method involves blending skim milk curd and full-fat curd post whey drainage, resulting in an acceptable low-fat Cheddar cheese (Fenelon et al., 1999). Selection of starter and adjunct cultures. Starter cultures hold significant importance in cheese production due to their impact on metabolism, which influences texture and flavor development. Choosing the right starter culture is crucial for achieving desired flavor and texture in full-fat cheeses. However, in the context of producing low-fat cheeses prone to developing off-flavors, the selection of an appropriate starter culture becomes even more critical. A starter culture suitable for producing high-quality standard fat Cheddar might not yield satisfactory results for a low-fat variant (Johnson & Chen, 1991). In the creation of low-fat cheeses, slower acid-producing starter cultures are preferable. This choice is essential as the controlled development of acid plays a pivotal role in minimizing the emergence of off-flavors. With reduced and low-fat cheeses containing higher moisture levels, maintaining a gradual acid development process during cheese production becomes vital. This approach ensures a robust buffering capacity in the cheese, preventing the formation of overly intense acid flavors.

Ardö (1997) suggests that when producing low-fat cheese with different temperature profiles than ordinary fat content cheese, starting cultures should be chosen based on their temperature sensitivity, autolytic characteristics, and proteolytic/peptidolytic activities. It is advised to induce autolysis at a lower temperature for decreased fat cheese. Recent investigations have shown that a harmonious interplay between autolysis and active cultures during ripening produces chemicals that are not typically present in milk or milk fat (Ganesan & Weimer, 2004). To enhance the flavor profile of reduced and low-fat cheeses, it is suggested to combine starter cultures with flavor adjunct cultures. Weimer et al. (1997) propose determining mixtures of cultures that provide the most favourable flavor profile for these cheeses. Culture suppliers now offer specially designed starter cultures for manufacturing reduced and low-fat cheeses. These cultures are characterized by slow growth, gradual acid development, and low proteolytic activity (Katsiari et al., 2002).

12.3.6 Use of Buttermilk and Surfactants

Madsen et al. were the first to record the use of buttermilk to improve the consistency of skim milk cheese (1966). According to Vanderghem et al. (2010), there are a lot of surface-active components in buttermilk that come from the milk fat globule membrane. When buttermilk was homogenized with cream and added, Mayes et al. (1994) found a little improvement. Nonetheless, they came to the conclusion that the impact wasn't big enough to warrant major financial investment. The use of buttermilk in cheesemaking improved the texture and body of reduced-fat Mozzarella cheese, according to Poduval and Mistry (1999), and it also boosted the formation of free oil during storage. Madsen et al. were the first to record the use of buttermilk to improve the consistency of skim milk cheese (1966). According to Vanderghem et al. (2010), many surface-active components in buttermilk come from the milk fat globule membrane. When buttermilk was homogenized with cream and added, Mayes et al. (1994) found a little improvement. Nonetheless, they came to the conclusion that the impact wasn't big enough to warrant major financial investment. The use of buttermilk in cheesemaking improved the texture and body of reduced-fat Mozzarella cheese, according to Poduval and Mistry (1999), and it also boosted the formation of free oil during storage.

Adding up to 5% ultrafiltered (UF) buttermilk to cheese milk improved the reduced-fat Cheddar's moisture retention. Insufficient curd fusion occurred when supplementation amounts exceeded 5% (Mistry et al., 1996). Turcot et al. (2002) showed that the preservation of almost half of the phospholipids from ultra-filtered buttermilk in the cheese resulted in a rise in moisture content. Using surfactants like lecithin before adding rennet resulted in Cheddar cheese with reduced fat content, a more porous structure, increased moisture levels, and a lighter color. However, this process had a detrimental impact on the flavor (Drake et al., 1998; Drake & Swanson, 1995). Amphiphilic materials may limit whey loss from curd by covering hydrophobic areas on para casein strands in cheese, hence decreasing the primary force that causes matrix shrinkage. Sipahioglu et al. (1999) produced a low-fat Feta cheese using tapioca starch and lecithin that closely resembled the flavor of the full-fat version. Dabour et al. (2006) found no alterations in microstructure or enhancements in texture while using lecithin to produce reduced-fat cheese.

12.3.7 Use of Selected Starters, Adjunct Starters, and Starter Stimulants

Mistry and Kasperson (1998) assert that using initial cultures significantly impacts the development of structure and flavor in reduced-fat cheese. Due to variations in settings and conditions, cultures suitable for full-fat cheese production may not be optimal for producing high-quality low-fat cheese. Choosing the appropriate beginning is crucial (Banks, 2004). Broadbent et al. (2013) discovered that reducing the fat concentration in Cheddar cheese altered the bacterial composition, without affecting the predominant bacterial strains. Biological variances in the aging process of full-fat and low-fat cheeses need additional starters to enhance the flavor of low-fat cheese. Skeie et al. (2013) used *L. casei* (TINE36) or *L. plantarum* (TINE18) as additives to enhance the flavor of low-fat cheese. Di Cagno et al. (2013) used microparticulated WPC, *Streptococcus thermophilus* (ST446), *L. plantarum* (LP), and *Lactobacillus rhamnosus* (LRA) to produce low-fat Caciotta-type cheese with improved texture and increased amounts of beneficial volatile compounds. Agrawal and Hassan (2007) used bacteria producing EPS to create low-fat Cheddar cheese with reduced acidity and increased moisture content using UF milk with a low concentration factor. Yeast extract, rich in amino acids and peptides, is crucial for promoting the growth of lactobacilli due to its supportive properties. Shakeel-Ur-Rehman et al. (2003) discovered that incorporating yeast extract into reduced-fat Cheddar cheese curd during salting boosted the population of nonstarter lactic acid bacteria (NSLAB) in the first ripening phase. This increased the level of unbound amino acids without causing the cheese to become more acidic. Abdel-Rafee et al. (2002) reported a similar occurrence with Domiati cheese.

Unwanted lactic acid bacteria that pollute the milk and dairy surroundings cause a second group of non-starter bacteria to grow in the curd when commercial cheese is made. NSLAB was first found in small amounts in factory-made Cheddar cheese. The population of these bacteria grows after about 3 months of maturing and stays at that level during the ripening process. Haque et al. (1997) and Fenelon et al. (2000) say that NSLAB counts are smaller in low-fat cheese than in full-fat cheese. There are no big changes in the NSLAB population, but there are big changes in the species and strain patterns as both full-fat and low-fat Cheddar cheeses ripen. There are several types of NSLAB in Cheddar cheese. The main ones are mesophilic *lactobacilli,* but there are also *Pediococcus* and *Leuconostoc spp.* It is possible to see differences in the NSLAB population in Cheddar cheeses made in different countries and even between production runs in the same plant. This variation could affect the changes in quality between batches of cheese. Non-starter lactic acid bacteria (NSLAB) can be left out of cheddar cheese while it ages, but they make it taste better and lessen the harshness and bitterness of some starting cultures. Because of this, there is rising interest in using certain extra strains of NSLAB to make both full-fat and low-fat Cheddars taste better. The taste of low-fat Cheddar got better when the right kinds and amounts of extra lactobacilli were added. After 6 months, pediococci helped make the aged Cheddar taste sharp, while micrococci added strong, unpleasant brothy flavors.

In 2002, Fenelon and his colleagues did a study that compared six different starter/adjunct pairs used to make low-fat Cheddar cheese. Each starter system had a base of Lactococcus lactis ssp. cremoris. In some tests, there was also an extra culture of *Lactococcus lactis ssp. lactis*, which made the medium more acidic. Different mixtures of *Lactobacillus helveticus, Lactobacillus helveticus/casei,*

Leuconostoc cremoris/Lactococcus lactis var diacetylactis/Streptococcus thermophilus/Lactobacillus helveticus were used in the adjunct culture mixes. The choice of starter or auxiliary had no major effect on primary proteolysis, but it did change the range of free amino acid levels. In order, glutamic acid, leucine, phenylalanine, valine, lysine, serine, and proline were the amino acids that were found in the most amounts. When acid-producing starters (*Lactococcus lactis* ssp. *lactis/ Lactococcus lactis* ssp. *cremoris*) and *Lactobacillus helveticus* were used to make cheese, it had higher levels of free amino acids and higher concentrations of low molecular weight peptides. At 90 and 180 days of ripening, the cheese had better flavor scores. Using *Lactobacillus casei* (7a) with sure starters also made a clean acid, non-bitter, milky, buttery, low-fat Cheddar that was mild in taste but consumers liked it (Midje et al., 2000). In this case, the starting culture mix had *Lactococcus lactis* ssp. *cremoris* (SK11), *Lactococcus lactis* ssp. *lactis biovar diacetylactis* (JV1), and *Lactobacillus casei* (7A). These strains were chosen to produce different flavors.

Weimer and colleagues successfully introduced *Brevibacterium linens* into low-fat Cheddar cheese in 1997. This differed from its typical usage in Trappist-style cheeses. The enhanced methanethiol synthesis by these cultures improved the flavor of Cheddar cheese (Ferchichi et al., 1985; Dias & Weimer, 1998a, b). Weimer et al. (1997) stated that the taste adjunct culture was effective only when combined with the beginning culture. *Brevibacterium linens* BL2 and *lactobacilli* exhibited distinct flavor preferences based on the specific starting culture used. Tungjaroenchai et al. (2001) assessed four adjunct cultures differing in aminopeptidase activity to gauge their influence on the flavor and texture of reduced-fat Edam cheese (20% fat). Although *Lactococcus lactis ssp. diacetylactis* exhibited higher aminopeptidase activity than *Lactobacillus helveticus* (LH 212), *Lactobacillus reuteri*, and *Brevibacterium linens* (BL2), cheeses containing *Lactobacillus helveticus* showcased the highest levels of free amino acids. Notably, beneficial texture effects were observed with *Lactobacillus. helveticus* (LH 212) and *Lactobacillus reuteri*. Katsiari et al. (2002) significantly improved the flavor and texture of a low-fat (9%) ewes' milk Kefalograviera-type cheese by selecting a commercial culture capable of producing acetate, diacetyl, and acetoin from citrate fermentation. This choice of commercial cultures enhanced the body, texture, and flavor intensity of the low-fat, high-moisture Kefalograviera-type cheese compared to the standard starter used in full-fat cheese production. Ryhänen et al. (2001) created a low-fat cheese containing 20% fat with unique bioactive characteristics, different from traditional full-fat cheeses. The starting culture employed in this invention consisted of a 12-strain combination of *Lactococci, Leuconostoc* sp., Propionibacterium, lactobacilli, Lactobacillus acidophilus, and Bifidobacterium spp. The probiotic bacteria Lactobacillus acidophilus and Bifidobacterium spp. were viable at a concentration of 10^6 CFU/g for 7 months, achieving the necessary threshold for a probiotic effect.

12.3.8 Attenuated Lactobacilli

Attenuated cultures used in cheese-making increase the bacterial enzyme variety in the cheese curd without interfering with the starter's acidification process. These cultures lack the ability to produce lactic acid yet provide active enzymes throughout the cheese's maturity. Attenuation procedures include heat treatment, freezing, spray drying, freeze drying, lysozyme treatment, or solvent treatment, as documented by Klein and Lortal (1999) and Johnson et al. (1998).

In the context of cheese-making, *Lactobacillus helveticus* attenuated cultures were employed by Ardö et al. (1997) to enhance flavor development and mitigate bitterness in a rindless, low-fat cheese. Skeie et al. (1995) improved the flavor of low-fat Gouda by incorporating heat-attenuated lactobacilli, also observing a salt-dependent debittering effect. Additionally, Johnson et al. (1998) compared the impact of different attenuation methods (spray drying, freeze drying, freezing) on reduced-fat Cheddar cheese, finding that each enhanced flavor development, albeit resulting in a sweet, nutty flavor not typical of Cheddar. However, cheeses made with freeze-dried or frozen attenuated cultures exhibited pronounced and unpleasant off-flavors.

Kebary et al. (1996) demonstrated the superiority of using freeze-shocked or heat-shocked lactobacilli in low-fat Ras cheese, resulting in a higher-quality cheese. Despite generally improving flavor, the use of attenuated cultures yielded mixed results. Notably, these cultures remain limited in availability for commercial sale.

12.4 Technologies to Improve Lower Fat Cheese

Various methods have been utilized to produce lower-fat cheese, aiming to overcome challenges related to yield, taste, and functionality. These approaches involve alterations in cheesemaking processes, the use of fat replacers and additives, specialized cultures, or a combination of these strategies.

A patented technique involved removing fat through centrifugation from fully-fat shredded mature Cheddar cheese, followed by pressing to reconstruct the cheese into a 50% reduced-fat version (Barbano, 2004; Nelson & Barbano, 2004). Amelia et al. (2013) developed a formula for low-fat Cheddar cheese with 6% fat, 28% protein, and 1.2% salt by blending reduced-fat Cheddar cheese, micellar casein concentrates (MCC), salt, and water. The MCC was derived from skim milk via ultrafiltration and microfiltration to yield a 10% protein concentrate.

Amelia et al. (2013) found that low-fat Cheddar cheese with a protein content below 28% had a mushy structure. Deegan et al. (2014) found that utilizing low-pressure homogenization at 10 MPa might enhance the quality of reduced-fat Finnish Emmental cheese. Nevertheless, this may not be the optimal technique for melting cheese. In 2013, Karaman and Akalın utilized a single-stage cream homogenization process at 20 MPa to create uniform skim milk with 0.75% fat for the production of low-fat Turkish white cheese. Skeie et al. (2013) used a two-step process to improve the overall quality of low-fat cheese. Small whey proteins and buttermilk were added to the cheese milk to enhance its creaminess. Particular strains of Lactobacillus bacteria were used to enhance the flavour of Norvegia, a Dutch-style cheese containing 10% fat. The findings showed that the extra Lactobacillus ssp. helped alter the texture of the cheese by decreasing hardness, possibly due to enhanced proteolysis. Ozturk et al. (2013) used high hydrostatic pressure (HHP) to alter the texture of low-fat cheese. The changes in texture and rheology caused by HHP differed based on the cheese's age. At first, using HHP caused cheese to melt more slowly and have a softer texture. After 1 month, there were slight variations in texture or rheology between the cheeses subjected to HHP and those that were not.

12.5 Effect of Structure on Functionality of Rennet-Curd Cheese

The functional properties of cheese refer to its ability to perform specific functions, such as deforming during chewing for easy swallowing or melting when heated for various applications. The para-casein network within cheese provides resistance to deformation, influencing texture and hardness. Milk fat, predominantly solid at low temperatures, contributes to elasticity and rigidity. As temperature increases, solid fat decreases, impacting cheese rigidity. Fat acts as a lubricant, reducing fracture force and hardness in deforming cheese. Adding butter to low-fat Mozzarella reduces hardness, while free fat released during heating facilitates the transition from a solid to a fluid state. High-pressure homogenization stabilizes fat against coalescence, reducing flowability and fluidity in heated Cheddar cheese (Guinee, 2016; Lopez et al., 2006; Wadhwani et al., 2011; Guinee et al., 2000b).

12.5.1 Effect of Fat Reduction on the Structure-Function–Function Relationships of Rennet-Curd Cheeses

Cheese primarily comprises fat, protein, and moisture, accounting for more than 95% of its total weight. Decreasing fat intake or adjusting the protein-to-fat ratio increases protein and moisture levels. However, this results in a reduction in the moisture-to-protein ratio, as shown by a decrease in moisture-in-nonfat substances (MNFS). Reducing fat content increases the volume percentage of the para-casein network, leading to a more cohesive microstructure. Cheese becomes harder, less deformable, and more elastic as a result of higher protein content and lower MNFS. Heated cheese becomes less flowable and extensible as its fat content decreases, becoming more transparent and less opaque (Fenelon & Guinee, 2000; Guinee et al., 2000a, b; Henneberry et al., 2015).

12.5.2 Structural Attenuation to Counteract the Effects of Fat Reduction

Various techniques are employed to enhance the texture and functionality of reduced-fat cheeses. These include:

1. In situ denaturation of whey protein through high heat treatment (HHT) of cheese milk.
2. Addition of microparticulated whey proteins to mimic fat globules and modify the para-casein network.
3. Reducing calcium-induced cross-linking of para-casein molecules in the network.
4. "Diluting" the casein network by reducing calcium levels or increasing hydration.

Reducing calcium levels notably affects cheese texture by lowering hardness, increasing flowability, and reducing chewiness. High heat treatment of milk leads to whey protein denaturation, affecting cheese gelation and firmness. Microparticulated whey proteins act as fillers, enhancing moisture retention and altering cheese texture. Diluting the casein network by reducing fat content increases moisture but decreases the moisture-to-protein ratio, affecting cheese hardness and chewiness. These methods aim to alter casein aggregation and minimize syneresis in reduced-fat cheeses (Chevanan et al., 2006; Henneberry et al., 2015).

12.6 Proteolysis in Low-Fat Cheddar Cheese

According to Banks et al. (1989), the presence of fat in cheese does not directly impact proteolysis. However, removing fat from milk results in a more rigid casein network, which indirectly affects the ripening process of low-fat cheese. Drake et al. (2010) found similar compounds in both low-fat and full-fat Cheddar cheeses during ripening but observed significant differences in key odorants. This suggests that biochemical changes and alterations in the cheese matrix influence the release of these important odorants.

Fenelon and Guinee (2000) found that decreasing the fat content of cheese results in notable increases in pH, moisture, and protein levels, along with a decrease in non-protein nitrogen fractions (MNFS). The alterations significantly impact the first breakdown of proteins. Constraints on flavour and texture development in low-fat cheese are due to many causes, such as the hindrance of proteolysis. The issues include insufficient breakdown of casein, autolysis of starting bacteria that does not match the need, leading to an amino acid deficit, and uncontrolled growth of non-starter lactic acid bacteria (NSLAB) (Ardö, 1997).

Cheese texture development depends on the degradation of αs1-casein (Mistry, 2001). The hardness of low-fat cheese is attributed to the incomplete breakdown of casein, namely αs1-casein, which happens more slowly than in full-fat cheese

(Fenelon & Guinee, 2000). The slow breakdown of αs1-casein as fat content decreases corresponds to a reduction in the ratio of remaining chymosin activity to protein in the cheese.

12.7 Types of Fat Replacers of Cheese

Fat replacers can be categorized into two types: fat substitutes and fat mimetics. Fat substitutes are molecules that emulate the physical and functional traits of traditional fat molecules, like triglycerides.

These substitutes can directly take the place of conventional fat molecules in foods, often as synthetic compounds that offer no energy calories or as structured lipid molecules providing reduced energy calories. They effectively maintain the taste and texture of foods, remaining heat stable and suitable for high-temperature cooking and frying. However, they might not replicate the taste properties of fat since fat itself contributes to the flavor of foods and acts as a carrier for other fat-soluble flavor compounds. On the other hand, fat mimetics imitates some organoleptic and physical characteristics of conventional fats but cannot replace fats on a weight-to-weight basis. Typically derived from proteins or carbohydrates and sometimes modified, these compounds contribute energy ranging from 0 to 4 kcal/g. Fat mimetics are generally unsuitable for high-temperature applications like frying due to susceptibility to denaturation or caramelization. Mimetics are often water-soluble and polar chemicals, unlike fats, and hence cannot perform some non-polar activities of fats, such as transporting lipid-soluble flavors. Their polar nature helps bind water, creating a creamy and lubricating sensation in meals akin to full-fat goods.

12.7.1 Fat Substitutes

Various fat substitutes have been created and are accessible in the market, while others are still in the developmental stage and not yet available commercially.

12.7.1.1 Sucrose Fatty Acid Polyesters

Olestra, produced by Procter and Gamble, is a sucrose polyester fat substitute composed of various esters of fatty acids and sucrose. Its physicochemical properties and applications in food depend on the fatty acids attached to sucrose. Synthesized from fatty acid methyl esters, it undergoes transesterification or interesterification under specific conditions. Olestra, approved by the FDA in 1996, can replace up to 100% of fat in savory snacks and is suitable for frying. While it mimics conventional fats in many aspects, it lacks caloric value as it passes through the digestive tract without being absorbed. Despite potential gastrointestinal side-effects, the

FDA rescinded the requirement for Olestra-containing foods to carry labels indicating such effects in 2003. Though it might impact carotenoid absorption, it doesn't affect other nutrients' absorption. FDA assessments deem Olestra safe, but consumer groups in the USA oppose its use. However, some argue that Olestra could aid obese individuals, reduce heart disease risk, and benefit colon cancer patients by substituting dietary fats.

12.7.1.2 Structured Lipids

Reduced-calorie fat substitutes like Caprenin and Salatrim blend short-chain and long-chain fatty acids esterified to glycerol. Caprenin, composed of caprylic, capric, and behenic fatty acids, provides only 5 kcal/g due to its unique fatty acid composition and poor absorption of one of its components. Salatrim, developed by Nabisco Foods Group, combines short-chain and long-chain fatty acids, resulting in lower caloric content (5 kcal/g) than regular fats. These substitutes exhibit properties akin to conventional fats like cocoa butter, finding applications in various food products such as confectionery, dairy items, fillings, and savoury dressings, owing to their unique functional and physical characteristics.

12.7.1.3 Other Fat Substitutes

Several patented fat substitutes, such as dialkyl dihexadecyl malonate (DDM) by Frito Lay, Inc., and esterified propoxylated glycerols (EPGs) by ARCO Chemical Company, have not yet received regulatory approval for commercial use. DDM, an ester compound, acts as a non-caloric fat substitute as it remains undigested and unabsorbed. On the other hand, EPGs are produced through the esterification of fatty acids to a polyether polyol derived from glycerol and propylene oxide. These substitutes, resistant to lipase, offer various functional properties and find potential applications in diverse food products like frozen desserts, salad dressings, baked goods, and spreads, and are suitable for cooking and frying.

12.7.2 Fat Mimetics

Fat mimetics imitate the sensory and textural qualities of conventional fat molecules present in food. They may be categorised into protein-based fat mimetics and carbohydrate-based fat mimetics.

12.7.2.1 Protein-Based Fat Mimetics

Protein-based fat mimetics are a suitable substitute for fats in dairy products, salad dressings, frozen desserts, and spreads. They are digested like proteins with a caloric value of 4 kcal/g but possess antigenic properties. Not ideal for frying due to protein denaturation, they work well in baking and similar cooking methods. These mimetics, derived from proteins like egg, milk, whey, soy, or wheat, are created through microparticulation, forming small gel particles that mimic fat's creamy texture. Simplesse® is one such mimetic made from whey protein concentrate and has been used in various food products since the early 1990s. It enhances creaminess, viscosity, and moisture retention while inhibiting water separation. Similar products include Dairy Lo from Pfizer and K-Blazer from Kraft Foods, both contributing to reduced-fat ice creams and cheeses.

12.7.2.2 Carbohydrate-Based Food Mimetics

Carbohydrates may effectively replace fats in meals, either partly or completely. Digestible carbohydrates supply 4 kcal/g, but non-digestible complex carbohydrates and hydrocolloid gums have negligible caloric content. Carbohydrate-based fat mimetics examples include gums, maltodextrins, polydextrose, cellulose derivatives, starch derivatives, and oat flour derivatives. The mimetics form a gel-like structure that increases thickness and texture similar to full-fat goods by stabilising water. They are not appropriate for frying, raise water activity in meals (which may affect shelf life by promoting microbial development), and affect flavour in varying ways depending on their type—gum-based mimetics usually do not change flavour, but starch- and cellulose-based ones often decrease flavour intensity.

12.7.2.3 Gums

Gums, high molecular weight, negatively charged carbohydrate hydrocolloids, function as stabilizers and gelling agents in food systems at low concentrations (0.1–0.5%). Varieties like xanthan, guar, locust bean, gum Arabic, carrageenan, and pectins are effective fat replacers alone or in combinations. They find diverse applications in dairy, dressings, soups, sauces, ice cream, icings, glazes, and baked goods. Selection considerations for these gums as fat mimetics encompass their gel's rheological properties, susceptibility to temperature and shearing forces, and compatibility with other ingredients.

12.7.2.4 Maltodextrins and Dextrins

These are unsweetened starch hydrolysates obtained from maize, potato, oat, rice, wheat, or tapioca starch with a dextrose equivalent (DE) below 20, indicating little sugar content. They improve the thickness, texture, and fullness in many food items.

A 20–35% aqueous solution of these compounds may replace oil, with 1 g of a 25% solution substituting 1 g of oil and lowering calorie content by 8 kcal/g. They act as fat substitutes in products such as table spreads, sour cream, frozen desserts, frostings, and salad dressings.

12.7.2.5 Polydextrose

Polydextrose, a glucose, sorbitol, citric, or phosphoric acid polymer, is a minimally metabolizable substance offering just 1 kcal/g. Primarily employed as a bulking agent in foods, it also aids in fat reduction. However, higher consumption levels may prompt a laxative effect. Its versatility extends to various food products like baked goods, frozen dairy treats, salad dressings, puddings, and frostings.

12.7.2.6 Cellulose

Various types of cellulose and cellulose compounds may mimic fat and are often combined with hydrocolloids. Microcrystalline cellulose (MCC) is a calorie-free fat substitute that mimics the properties of fat in water, such as body, flavour, and viscosity. Approximately 60–70% of the microcrystals are less than 0.2 mm long, making them water insoluble. When combined with additional substances such as gums, these microcrystals bind together to form a network that provides the product with a smooth texture and a rich consistency similar to full-fat goods. Powdered cellulose, methylcellulose, and hydroxypropyl methylcellulose are alternative forms of cellulose used as fat alternatives in baked products, frozen delights, sauces, and salad dressings.

12.7.2.7 Starch

Modified starches from potato, maize, oat, rice, wheat, and tapioca are efficient fat substitutes, improving creaminess and moisture retention in food products. These modified starches, produced by chemical and/or physical processes, provide economical choices readily accessible in the market. However, its durability when exposed to freeze/thaw cycles, heat, acid, and shear may not be enough for specific uses. Furthermore, they may somewhat reduce the authentic taste of the included dishes. They are fat substitutes in high-moisture dishes such as salad dressings, sauces, baked products, frostings, and sausages.

12.7.2.8 Oatrim

Oatrim is a fat mimetic derived from oat flour through partial enzymatic hydrolysis, offering a mouthfeel akin to triglycerides. Comprising 5% β-glucan, soluble fiber, and amylodextrin, it exists as a dry powder (4 kcal/g) or a gel hydrated with three parts water (1 kcal/g). Resilient to high temperature-short time (HTST) processing and retorting but unsuitable for frying, it has been utilized to create fat-free, cholesterol-free milk with a taste resembling 2% fat milk. Its versatility extends to inclusion in confectionery, frozen desserts, baked goods, and meat products.

12.8 Flavor Improvement in Low-Fat Cheese

Reduced-fat cheese lacks typical Cheddar flavors, showing slow progression and a shift towards meaty-brothy or burnt-brothy notes due to elevated levels of specific compounds during ripening. Studies by Fenelon et al. (2000) and Drake et al. (2010) identified compounds like furanone, 1-octen-3-one, and phenoxyethanol, attributing them to the altered flavors in low-fat Cheddar. These compounds create significant differences in odorant concentrations between full-fat and low-fat cheeses, influenced by compound variations and matrix disparities affecting compound release. Achieving satisfactory and consistent flavor remains the primary challenge for low-fat cheeses. While some reduced-fat cheeses (up to 50% reduction) mirror full-fat counterparts in flavor, further fat reduction widens the flavor quality gap. Emulating the distinct flavors of aged Cheddar and Gouda in low-fat versions is intricate due to changes in the cheese environment resulting from altered chemical compositions in manufacturing processes.

Low-fat versions of certain cheeses, like mould-ripened or surface-ripened varieties, are easier to replicate as their flavours mainly stem from deliberately added microflora and enzymes. In contrast, achieving desired flavors in low-fat Cheddar involves complex interactions of microbial metabolism, enzymatic reactions, and altered chemical environments. Two hypotheses explain the limited flavor development in low-fat cheeses. One emphasizes the inadequate formation of desired flavor compounds due to the altered chemical environment, impacting microbial metabolism and enzymatic activity. The other highlights are inhibiting necessary enzymatic and microbiological activity and favouring undesirable flavour components. Factors like pH, fat, and serum impact how flavors are perceived in high-moisture, low-fat cheese, where undesirable flavors may be more detectable. Researchers are exploring methods involving specific additives, enzymes, and ripening conditions to improve flavor development in low-fat cheese, aiming to address the shortcomings in enzymatic activity. Success in achieving solid and aged flavors in low-fat cheeses primarily occurs with modest fat reductions. In contrast, higher reductions pose more significant challenges due to accelerated adverse reactions from high moisture, low salt, and low acid content. This often results in less flavourful, mild cheeses rarely aged beyond 6 months in commercial settings due to difficulties in balancing ageing without compromising texture or flavour quality.

12.9 Conclusion

The production of low-fat cheese involves a complex interplay of factors, including milk standardization, acidification, curd washing, cooking temperature, salting, starter culture selection, and the use of fat replacers. By carefully manipulating these variables, cheesemakers can produce low-fat cheese with desirable texture, flavour, and functionality. While significant advancements have been made, ongoing research and innovation are essential to further optimize the production process and develop low-fat cheese that can rival the taste and texture of full-fat varieties.

References

Abdel-Rafee, S., Okasha, A. I., & Mehana, M. Y. (2002). Fresh ripened soft cheese. *Egyptian Journal of Food Science, 30*(2), 331–338.

Agrawal, P., & Hassan, A. N. (2007). Ultrafiltered milk reduces bitterness in reduced-fat Cheddar cheese made with an exopolysaccharide-producing culture. *Journal of Dairy Science, 90*(7), 3110–3117.

Amelia, I., Drake, M., Nelson, B., & Barbano, D. M. (2013). A new method for the production of low-fat Cheddar cheese. *Journal of Dairy Science, 96*(8), 4870–4884.

Anderson, D. L., Mistry, V. V., Brandsma, R. L., & Baldwin, K. A. (1993). Reduced fat Cheddar cheese from condensed milk. 1. Manufacture, composition, and ripening. *Journal of Dairy Science, 76*(10), 2832–2844.

Ardö, Y. (1997). Flavour and texture in low-fat cheese. In *Microbiology and biochemistry of cheese and fermented milk* (pp. 207–218). Springer.

Banks, J. M. (2004). The technology of low-fat cheese manufacture. *International Journal of Dairy Technology, 57*(4), 199–207.

Banks, J. M., Brechany, E. Y., & Christie, W. W. (1989). The production of low fat Cheddar-type cheese. *International Journal of Dairy Technology, 42*(1), 6–9.

Banks, J. M., Hunter, E. A., & Muir, D. D. (1993). Sensory properties of low fat Cheddar cheese: Effect of salt content and adjunct culture. *International Journal of Dairy Technology, 46*(4), 119–123.

Barbano, D. (2004). U.S. Patent No. 6,808,735. U.S. Patent and Trademark Office.

Broadbent, J. R., Brighton, C., McMahon, D. J., Farkye, N. Y., Johnson, M. E., & Steele, J. L. (2013). Microbiology of Cheddar cheese made with different fat contents using a Lactococcus lactis single-strain starter. *Journal of Dairy Science, 96*(7), 4212–4222.

Buch Kristensen, J. M. (1999). *Cheese technology—A north European approach* (p. 109). International Dairy Books/Danish Dairy Board.

Chevanan, N., Muthukumarappan, K., Upreti, P., & Metzger, L. E. (2006). Effect of calcium and phosphorus, residual lactose and salt-to-moisture ratio on textural properties of cheddar cheese during ripening. *Journal of Texture Studies, 37*(6), 711–730.

Dabour, N., Kheadr, E., Benhamou, N., Fliss, I., & LaPointe, G. (2006). Improvement of texture and structure of reduced-fat Cheddar cheese by exopolysaccharide-producing lactococci. *Journal of Dairy Science, 89*(1), 95–110.

Dalgleish, D. G., & Law, A. J. (1988). pH-induced dissociation of bovine casein micelles. I. Analysis of liberated caseins. *Journal of Dairy Research, 55*(4), 529–538.

Deegan, K. C., Holopainen, U., McSweeney, P. L., Alatossava, T., & Tuorila, H. (2014). Characterisation of the sensory properties and market positioning of novel reduced-fat cheese. *Innovative Food Science & Emerging Technologies, 21*, 169–178.

Dias, B., & Weimer, B. (1998a). Conversion of methionine to thiols by lactococci, lactobacilli, and brevibacteria. *Applied and Environmental Microbiology, 64*(9), 3320–3326.

Dias, B., & Weimer, B. (1998b). Purification and characterization of methionine gamma-lyase from *Brevibacterium linens* BL2. *Applied and Environmental Microbiology, 64*(9), 3327–3331.

Drake, M. A., Boylston, T. D., & Swanson, B. G. (1996). Fat mimetics in low-fat Cheddar cheese. *Journal of Food Science, 61*(6), 1267–1271.

Drake, M. A., Chen, X. Q., Gerard, P. D., & Gurkin, S. U. (1998). Composition and quality attributes of reduced-fat cheese as affected by lecithin type. *Journal of Food Science, 63*(6), 1018–1023.

Drake, M. A., Miracle, R. E., & McMahon, D. J. (2010). Impact of fat reduction on flavor and flavor chemistry of Cheddar cheeses. *Journal of Dairy Science, 93*(11), 5069–5081.

Farkye, N. Y., Prasad, B. B., Rossi, R., & Noyes, O. R. (1995). Sensory and textural properties of Queso Blanco-type cheese influenced by acid type. *Journal of Dairy Science, 78*(8), 1649–1656.

Fenelon, M. A., & Guinee, T. P. (2000). Primary proteolysis and textural changes during ripening in Cheddar cheeses manufactured to different fat contents. *International Dairy Journal, 10*(3), 151–158.

Fenelon, M. A., Guinee, T. P., & Reville, W. J. (1999). Characteristics of reduced-fat Cheddar prepared from the blending of full-fat and skim cheese curds at whey drainage. *Milchwissenschaft, 54*(9), 506–510.

Fenelon, M. A., O'Connor, P., & Guinee, T. P. (2000). The effect of fat content on the microbiology and proteolysis in Cheddar cheese during ripening. *Journal of Dairy Science, 83*(10), 2173–2183.

Ferchichi, M., Hemme, D., Nardi, M., & Pamboukdjian, N. (1985). Production of methanethiol from methionine by *Brevibacterium linens* CNRZ 918. *Microbiology, 131*(4), 715–723.

Ganesan, B., & Weimer, B. C. (2004). Role of aminotransferase IlvE in production of branched-chain fatty acids by *Lactococcus lactis subsp. lactis*. *Applied and Environmental Microbiology, 70*(1), 638–641.

Guinee, T. P. (2016). Protein in cheese and cheese products: Structure-function relationships. In *Advanced dairy chemistry: Volume 1B: Proteins: Applied aspects* (pp. 347–415). Springer.

Guinee, T. P., Auty, M. A., & Fenelon, M. A. (2000a). The effect of fat content on the rheology, microstructure and heat-induced functional characteristics of Cheddar cheese. *International Dairy Journal, 10*(4), 277–288.

Guinee, T. P., Auty, M. A. E., Mullins, C., Corcoran, M. O., & Mulholland, E. O. (2000b). Preliminary observations on effects of fat content and degree of fat emulsification on the structure-functional relationship of Cheddar-type cheese. *Journal of Texture Studies, 31*(6), 645–663.

Haque, Z. U., Kucukoner, E., & Aryana, K. J. (1997). Aging-induced changes in populations of lactococci, lactobacilli, and aerobic microorganisms in low-fat and full-fat Cheddar cheese. *Journal of Food Protection, 60*(9), 1095–1098.

Henneberry, S., Wilkinson, M. G., Kilcawley, K. N., Kelly, P. M., & Guinee, T. P. (2015). Interactive effects of salt and fat reduction on composition, rheology and functional properties of mozzarella-style cheese. *Dairy Science & Technology, 95*, 613–638.

Johnson, M. E., & Chen, C. (1991, March). *Making quality reduced-fat cheese*. In Low-fat cheddar seminar.

Johnson, M. E., & Chen, C. M. (1995). Technology of manufacturing reduced-fat Cheddar cheese. In *Chemistry of structure-function relationships in cheese* (pp. 331–337). Springer US.

Johnson, M. E., Steele, J. L., Broadbent, J., & Weimer, B. C. (1998). Manufacture of Gouda and flavour development in reduced-fat Cheddar cheese. *Australian Journal of Dairy Technology, 53*(2), 67.

Katsiari, M. C., Voutsinas, L. P., & Kondyli, E. (2002). Improvement of sensory quality of low-fat Kefalograviera-type cheese by using commercial special starter cultures. *Journal of Dairy Science, 85*(11), 2759–2767.

Kebary, K. M. K., Khader, A. E., Zedan, A. N., & Mahmoud, S. F. (1996). Accelerated ripening of low fat Ras cheese by attenuated lactobacilli cells. *Food Research International, 29*(8), 705–713.

Kelly, M., Fox, P. F., & McSweeney, P. L. H. (1996). Effect of salt-in-moisture on proteolysis in Cheddar-type cheese. *Milchwissenschaft, 51*(9), 498–501.

Klein, N., & Lortal, S. (1999). Attenuated starters: An efficient means to influence cheese ripening—A review. *International Dairy Journal, 9*(11), 751–762.

Kosikowski, F. V., & Mistry, V. V. (1997). *Cheese and fermented milk foods* (3rd ed.). F. V. Kosikowski, LLC.

Laloy, E., Vuillemard, J. C., El Soda, M., & Simard, R. E. (1996). Influence of the fat content of Cheddar cheese on retention and localization of starters. *International Dairy Journal, 6*(7), 729–740.

Lawrence, R. C., Creamer, L. K., & Gilles, J. (1987). Texture development during cheese ripening. *Journal of Dairy Science, 70*(8), 1748–1760.

Lopez, C., Briard-Bion, V., Camier, B., & Gassi, J. Y. (2006). Milk fat thermal properties and solid fat content in emmental cheese: A differential scanning calorimetry study. *Journal of Dairy Science, 89*(8), 2894–2910.

Lucey, J. A., & Fox, P. F. (1993). Importance of calcium and phosphate in cheese manufacture: A review. *Journal of Dairy Science, 76*(6), 1714–1724.

Madsen, F. M., Reinbold, G. W., & Clark, W. S., Jr. (1966). Comparison of fresh buttermilk, reconstituted buttermilk and reconstituted skimmilk for the manufacture of low-fat cheese. *Manufactured Milk Products Journal, 57*(10), 18–22.

Mayes, J. J., Urbach, G., & Sutherland, B. J. (1994). Does addition of buttermilk affect the organoleptic properties of low-fat Cheddar cheese? *Australian Journal of Dairy Technology, 49*(1), 39.

McGregor, J. U., & White, C. H. (1990). Optimizing ultrafiltration parameters for the development of a lowfat cheddar cheese. *Journal of Dairy Science, 73*(2), 314–318.

Metzger, L. E., Barbano, D. M., & Kindstedt, P. S. (2001). Effect of milk preacidification on low fat Mozzarella cheese: III. Post-melt chewiness and whiteness. *Journal of Dairy Science, 84*(6), 1357–1366.

Midje, D. L., Bastian, E. D., Morris, H. A., Martin, F. B., Bridgeman, T., & Vickers, Z. M. (2000). Flavor enhancement of reduced fat cheddar cheese using an integrated culturing system. *Journal of Agricultural and Food Chemistry, 48*(5), 1630–1636.

Mistry, V. V. (2001). Low fat cheese technology. *International Dairy Journal, 11*(4–7), 413–422.

Mistry, V. V., & Kasperson, K. M. (1998). Influence of salt on the quality of reduced fat Cheddar cheese. *Journal of Dairy Science, 81*(5), 1214–1221.

Nelson, B. K., & Barbano, D. M. (2004). Reduced-fat Cheddar cheese manufactured using a novel fat removal process. *Journal of Dairy Science, 87*(4), 841–853.

Ozturk, M., Govindasamy-Lucey, S., Jaeggi, J. J., Johnson, M. E., & Lucey, J. A. (2013). The influence of high hydrostatic pressure on regular, reduced, low and no salt added Cheddar cheese. *International Dairy Journal, 33*(2), 175–183.

Poveda, M., Arnold, M., & Farkye, N. (2013). Effects of chelating agents on texture of low-fat Cheddar cheese. *Journal of Dairy Science, 96*(E-Suppl. 1), 304.

Rodriguez, J. (1998). Recent advances in the development of low-fat cheeses. *Trends in Food Science & Technology, 9*(6), 249–254.

Ryhänen, E. L., Pihlanto-Leppälä, A., & Pahkala, E. (2001). A new type of ripened, low-fat cheese with bioactive properties. *International Dairy Journal, 11*(4–7), 441–447.

Sipahioglu, O., Alvarez, V. B., & Solano-Lopez, C. (1999). Structure, physico-chemical and sensory properties of feta cheese made with tapioca starch and lecithin as fat mimetics. *International Dairy Journal, 9*(11), 783–789.

Skeie, S., Alseth, G. M., Østlie, H., Abrahamsen, R. K., Johansen, A. G., & Øyaas, J. (2013). Improvement of the quality of low-fat cheese using a two-step strategy. *International Dairy Journal, 33*(2), 153–162.

St-Gelais, D., Roy, D., & Audet, P. (1998). Manufacture and composition of low fat Cheddar cheese from milk enriched with different protein concentrate powders. *Food Research International, 31*(2), 137–145.

Tungjaroenchai, W., Drake, M. A., & White, C. H. (2001). Influence of adjunct cultures on ripening of reduced fat Edam cheeses. *Journal of Dairy Science, 84*(10), 2117–2124.

Turcot, S., St-Gelais, D., & Turgeon, S. L. (2002). Affinage de fromages allégés de type Cheddar fabriqués à partir de laits enrichis en phospholipides. *Le Lait, 82*(2), 209–223.

Vanderghem, C., Bodson, P., Danthine, S., Paquot, M., Deroanne, C., & Blecker, C. (2010). *Milk fat globule membrane and buttermilks: From composition to valorization*. Base.

Wadhwani, R., McManus, W. R., & McMahon, D. J. (2011). Improvement in melting and baking properties of low-fat Mozzarella cheese. *Journal of Dairy Science, 94*(4), 1713–1723.

Weimer, B., Dias, B., Ummadi, M., Broadbent, J., Brennand, C., Jaegi, J., Johnson, M., Milani, J., Steele, J., & Sisson, D. V. (1997). Influence of NaCl and pH on intracellular enzymes that influence Cheddar cheese ripening. *Le Lait, 77*(3), 383–398.

Poduval, V. S., & Mistry, V. V. (1999). Manufacture of reduced fat Mozzarella cheese using ultrafiltered sweet buttermilk and homogenized cream. Journal of Dairy Science, 82(1), 1-9.

Mistry, V. V. (2001). Low fat cheese technology. International dairy journal, 11(4-7), 413-422.

Di Cagno, R., Coda, R., De Angelis, M., & Gobbetti, M. (2013). Exploitation of vegetables and fruits through lactic acid fermentation. Food microbiology, 33(1), 1-10.

Shakeel-ur-Rehman, S. U. R., Farkye, N. Y., Considine, T., Schaffner, A., & Drake, M. A. (2003). Effects of standardization of whole milk with dry milk protein concentrate on the yield and ripening of reduced-fat cheddar cheese.

Skeie, S., Narvhus, J., Ardö, Y., & Abrahamsen, R. K. (1995). Influence of liposome-encapsulated Neutrase and heat-treated lactobacilli on the quality of low-fat Gouda-type cheese. Journal of dairy research, 62(1), 131-139.

Chapter 13
Low Sodium Cheese

Rajat Singh and Bazilla Gayas

13.1 Introduction

Cheese has been a part of human diets for thousands of years. Cheese, originally a concentrated milk product, had a lengthy shelf life in ancient times. Cheese was an ideal food for our labouring ancestors because of its high protein and fat content; it provided them with energy and nutrients. Over the past 20 years, study into lowering the amount of sodium in cheese has grown and now includes a wide range of fields. The goal is to lower the amount of sodium in cheese without affecting its quality or safety. Table salt is the main source of sodium in cheese, along with the natural amount of salt in cheese (Bansal & Mishra 2020). Salt contains sodium, which is the predominant constituent in extracellular fluids and facilitates the transportation of nutrients. Intake of this substance is crucial, since it plays a vital role in regulating blood pressure, transporting water inside cells, maintaining osmotic pressure, and transmitting nerve impulses. Cheese is a significant contributor to dietary salt intake (DSI) among various dairy products. In the United States, the DSI of cheese ranges from 3.8% to 11% (CDC, 2012), while in Germany it is 10% (Bisig, 2017), in Switzerland it is 7.5% (Bisig, 2017), in Australia it is 4–7% (Charlton et al., 2010; Grimes et al., 2011), in Canada it is 3.2% to 5.4% (Fischer et al., 2009), in the United Kingdom it is 4% (Hashem et al., 2014), and in New Zealand it is 2.8% (Ministry of Health, 2003). Cheese is a constituent of multi-component dishes like pizzas, pastas, and quesadillas, where reducing the overall sodium can be done by gradually decreasing the salt concentration of each individual component. Therefore, there is a desire to produce low- and reduced-sodium cheeses in the

R. Singh (✉)
Department of Food Engineering, National Institute of Food Technology Entrepreneurship and Management (NIFTEM-K), Sonepat, Haryana, India

B. Gayas
Department of Food Technology, Chandigarh university, Chandigarh, Punjab, India

S. Mansha Rafiq (ed.), *Technological Advances and Trends in Cheese Making*,
https://doi.org/10.1007/978-3-031-94476-5_13

United States. Nevertheless, this poses difficulties for the conventional cheese production process.The use of strategies to decrease the amount of NaCl in cheese may be dated back to the 1980s, when the initial strategy was reducing the rate at which sodium chloride was added. Since 1980, several technologies have been created, but the presence of low-sodium cheeses (LSCs) in stores is still quite limited due to negative effects on consumer acceptance and the cheese's quality. Despite over 40 years of global research, scientists are still striving to find methods to decrease the salt content in cheese while maintaining its normal quality attributes. One way to decrease the amount of sodium in cheese is by reducing the NaCl concentration. However, this approach may pose challenges in terms of safety and microbiological quality if the rate of NaCl addition is decreased (Johnson et al., 2009).

In addition to safety concerns, reduced salt cheeses are also associated with increased moisture content, water activity, glycolysis, and degree of proteolysis, as well as decreased firmness (Rulikowska et al., 2013; McMahon et al., 2014; Murtaza et al., 2014). Hence, the primary strategy for decreasing the salt content in cheese primarily revolves around substituting sodium salt with potassium (K), calcium (Ca), or magnesium (Mg) salts. However, substituting NaCl, either partially or completely, in cheese presents a substantial difficulty due to the impact of K, Ca, and Mg ions on different biochemical alterations in the cheese structure as shown in Table 13.1.

13.2 Sodium Chloride and Public Health

According to Kaplan (2000), it is advised that individuals consume roughly 2.4 g of sodium (Na) or 6 g of sodium chloride (NaCl) day, which can be naturally found in food. Elevated consumption of dietary sodium significantly raises blood pressure, so dramatically augmenting the likelihood of stroke, heart disease, cardiac failure, and kidney disease (Doyle & Glass, 2010). As to the FDA, any food with less than 140 mg of salt per serving would be classified as low sodium. Decreasing the amount of salt in one's diet has a clear and gradual impact on reducing blood pressure in specific groups of people (He & MacGregor, 2010). Approximately 75% of the salt in our diet comes from processed foods that have been enhanced with sodium chloride, while just 5% of the sodium is naturally present (US Department of Agriculture, 2010). The US dietary guidelines establish the recommended minimum consumption level for salt at 500 mg per day and the maximum intake level at 2300 mg per day (US Department of Agriculture, 2010). Canada, the United Kingdom, and Australia, which have lower average sodium intakes, have established guidelines proposing adequate intake levels ranging from 460 to 1500 mg/day and upper intake limits ranging from 2200 to 2400 mg/day (NHMRC, 2006; Health Canada, 2010; Wyness et al., 2012; WHO, 2013). The projected salt intake in India exceeds the recommended level of 2000 mg sodium (5 g salt) per day given by the World Health Organisation. In order to meet these recommended amounts of dietary intake, there is a growing need for foods that have lower salt content. Sodium

Table 13.1 Outcome of various salt reduction strategies on organoleptic properties of LSCs

Sr. no.	Salt reduction strategies	Cheese type	References	Outcome
1	Decreasing the rate of NaCl addition	Cheddar	Rulikowska et al. (2013)	The reduction of NaCl resulted in a pronounced bitterness and a diminished level of flavour development
2		Cheddar		Gradual flavour drop with 50% NaCl reduction during ripening
3		Mozzarella	Arboatti et al. (2014) and Ganesan et al. (2014)	Cheese liking scores were considerably affected by salt reduction compared to control
	Salt substitution			
5		Halloumi cheese	Kamleh et al. (2012)	Acceptability was same, but sensory scores decreased with KCl substitution
6		Processed cheese	Patel et al. (2013)	Cheese prepared by partial substation of NaCl with KCl (1:1) had high bitterness, low saltiness, and a soft body
7	Salt substitution with bitter blocker or flavor enhancer	Cheddar	Grummer et al. (2013)	Some flavour enhancers improved cheese's sensory qualities, while others degraded them
8		Cheddar	Khetra et al. (2016)	Cheese with 75% NaCl replacement, 2 g/L HVP, and 300 mg/L AMP had comparable sensory qualities to the control
9	Process modification	Cheddar	Ozturk et al. (2015)	HHP treatment of LSCs with UF retentate did not significantly change bitterness compared to the control
10		Cheddar	Bansal and Mishra (2020)	Low-sodium cheese made by adding protein to cheese milk with UF retentate was better than cheese without it
11		Feta	Karimi et al. (2012)	UF feta cheese with 25% salt replacement produces the highest sensory acceptability

chloride (NaCl), commonly referred to as table salt, is a traditional food additive that is widely utilised to enhance the shelf life and taste of various food products. Consequently, food scientists are striving to reduce the sodium content in food by either reducing the rate at which sodium chloride (NaCl) is added or by partially or fully substituting NaCl with other appropriate salts such as potassium chloride (KCl), calcium chloride ($CaCl_2$), or magnesium chloride ($MgCl_2$) (Bansal and Mishra (2020). One such food commodity is cheese. Retail cheeses in the US have different salt levels, with average values ranging from 615 mg/100 g in Cheddar cheese to 1242 mg/100 g in process cheese (Agarwal et al., 2011). In the United

States, the average amount of cheese consumed per person each year is 15.3 kg (CDIC, 2013). This translates to a daily consumption of 42 g of cheese. The consumption of different types of cheese gives a salt content ranging from 250 to 500 mg, with the higher end of the spectrum meeting the recommended intake limits.

13.3 Cheese

Cheese is rich in proteins, bioactive peptides, amino acids, lipids, fatty acids, vitamins, minerals and includes a lot of saturated and trans-fatty acids, yet it has not been linked to any disease. Cheese's essential amino acids boost body growth. Conjugated linoleic acid and sphingolipids in cheese may fight cancer. Cheese is known for its calcium content, which promotes strong bones and teeth. It also lowers blood pressure and aids weight loss when paired with low-calorie diets. Natural cheese has 40–800 mg/100 g of salt, while processed cheese has a comparable range. Maximum concentration is 1500 mg/100 g (El-Bakry, 2012). Table 13.2 shows the average salt content of popular cheeses.

The addition of emulsifying salts based on sodium during the cooking process of processed cheeses results in an enhanced sodium content in these cheeses. Given this, decreasing the amount of sodium chloride in cheese could potentially enhance health, social, and public spending results. In response to this directive, numerous global health organisations are currently developing legislation to efficiently decrease the magnitude of NaCl in foods (Webster et al., 2014; Trieu et al.. 2015). In addition, the WHO has set a worldwide goal of reducing salt chloride intake by 30% by the year 2025.

Table 13.2 Sodium content of some cheese varieties

Sr. no.	Cheese varieties	References	Quantity (mg/100 g)
1	Parmesan cheese	Doyle and Glass (2010)	1602
2	Roquefort	Akan et al. (2017)	1600
3	Grated hard cheese	Agarwal et al. (2011)	1530
4	Feta	Akan et al. (2017)	1400
5	Coalho cheese	Matera et al. (2018)	570 to 1093
6	Brie, Beaufort, and Tomme cheese	Akan et al. (2017)	600 to 800
7	Cheddar, Colby, Jack and Mozzarella cheese	Agarwal et al. (2011)	668
8	Cream cheese	Agarwal et al. (2011)	408
9	Cottage cheese	Agarwal et al. (2011) and Cruz et al. (2011)	215 to 347
10	Swiss cheese	Doyle and Glass (2010)	192

13.4 Sodium Less Cheeses

Lowering the amount of sodium in cheeses is a difficult endeavour since salt plays multiple important roles that enhance the quality and safety of the product (Rulikowska et al., 2013). In contrast to other foods, merely adjusting the components is not enough to reduce the salt level in cheese. Doing so would result in an increase in moisture content, water activity (aw), the rate of lactose metabolism, and the intensity of proteolytic events. Equally significant is the occurrence of a reduced rate of bacterial cell autolysis, which has the potential to disrupt the normal ageing process of cheese (McCarthy et al., 2015). Consequently, cheeses that have less salt may exhibit sensory flaws, altered physical qualities, or be more prone to decomposition or structural issues over time. Due to the various types of cheese that can be made, each with distinct physical, chemical, and microbiological properties, reducing salt content may have specific negative effects on the product, as mentioned above. Simultaneously, many ways can be employed to decrease the sodium level in cheeses based on their specific typology. Several approaches to reduce sodium (Na) content in soft, semi-hard, hard, and mold-ripened cheeses.

To effectively choose low-sodium cheese, it is advisable to carefully examine the nutrition label on the cheese packaging and ensure that the sodium level is relatively low. The 2020–2025 Dietary Guidelines for Americans advise individuals to limit their daily sodium intake to 2300 mg. Some experts recommend an even lower consumption, around 1500 mg, particularly for individuals with high blood pressure, kidney disease, or other health conditions that can be negatively affected by excessive salt consumption. Certain types of cheese might have a high salt content, causing you to meet your recommended sodium intake rapidly. According to the USDA, a 1/4 cup of feta cheese has 428 mg of sodium, which is approximately 1/3 of the recommended 1500 mg limit for individuals with specific medical conditions. Incorporating cheese into one's diet is associated with a variety of potential health advantages. According to a 2019 study published in Advances in Nutrition, consuming dairy products like cheese can contribute to maintaining a healthy body weight. This is likely because dairy provides essential nutrients such as protein, calcium, and vitamin B_{12}, and its combination of carbohydrates, fats, and proteins is fulfilling. In addition, a 2022 Nutrients analysis reveals that dairy products, particularly various types of cheese, are rich in vitamin K_2, a nutrient synthesised by bacteria. According to a 2023 review in Food Science & Nutrition, consuming enough of this vitamin has been proven to reduce the development of dangerous calcium deposits in the heart and is crucial in the prevention of cardiovascular disease and Type 2 diabetes. However, there were several approaches to reduce sodium content in soft, semi-hard, hard, and mold-ripened cheeses.

13.5 Soft Cheeses

Cheeses are commonly recognised as a food group that is rich in sodium (Na), despite the fact that soft cheeses may include a varying range (0.7–2.3%) of sodium chloride (NaCl), which can have varied effects (Bansal & Mishra, 2020; Dugat-Bony et al., 2016). Salt is a significant hindrance in managing bacterial growth in cheese. Soft cheeses are generally characterised by their low fat and high moisture levels. Hence, decreasing the salt content in these cheeses could potentially have adverse effects on the security and longevity of the products by decreasing the ratio of salt to moisture (Bishop & Smukowski, 2006). To minimise the sodium content in soft cheeses, one can either decrease the rate at which NaCl is added or substitute it with other chloride salts as KCl, $CaCl_2$, or $MgCl_2$ (Loudiyi & Aït-Kaddour, 2019). It is important to thoroughly assess the reduction of salt or the substitution of salt with other ingredients in soft cheeses, since the specific physical and chemical properties of these cheeses might promote the growth of harmful bacteria and germs that cause spoiling. Choosing the right LAB starter culture is crucial for ensuring the safety of these cheeses. This is because the LAB starter culture helps produce organic acids, particularly lactic acid, as well as other substances derived from secondary bacterial metabolism, such as hydrogen peroxide, enzymes, or bacteriocins. These substances have a competitive effect against harmful microorganisms, thus protecting the cheese (Fox et al., 2004; García-Díez & Saraiva, 2021). In addition, substituting NaCl with KCl can have beneficial effects. For instance, when 50% of the brine solution was replaced with KCl at a reduced concentration of 7.5%, it improved the survival of probiotic bacteria that were added. In addition, this replacement resulted in a substantial augmentation of the liberation of four crucial amino acids: phenylalanine, tryptophan, valine, and leucine (Gandhi & Shah, 2016). A natural ingredient was evaluated to inhibit the formation of moulds that can lead to spoiling in low-sodium perishable soft cheeses. Prior to the use of rennet, *Nigella sativa* oil was introduced to pasteurised milk. Following a storage period of 14 days, the cheese exhibited antifungal effects against *Candida albicans* and *Aspergillus parasiticus*, (Abdel-Latif et al., 2021). The enzymatic activity in mozzarella is affected by the concentration of ions, such as Na+ and K+. Nevertheless, the proteolysis level in mozzarella with low sodium concentration remained relatively stable, possibly due to its limited shelf-life (Arboatti et al., 2014). Previous studies have also shown similar results when potassium chloride (KCl) was used instead of sodium chloride (NaCl) in various types of soft cheeses, including Halloumi, Feta, and white cheeses (Güven et al., 2008; Katsiari et al., 2001; Güven & Karaca, 2001). According to sensory analysis, a study comparing mozzarella and cottage cheese made with 30–35% less NaCl did not find any significant variations in overall acceptability. However, reducing the salt content at greater rates had a detrimental impact on sensory characteristics (Ganesan et al., 2014). A different method was used to decrease the sodium content in white cheese. This involved creating a brine solution with hydrocolloids at concentrations of 8% and 12%. The results were encouraging, particularly when carrageenan and gelatin were used instead of guar or xanthan gum. The latter two negatively affected the flavour, texture, and appearance of the cheese (Cankurt, 2019).

13.6 Semi-hard and Hard Cheeses

Hystead et al. (2013) put several *Listeria monocytogenes* strains into salted, low-Na cheddar cheese at various ages. In the 33rd week of ageing, *L. monocytogenes* survived better in cheeses with less sodium, but otherwise there were no significant differences. An increased degradation of proteins raises pH, increasing unbound amino acids. This can improve pathogen stress tolerance. When deliberately contaminated with *Listeria innocua*, Edam cheese with Na, K substitution did not change its microbiological composition (Luzzi et al., 2021). *Salmonella spp.* was shown to survive for up to 90 days in low-salt Cheddar cheese held at temperatures of 4 °C or 10 °C, and for up to 30 days at a temperature of 21 °C (Silva Ferrari et al., 2016). Multiple endeavours were undertaken to decrease the NaCl content in Cheddar cheese, revealing divergent and inconclusive outcomes. Substituting more than 30% of NaCl with $MgCl_2$, KCl, or $CaCl_2$ in the employment process resulted in flaws such as changes in texture, the emergence of a metallic flavour, and a decline in sensory characteristics, particularly due to an unpleasant aftertaste (Fox et al., 2004; Bae et al., 2017; Hoffmann et al., 2020). Decreasing the amount of salt in hard cheeses can result in increased microbial development and proteolysis, which can impact the hardness and gumminess of the cheeses produced (Sheibani et al., 2015). However, other experiments have successfully produced high-quality Cheddar cheeses with minimal sodium content (Ganesan et al., 2014). Optimisation of processing conditions can mitigate the deleterious effects of salt reduction or replacement in Cheddar-type cheeses. These include lowering curd residual lactose, changing rennet dosage, adding adjunct cultures/starter, or employing ultra-filtered retentate in milk. These optimisations are needed since cheeses' salt balance is fragile (Bae et al., 2017). A 50% salt reduction in Cheddar cheese's physical, sensory and chemical properties showed a gradual loss of taste.

Moisture management in cheese making is achieved by lowering the size of the curd grains and increasing the cooking time at 40 °C. This has been found to be the most successful technological process for producing Cheddar cheeses with recovered textural qualities. Replacing NaCl with KCl at 30 or 50% in a Grana-type hard cheese allowed for the production of similar cheeses, with no significant variations in terms of chemical composition, microbiological counts, and sensory evaluations (Tidona et al., 2019).

13.7 Mold-Ripened Cheeses

Mold-ripened cheeses can be categorised into two groups depending on the manner in which the moulds are dispersed as an additional starter during the cheese-making procedure. (i) Surface-ripened cheeses are produced by applying the inoculum onto the rind prior to the ripening of the cheese. (ii) Blue-veined cheeses are produced by introducing spores of particular mould species directly into the milk during the cheese production process. Surface-ripened cheeses have 1.5–2.3% salt. However, the rind has more salt early in ripening. This creates an acidic, salty environment that promotes bacteria growth. Washed-rind cheeses like Taleggio, Gruyere, Epoisses, Tilsit, and Limburger have moulds on their surfaces. (Wolfe et al., 2014; Irlinger et al., 2015). Reducing salt in surface-ripened cheeses may affect the salt-tolerant microbial ecology, causing spoilage. P. camemberti rapidly increased in 0.8% NaCl-salted Camembert cheese, resulting in unsatisfactory and uneven results (Fox et al., 2004). According to Dugat-Bony et al. (2016), purposely putting *Pseudomonas fragi* onto surface-ripened cheeses with a reduced salt concentration (1.3%) increased its proliferation compared to the control group (1.8%). Camembert cheese made with 20% less NaCl and partially replaced with KCl had no effect on its physico-chemical properties, microbiological diversity of the adjunct starter culture (*Geotrichum candidum and P. camemberti*, 1:2), spoiling bacteria, or sensory acceptance (Galli et al., 2016). With decreasing NaCl concentration, *G. candidum* increased significantly in Camembert cheese, causing uneven mould growth, according to RNA-seq study. The study did not examine this, but lowering salt could allow *G. candidum* to outperform P. camemberti, producing cheeses with a higher ammoniacal taste and bitterness. This is due to the significant proteolytic activity associated with *G. candidum* (Galli et al., 2016), or it could result in a surface defect known as "toad skin" (Dugat-Bony et al., 2019). Salted blue-veined cheeses contain 3–5% NaCl. They are often seasoned with dried salt. Mold-ripened cheeses mature due to Penicillium roqueforti spores, which grow with 1% NaCl. This study studied salt's effect on lipolysis. The results demonstrated that salt dosage affects lipolysis. When the NaCl content ranged from 0.5% to 3%, there were no significant differences in lipolysis. However, increasing the NaCl level to approximately 4% to 6% resulted in an increase in lipolysis. At higher concentrations of salt, lipolysis tended to decrease (Loudiyi & Aït-Kaddour, 2019).

13.8 Naturally Lower-Sodium Cheeses

13.8.1 Cottage Cheese

When served alongside poached eggs and whole-wheat toast, cottage cheese has become the go-to choice for enhancing a nutritious breakfast. Nevertheless, conventional cottage cheese may have a high sodium content. Therefore, it is advisable to choose the no-salt-added options. These kinds often do not include any additional preservatives compared to salt-added cottage cheese.

13.8.2 Ricotta

This versatile cheese, ranging from lasagna to manicotti, possesses a delicate and ethereal texture that well balances the richness of various Italian meals. Due to its high moisture content and the fact that it is consumed fresh, ricotta does not need to be preserved or have its moisture reduced with the addition of salt. "Furthermore, the presence of ricotta in a dish is not essential for enhancing its flavour, as the other ingredients typically contribute sufficient flavour,"

13.8.3 Cream Cheese

Bagels with cream cheese are a delectable choice for breakfast, and it is noteworthy that this spread is also low in salt. Cream cheese is a type of cheese with a high moisture content that is spread, stored in a refrigerator, and consumed when still fresh. "Therefore, salt is not required for extracting moisture from the curd, nor is it necessary for preserving the cheese during the distribution process."

13.8.4 Parmesan

Similar to cream cheese, the amount of Parmesan cheese you typically sprinkle on a bowl of pasta is limited to a few grated spoon full. As a result, you consume less salt compared to other types of hard cheeses. However, when comparing equal weights, it may not be low in salt. The recommended serving size for Parmesan cheese by the FDA is 5 g. In contrast, most other cheeses have a serving size of 30 g and c8ontain approximately 190 mg of sodium. The portion size is limited because Parmesan is typically used as a seasoning rather than a main component of a dish.

13.8.5 Fresh Mozzarella

It is important to choose fresh mozzarella, specifically the sort commonly found in Italy. This form of mozzarella has a high moisture content and is a fresh cheese that does not need salt to eliminate moisture or aid in ageing. The mozzarella that Americans usually think of is often sold in shredded form for pizza and is a cheese that is firmer compared to the Italian variety. American mozzarella exhibits a reduced moisture content and an extended duration of freshness. Salt is essential for extracting moisture, maintaining the texture of the cheese, and contributing to its flavour profile.

13.8.6 Swiss

The most sacred cheese is typically minimal in salt. The slower production process enables the removal of moisture with less salt used. In addition, it includes propionic acid, which is a naturally occurring short-chain saturated fatty acid. This acid helps in preserving the cheese as it ages, as stated by the NIH.

13.8.7 Monterey Jack

Although Jack cheese still has a considerable sodium content, when compared to cheddar, Jack cheese has a higher moisture content and is usually matured for a shorter period of time, resulting in a reduced requirement for salt. Furthermore, many varieties of Jack cheese are enhanced with the addition of peppers, thereby concealing the potential lack of flavour that may be present in low-sodium products. Table 13.3 is representing nutritional value of different naturally lower-sodium cheeses in each serving (Marissa Miller, 2019).

13.9 Composition and Therapeutic Significance

The composition of cheese is mostly influenced by the milk's quality and the particular manufacturing procedures applied. There may not be a direct correlation between the presence of NaCl and the nutritional makeup of cheese. However, when

Table 13.3 Nutritional value of different naturally lower-sodium cheeses per serving)

Sr .no.	Naturally lower-sodium cheeses	Calories	Fat (g)	Carbs (g)	Sodium (mg)	Sugar (g)	Fiber (g)	Protein (g)
1	Cottage cheese	81	1 g (1 g saturated)	3	15	3	0	14
2	Ricotta	171	10 g (6 g saturated)	6	123	0	0	14
3	Cream cheese	31	3 g (1.5 g saturated)	1	48	1	0	1
4	Parmesan	40	3 g (1.5 g saturated)	0	121	0	0	4
5	Fresh mozzarella	85	6 g (4 g saturated)	1	138	0	0	6
6	Swiss	50	1 g (1 g saturated)	1	56	0	0	8
7	Monterey jack	53	4 g (2 g saturated)	0	133	0	0	5

NaCl is reduced or substituted, it might affect the expulsion of whey and the enzymatic activity of the cheese. These changes can potentially alter the composition and therapeutic benefits of the cheese. Table 13.4 illustrates the effect of reducing NaCl on the composition and therapeutic value of LSCs. The content and quality of cheese are significantly affected by the rate and amount of syneresis. This process determines the loss of nutrients in the whey and the speed at which glycolysis, proteolysis, and lipolysis occur during ripening (Daviau et al., 2000; Dejmek & Walstra, 2004; Everard et al., 2008). The decrease or replacement of NaCl is inversely related to the syneresis (Johnson et al., 2009; Lu & McMahon, 2015), which leads to variations in moisture, fat, and protein concentrations in Low Salt Cheeses (LSCs) compared to cheeses salted at the usual level. (Fox et al., 2000; Guinee & Fox, 2004; Grummer et al., 2013). It is generally observed that an increase in moisture leads to a decrease in the levels of fat and protein.

Rulikowska et al. (2013) conducted an experiment to analyse the composition of Cheddar cheeses. The experiment involved adding different amounts of NaCl (0.50%, 1.25%, 1.80%, 2.25%, 2.50%, and 3.0% w/w) during the preparation of the cheeses. The investigators found that decreasing NaCl addition increased moisture and decreased fat, protein, and ash. The researchers found no statistically significant differences in Cheddar cheese composition when salted 1.80% to 3.0% w/w NaCl. Cheeses salted at 0.5% and 1.25% changed composition more than the other samples.

Table 13.4 Effects of reducing or substituting salt on the composition and therapeutic characteristics of LSCs: scientific study findings

Cheese type	References	Outcome
Cheddar	Rulikowska et al. (2013)	Rise in moisture content resulted in a decrease in fat, protein, and ash amounts, indicating a drop in the rate of NaCl addition
Akawi	Gandhi and Shah (2016)	Release of essential amino acids (EAAs) from cheese extracts held in a decreased sodium chloride (NaCl) brine was enhanced, although the activity of peptides remained mostly unchanged compared to the control
Cheddar, Mozzarella Kefalograviera, Feta Halloumi, Akawi	Ayyash and Shah (2010, 2011b) and Ayyash et al. (2012)	Study found no significant variations in the moisture content and nutritional composition of cheeses when NaCl was partially substituted with KCl
Akawi	Ayyash et al. (2012)	ACE-inhibitory action remained unaffected when NaCl was partially replaced with KCl
Cheddar	Rulikowska et al. (2013)	Cheeses salted at 0.5% or 1.25% NaCl had a notably greater content of free amino acids (FAAs) compared to cheeses salted at 2.5% or 3.0% NaCl
Low moisture Mozzarella cheese	Ayyash and Shah (2011b)	Levels of soluble calcium (Ca) and phosphorus (P) in all tested cheeses showed a considerable rise after storage
Akawi, Nabulsi	Ayyash and Shah (2011a) and Ayyash et al. (2012)	Concentration of calcium (Ca) and phosphorus (P) exhibited a substantial reduction during the course of the storage duration.

Rowney et al. (2004) conducted a study on Mozzarella cheese to investigate how varying quantities of NaCl and different salting methods affect the composition of the cheese. The scientists noticed a similar trend in moisture content for RSCs as that of Cheddar cheese, with LSCs having higher moisture content compared to the norm. Cheeses lose moisture due to osmosis and casein shrinkage or expansion in the presence of sodium ions. Since cheese is rich in Ca, Mg, and P, LSC mineral composition has been extensively studied.

Schroeder et al. (1988) calculated Cheddar cheese Na, K, and Ca values by varying salt addition from 0.5% to 2.0%. The authors used atomic absorption spectrophotometry to show that the experimental Cheddar cheeses had no significant K and Ca changes. In a similar vein, Rulikowska et al. (2013) observed a statistically minor decrease in Cheddar cheese Ca concentration when NaCl decreased. However, Cheddar cheese salted at 0.5% had a significantly different calcium content than cheese salted at 3.0%.Several studies have studied the effects of lowering salt and/or substituting it with KCl on cheese composition, but nothing is known about its therapeutic benefits. It is crucial to investigate the medicinal qualities of LSCs in order to confirm the health benefits associated with these cheeses.

13.10 Physicochemical Properties

Cheese quality is assessed by physical and chemical qualities throughout manufacture and ripening. Cheese quality is influenced by factors such as water activity (a_w) and pH which affect sensory, nutritional, textural, rheological, functional, and microbiological qualities. These characteristics are interconnected. Therefore, studying these characteristics is crucial for creating high-quality low-sodium cheeses/RSCs.

13.10.1 Water Activity (a_w)

Salt to Moisture ratio (S/M) and water activity affect the growth of starter and non-starter lactic acid bacteria (SLAB and NSLAB), which affect cheese quality and safety. Each cheese has ideal S/M. Deviation from the optimal value causes bacterial growth and unregulated enzymatic activity, lowering cheese quality. It is well known that cheese aw decreases with S/M (Rulikowska et al., 2013; Khetra et al., 2019). Rulikowska et al. (2013) examined Cheddar cheese water activity. They observed that lowering NaCl increased water activity. Water activity in different cheeses was tested after substituting NaCl with KCl.

13.10.2 PH

The pH of the cheese matrix plays a vital role in controlling the growth of cheese microflora and the activity of enzymes during ripening also affects the absorption and diffusion of salt as well as the expulsion of whey Hence, assessing the impact of reducing salt content and/or substituting it on pH levels is crucial for maintaining the overall quality and safety of RSCs. In their study, Ganesan et al. (2014) found that the pH of Mozzarella cheese remained constant regardless of the rate at which NaCl was added. However, the pH of low sodium Cheddar cheeses decreased from 5.3 to 5.0–5.1 in comparison to control cheeses. Arboatti et al. (2014) discovered that there was no notable disparity in pH levels between full salt Mozzarella cheese and Mozzarella cheese with reduced salt content of 35% and 60%. Upreti and Metzger (2007) determined that the pH of cheese is affected by various elements, such as the proportion of calcium and phosphorus in the colloidal and serum phase, the remaining lactose, and the S/M ratio. The debate on pH suggests that different salt reduction tactics may have beneficial or negative effects depending on the type of cheese. Salt-reduction techniques should prioritise maintaining the same pH change rate in RSCs as the control, as it is the most important element affecting cheese quality and safety. Further research is needed to assess the impact of Ca, Mg, phosphate, citrate, and other ions on cheese pH, which can significantly impact quality and safety.

13.10.3 Texture and Rheology

Rheology and texture assessment are essential for assessing cheese quality during manufacturing, ageing, and consumption. Salt reduction affects cheese's manipulation, packaging, and mastication, therefore studying its rheological and textural effects is crucial. These features are important quality factors for manufacturers and customers. Typically, low salt concentration produces weak, soft, and doughy cheeses. Conversely, cheeses that have a higher percentage of salt are typically characterised by being compact and fragile, with a parched and firm consistency. (Guinee & Fox, 2004). Similar results have been obtained in previous research for Gaziantep (Kaya, 2002) and Feta cheeses. Rulikowska et al. (2013) conducted a study on the textural characteristics of cheddar cheese for a period of 224 days, while making adjustments to the rate at which salt was added. An inverse association was discovered between the salt concentration, hardness, fracture stress, and fracture strain. Decreasing the amount of sodium chloride, increasing the breakdown of proteins, and reducing the salt-to-moisture ratio may result in a decrease in the textural characteristics of LSCs. Ganesan et al. (2014) examined the effects of NaCl concentrations (0.7%, 1.0%, 1.25%, 1.35%, and 1.8%) on Cheddar cheese texture properties during storage at 22 °C for 3 and 6 months. After 90 days of ripening, the investigators found no influence of salt on cheese resilience, cohesion, or hardness. Loudiyi et al. (2017) examined the impact of lowering NaCl concentration in Cantal-type cheese, with or without replacing it with KCl, on its rheological properties, including fracture stress (G′) and strain (G″). When heating the samples, the authors observed a drop in G′ and G″ values in the 0.5% and 1.0% NaCl-seasoned cheeses compared to the 2% NaCl-seasoned cheeses. However, cheeses made with 25% and 50% KCl replacement showed greater G′ and G″ values than the control.

13.10.4 Cheese Microflora and Safety

Salt inhibits the growth of spoilage-causing germs and controls the growth of beneficial microbes in cheese during storage. Cheese quality and safety are ensured by NaCl, acidic pH, and low temperature, which inhibit microorganism growth. NaCl's antimicrobial properties come from water activity (aw) and osmotic pressure. An increase in NaCl content decreases water activity (aw) and increases the osmotic gradient among bacterial cells and cheese liquid. This inhibits and disables bacteria. Since NaCl reduction or substitution affects cheese water activity (aw) and pH, microbiological features of Low Sodium Cheeses (LSCs) must be examined to ensure quality and safety. Table 6 shows how decreasing NaCl levels affects LSC microbiology. Hystead et al. (2013) conducted an experiment to examine the growth of *Listeria monocytogenes* in Cheddar cheese, comparing samples with and without potassium chloride (KCl) substitution. Cheddar cheeses with reduced sodium content (25% and 53% decrease) were created using two different types of Lactococcus lactis bacteria. The samples were deliberately contaminated with *Listeria monocytogenes* at four separate stages (2, 15, 22, and 33 weeks) during the ageing process. They were then stored for 27 and 50 days at temperatures of 4 °C and 12 °C, respectively. Following the inoculation process, which occurred 2 weeks after manufacturing, the authors observed that neither salt nor starting culture had any impact on the survival of *Listeria monocytogenes*.

Silva et al. (2018) evaluated the effects of replacing NaCl with KCl 1:1 and adding taste boosters (1% arginine, 1% yeast extract, and 1% oregano extract) on Prato cheese probiotic concentrations. The trial cheeses were stored at 4 °C for 60 days. Stored samples were analysed at 1st, 30th, and 60th days. The study found that replacing NaCl with KCl, with or without flavourings, significantly reduced *Lactobacillus casei* bacteria at the completion of storage ($p < .05$).

The above-presented data indicate that reducing NaCl has a beneficial impact on probiotic counts, while having an insignificant influence on pathogenic and spoilage-causing microbes in RSCs. This suggests that it is possible to manufacture RSCs (reduced-sodium cheeses) without sacrificing their safety. This discovery could serve as a catalyst for worldwide food safety authorities to establish regulations with the goal of minimising the NaCl levels in food items, hence reducing the intake of dietary sodium by individuals. It is important to acknowledge that harmful and spoilage-causing microbes can endure in LSCs, underscoring the necessity of implementing effective production processes to guarantee the stability and safety of LSCs throughout storage.

13.11 Conclusion

The dairy sector is responsible for developing technology to produce cheese with lower sodium content due to public health concerns associated with excessive sodium levels. Modifying the salt intake of individuals who are already accustomed to high consumption poses a challenge and requires multiple approaches. These approaches should include a gradual decrease in the salt content of products, as well as technological measures that enhance the perception of salt in foods where salt has already been reduced. To successfully implement any modifications in the cheese production process, a rigorous level of control is necessary. Additionally, further research is needed to determine the acceptable quality and quantity of salt for marketing cheese with reduced salt content in the market. The projected success in reducing sodium content in cheese can be achieved through improvements in rheological, structural, and sensory factors, together with targeted consumer promotions emphasising the added value of the product. It has been noted that lowering salt in products may help reduce sodium intake, but it does not ensure a drop in diseases associated with consuming such items without the additional support of healthy habits like regular exercise and weight management. Crystalline particles containing both KCl and lysine mono-hydrochloride could serve as a viable alternative to food-grade NaCl, as these two components are resistant to separation. Although certain goods have employed the tactic of combining NaCl with KCl to diminish sodium intake, there are only a limited number of sodium-free products available in the market. The advantages of consuming a diet that is low in sodium are further enhanced when accompanied by modifications in one's lifestyle.

References

Abdel-Latif, E. F., Abbas, K. A., Abdelmontaleb, H. S., & Hamdy, S. M. (2021). Nigella sativa oil: A promising prospective antifungal agent in the manufacture of low-salt soft cheese. *Italian Journal of Food Safety, 10*, 9862. https://doi.org/10.4081/ijfs.2021.9862

Agarwal, S., McCoy, D., Graves, W., Gerard, P. D., & Clark, S. (2011). Sodium content in retail Cheddar, Mozzarella, and process cheeses varies considerably in the United States. *Journal of Dairy Science, 94*(3), 1605–1615.

Akan, E., Yerlikaya, O., & Kinik, Ö. (2017). Importance of salt in dairy products and sodium reduction strategies in food and dairy products. *Agro Food Industry Hi-Tech, 28*(2), 60–62.

Arboatti, A. S., Olivares, M. L., Sabbag, N. G., Costa, S. C., Zorrilla, S. E., & Sihufe, G. A. (2014). The influence of sodium chloride reduction on physicochemical, biochemical, rheological and sensory characteristics of Mozzarella cheese. *Dairy Science & Technology, 94*(4), 373–386.

Ayyash, M. M., & Shah, N. P. (2010). Effect of partial substitution of NaCl with KCl on Halloumi cheese during storage: Chemical composition, lactic bacterial count, and organic acids production. *Journal of Food Science, 75*(6), C525–C529.

Ayyash, M. M., & Shah, N. P. (2011a). Proteolysis of low-moisture Mozzarella cheese as affected by substitution of NaCl with KCl. *Journal of Dairy Science, 94*(8), 3769–3777.

Ayyash, M. M., & Shah, N. P. (2011b). The effect of substituting NaCl with KCl on Nabulsi cheese: Chemical composition, total viable count, and texture profile. *Journal of Dairy Science, 94*(6), 2741–2751.

Ayyash, M. M., Sherkat, F., & Shah, N. P. (2012). The effect of NaCl substitution with KCl on Akawi cheese: Chemical composition, proteolysis, angiotensin-converting enzyme-inhibitory activity, probiotic survival, texture profile, and sensory properties. *Journal of Dairy Science, 95*(9), 4747–4759.

Bae, I., Park, J. H., Choi, H. Y., & Jung, H. K. (2017). Emerging innovations to reduce the salt content in cheese; effects of salt on flavor, texture, and shelf life of cheese; and current salt usage: A review. *Korean Journal for Food Science of Animal Resources, 37*, 793–798. https://doi.org/10.5851/kosfa.2017.37.6.793

Bansal, V., & Mishra, S. K. (2020). Reduced-sodium cheeses: Implications of reducing sodium chloride on cheese quality and safety. *Comprehensive Reviews in Food Science and Food Safety, 19*, 733–758. https://doi.org/10.1111/1541-4337.12524

Bishop, J. R., & Smukowski, M. (2006). Storage temperatures necessary to maintain cheese safety. *Food Protection Trends, 26*, 714–724.

Bisig, W. (2017). *The importance of salt in the manufacturing and ripening of cheese*. International Dairy Federation.

Canadian Dairy Information Centre (CDIC). (2013). *Global and retail consumption*. http://www.dairyinfo.gc.ca/index_e.php. Accessed 1 Aug 2013.

Cankurt, H. (2019). The effects of adding different stabilizers in brine on the physicochemical, sensory, microbiological and textural properties of white cheese. *Foods, 8*(4), E133. https://doi.org/10.3390/foods8040133

Centers for Disease Control and Prevention (CDC). (2012). Vital signs: Food categories contributing the most to sodium consumption United States, 2007–2008. *Morbidity and Mortality Weekly Report, 61*(5), 92–98.

Charlton, K., Yeatman, H., Houweling, F., & Guenon, S. (2010). Urinary sodium excretion, dietary sources of sodium intake and knowledge and practices around salt use in a group of healthy Australian women. *Australian and New Zealand Journal of Public Health, 34*(4), 356–363.

Cruz, A. G., Faria, J. A., Pollonio, M. A., Bolini, H. M., Celeghini, R. M., Granato, D., & Shah, N. P. (2011). Cheeses with reduced sodium content: Effects on functionality, public health benefits and sensory properties. *Trends in Food Science & Technology, 22*(6), 276–291.

Daviau, C., Pierre, A., Famelart, M. H., Goudédranche, H., Jacob, D., Garnier, M., & Maubois, J. L. (2000). Characterisation of whey drainage kinetics during soft cheese manufacture in relation with the physicochemical and technological factors, pH at renneting, casein concentration and ionic strength of milk. *Le Lait, 80*(4), 417–432.

Dejmek, P., & Walstra, P. (2004). The syneresis of rennet-coagulated curd. In P. L. H. McSweeney, P. F. Fox, T. M. Cogan, & T. P. Guinee (Eds.), *Cheese: Chemistry, physics and microbiology* (Vol. 1, 3rd ed., pp. 71–103). Elsevier.

Doyle, M. E., & Glass, K. A. (2010). Sodium reduction and its effect on food safety, food quality, and human health. *Comprehensive Reviews in Food Science and Food Safety, 9*(1), 44–56.

Dugat-Bony, E., Sarthou, A.-S., Perello, M.-C., de Revel, G., Bonnarme, P., & Helinck, S. (2016). The effect of reduced sodium chloride content on the microbiological and biochemical properties of a soft surface-ripened cheese. *Journal of Dairy Science, 99*(4), 2502–2511. https://doi.org/10.3168/jds.2015-10502

Dugat-Bony, E., Bonnarme, P., Fraud, S., Catellote, J., Sarthou, A. S., Loux, V., et al. (2019). Effect of sodium chloride reduction or partial substitution with potassium chloride on the microbiological, biochemical and sensory characteristics of semi-hard and soft cheeses. *Food Research International, 125*, 108643. https://doi.org/10.1016/j.foodres.2019.108643

El-Bakry, M. (2012). Salt in Cheese: A Review. *Current Research in Dairy Sciences, 4*(1), 1-5.

Everard, C. D., O'Callaghan, D. J., Mateo, M. J., O'Donnell, C. P., Castillo, M., & Payne, F. A. (2008). Effects of cutting intensity and stirring speed on syneresis and curd losses during cheese manufacture. *Journal of Dairy Science, 91*(7), 2575–2582.

Fischer, P. W., Vigneault, M., Huang, R., Arvaniti, K., & Roach, P. (2009). Sodium food sources in the Canadian diet. *Applied Physiology, Nutrition, and Metabolism, 34*(5), 884–892.

Fox, P. F., Guinee, T. P., Cogan, M. T., & McSweeney, P. L. H. (2000). Salting of cheese curd. In P. F. Fox (Ed.), *Fundamentals of cheese science* (pp. 153–168). Aspen Publication.

Fox, P. F., McSweeney, P. L. H., Cogan, T. M., & Guinee, T. P. (2004). *In "cheese chemistry, physics and microbiology"* (Vol. 1, 3rd ed.). Elsevier Academic Press.

Galli, B. D., Martin, J. G. P., da Silva, P. P. M., Porto, E., & Spoto, M. H. F. (2016). Sensory quality of camembert-type cheese: Relationship between starter cultures and ripening molds. *International Journal of Food Microbiology, 234*, 71–75. https://doi.org/10.1016/j.ijfoodmicro.2016.06.025

Gandhi, A., & Shah, N. P. (2016). Salt reduction in a model high-salt Akawi cheese: Effects on bacterial activity, pH, moisture, potential bioactive peptides, amino acids, and growth of human colon cells. *Journal of Food Science, 81*(4), H991–H1000. https://doi.org/10.1111/1750-3841.13260

Ganesan, B., Brown, K., Irish, D. A., Brothersen, C., & McMahon, D. (2014). Manufacture and sensory analysis of reduced- and low- sodium Cheddar and mozzarella cheeses. *Journal of Dairy Science, 97*, 1970–1982. https://doi.org/10.3168/jds.2013-7443

García-Díez, J., & Saraiva, C. (2021). Use of starter cultures in foods from animal origin to improve their safety. *International Journal of Environmental Research and Public Health, 18*, 2544. https://doi.org/10.3390/ijerph18052544

Grimes, C. A., Campbell, K. J., Riddell, L. J., & Nowson, C. A. (2011). Sources of sodium in Australian children's diets and the effect of the application of sodium targets to food products to reduce sodium intake. *British Journal of Nutrition, 105*(3), 468–477.

Grummer, J., Bobowski, N., Karalus, M., Vickers, Z., & Schoenfuss, T. (2013). Use of potassium chloride and flavor enhancers in low sodium Cheddar cheese. *Journal of Dairy Science, 96*(3), 1401–1418.

Guinee, T. P., & Fox, P. F. (2004). Salt in cheese: Physical, chemical and biological aspects. In P. L. H. McSweeney, P. F. Fox, T. M. Cogan, & T. P. Guinee (Eds.), *Cheese: Chemistry, physics and microbiology* (Vol. 1, 3rd ed., pp. 207–259). Elsevier Ltd.

Güven, M., & Karaca, O. B. (2001). Proteolysis levels of white cheeses salted and ripened in brines prepared from various salts. *International Journal of Dairy Technology, 54*, 29–33.

Güven, M., Cadun, C., Karaca, O., & Hayaloğlu, A. (2008). Influence of rennet concentration on ripening characteristics of Halloumi cheese. *Journal of Food Biochemistry, 32*, 615–627. https://doi.org/10.1111/j.1745-4514.2008.00187.X

Hashem, K. M., He, F. J., Jenner, K. H., & MacGregor, G. A. (2014). Cross-sectional survey of salt content in cheese: A major contributor to salt intake in the UK. *British Medical (Association) Journal (BMJ Open), 4*(8), e005051.

He, F. J., & MacGregor, G. A. (2010). Reducing population salt intake worldwide: From evidence to implementation. *Progress in Cardiovascular Diseases, 52*, 363–382.

Health Canada. 2010. *Dietary reference intakes tables*. http://www.hc-sc.gc.ca/fn-an/nutrition/reference/table/index-eng.php. Accessed 1 Aug 2013.

Hoffmann, W., Luzzi, G., Steffens, M., Clawin-Rädecker, I., Franz, C. M. A. P., & Fritsche, J. (2020). Salt reduction in film-ripened, semihard Edam cheese. *International Journal of Dairy Technology, 73*, 270–282.

Hystead, E., Diez-Gonzalez, F., & Schoenfuss, T. C. (2013). The effect of sodium reduction with and without potassium chloride on the survival of listeria monocytogenes in Cheddar cheese. *Journal of Dairy Science, 96*(10), 6172–6185. https://doi.org/10.3168/jds.2013-6675

Irlinger, F., Layec, S., Hélinck, S., & Dugat-Bony, E. (2015). Cheese rind microbial communities: Diversity, composition and origin. *FEMS Microbiology Letters, 362*, 1–11. https://doi.org/10.1093/femsle/fnu015

Johnson, M. E., Kapoor, R., McMahon, D. J., McCoy, D. R., & Narasimmon, R. G. (2009). Reduction of sodium and fat levels in natural and processed cheeses: Scientific and technological aspects. *Comprehensive Reviews in Food Science and Food Safety, 8*(3), 252–268.

Kamleh, R., Olabi, A., Toufeili, I., Najm, N. E. O., Younis, T., & Ajib, R. (2012). The effect of substitution of sodium chloride with potassium chloride on the physicochemical, microbiological, and sensory properties of halloumi cheese. *Journal of Dairy Science, 95*(3), 1140 1151.

Kaplan, N. M. (2000). The dietary guideline for sodium: Should we shake it up? No. *The American Journal of Clinical Nutrition, 71*(5), 1020–1026.

Karimi, R., Mortazavian, A. M., & Karami, M. (2012). Incorporation of Lactobacillus casei in Iranian ultrafiltered Feta cheese made by partial replacement of NaCl with KCl. *Journal of Dairy Science, 95*(8), 4209–4222.

Katsiari, M. C., Voutsinas, L. P., Alichanidis, E., & Roussis, I. G. (2001). Lipolysis in reduced sodium Kefalograviera cheese made by partial replacement of NaCl with KCl. *Food Chemistry, 72*, 193–197. https://doi.org/10.1016/S0308-8146(00)00218-1

Kaya, S. (2002). Effect of salt on hardness and whiteness of Gaziantep cheese during short-term brining. *Journal of Food Engineering, 52*(2), 155–159.

Khetra, Y., Kanawjia, S. K., & Puri, R. (2016). Selection and optimization of salt replacer, flavor enhancer and bitter blocker for manufacturing low sodium Cheddar cheese using response surface methodology. *LWT - Food Science and Technology, 72*, 99–106.

Khetra, Y., Kanawjia, S. K., Puri, R., Kumar, R., & Meena, G. S. (2019). Using taste-induced saltiness enhancement for reducing sodium in Cheddar cheese: Effect on physico-chemical and sensorial attributes. *International Dairy Journal, 91*, 165–171.

Loudiyi, M., & Aït-Kaddour, A. (2019). Evaluation of the effect of salts on chemical, structural, textural, sensory and heating properties of cheese: Contribution of conventional methods and spectral ones. *Critical Reviews in Food Science and Nutrition, 59*, 2442–2457. https://doi.org/10.1080/10408398.2018.1455637

Loudiyi, M., Karoui, R., Rutledge, D. N., Lavigne, R., Montel, M. C., & Ait-Kaddour, A. (2017). Contribution of fluorescence spectroscopy and independent components analysis to the evaluation of NaCl and KCl effects on molecular-structure and fat melting temperatures of Cantal-type cheese. *International Dairy Journal, 73*, 116–127.

Lu, Y., & McMahon, D. J. (2015). Effects of sodium chloride salting and substitution with potassium chloride on whey expulsion of Cheddar cheese. *Journal of Dairy Science, 98*(1), 78–88.

Luzzi, G., Brinks, E., Fritsche, J., & Franz, C. M. A. P. (2021). Effect of reduction of sodium content on the microbial ecology of Edam cheese samples. *AMB Express, 11*, 28. https://doi.org/10.1186/s13568-021-01188-7

Matera, J., Luna, A. S., Batista, D. B., Pimentel, T. C., Moraes, J., Kamimura, B. A., et al. (2018). Brazilian cheeses: A survey covering physicochemical characteristics, mineral content, fatty acid profile and volatile compounds. *Food Research International, 108*, 18–26.

McCarthy, C. M., Wilkinson, M. G., Kelly, P. M., & Guinee, T. P. (2015). Effect of salt and fat reduction on the composition, lactose metabolism, water activity and microbiology of Cheddar cheese. *Dairy Science & Technology, 95*, 587–611. https://doi.org/10.1007/s13594-015-0245-2

McMahon, D. J., Oberg, C. J., Drake, M. A., Farkye, N., Moyes, L. V., Arnold, M. R., et al. (2014). Effect of sodium, potassium, magnesium, and calcium salt cations on pH, proteolysis, organic acids, and microbial populations during storage of full-fat Cheddar cheese. *Journal of Dairy Science, 97*(8), 4780–4798.

Miller, M. (2019). The 7 best low-sodium cheeses, according to nutritionists. Women's Health. https://www.womenshealthmag.com/food/a19930240/salty-cheeses/Women'sHealth

Ministry of Health. (2003). Nutrition and the burden of disease: New Zealand 1997–2011. Wellington, New Zealand: Author.

Murtaza, M. A., Huma, N., Sameen, A., Murtaza, M. S., Mahmood, S., Mueen-ud-Din, G., & Meraj, A. (2014). Texture, flavor, and sensory quality of buffalo milk Cheddar cheese as influenced by reducing sodium salt content. *Journal of Dairy Science, 97*(11), 6700–6707.

National Health and Medical Research Council (NHMRC). (2006). *Nutrient reference values for Australia and New Zealand including recommended dietary intakes*. NHMRC.

Ozturk, M., Govindasamy-Lucey, S., Jaeggi, J. J., Johnson, M. E., & Lucey, J. A. (2015). Low-sodium Cheddar cheese: Effect of fortification of cheese milk with ultrafiltration retentate and high-hydrostatic pressure treatment of cheese. *Journal of Dairy Science, 98*(10), 6713–6726.

Patel, J. B., Patel, P. V., & Lee, E. J. (2013). The effect of potassium chloride as a salt replacer on the qualities of processed cheese. *Journal of Students Research, 13*, 313–329.

Rowney, M. K., Roupas, P., Hickey, M. W., & Everett, D. W. (2004). Saltinduced structural changes in 1-day old Mozzarella cheese and the impact upon free oil formation. *International Dairy Journal, 14*(9), 809–816.

Rulikowska, A., Kilcawley, K. N., Doolan, I. A., Alonso-Gomez, M., Nongonierma, A. B., Hannon, J. A., et al. (2013). The impact of reduced sodium chloride content on Cheddar cheese quality. *International Dairy Journal, 28*(2), 45–55. https://doi.org/10.1016/j.idairyj.2012.08.007

Schroeder, C. L., Bodyfelt, F. W., Wyatt, C. J., & McDaniel, M. R. (1988). Reduction of sodium chloride in Cheddar cheese: Effect on sensory, microbiological, and chemical properties. *Journal of Dairy Science, 71*(8), 2010–2020.

Sheibani, A., Ayyash, M. M., Shah, N. P., & Mishra, V. K. (2015). The effects of salt reduction on characteristics of hard type cheese made using high proteolytic starter culture. *International Food Research Journal, 22*, 2452–2459.

Silva Ferrari, I., de Souza, J. V., Ramos, C. L., da Costa, M. M., Schwan, R. F., & Dias, F. S. (2016). Selection of autochthonous lactic acid bacteria from goat dairies and their addition to evaluate the inhibition of Salmonella typhi in artisanal cheese. *Food Microbiology, 60*, 29–38. https://doi.org/10.1016/j.fm.2016.06.014

Silva, H. L., Balthazar, C. F., Esmerino, E. A., Neto, R. P., Rocha, R. S., Moraes, J., et al. (2018). Partial substitution of NaCl by KCl and addition of flavor enhancers on probiotic Prato cheese: A study covering manufacturing, ripening and storage time. *Food Chemistry, 248*, 192–200.

Tidona, F., Francolino, S., Ghiglietti, R., Locci, F., Carminati, D., Zambrini, A. V., et al. (2019). The impact of sodium chloride reduction on Grana-like cheese production. *The Journal of Dairy Research, 86*, 470–476. https://doi.org/10.1017/S0022029919000797

Trieu, K., Neal, B., Hawkes, C., Dunford, E., Campbell, N., RodriguezFernandez, R., et al. (2015). Salt reduction initiatives around the world–a systematic review of progress towards the global target. *PLoS One, 10*(7), e0130247.

Upreti, P., & Metzger, L. E. (2007). Influence of calcium and phosphorus, lactose, and salt-to-moisture ratio on Cheddar cheese quality: pH changes during ripening. *Journal of Dairy Science, 90*(1), 1–12.

U.S. Department of Agriculture and U.S. Department of Health and Human Services. (2010). Dietary Guidelines for Americans, (7th ed.). Washington, DC: U.S. Government Printing Office.

Webster, J., Trieu, K., Dunford, E., & Hawkes, C. (2014). Target salt 2025: A global overview of national programs to encourage the food industry to reduce salt in foods. *Nutrients, 6*(8), 3274–3287.

WHO. (2013). WHO issues new guidance on dietary salt and potassium. *Central European Journal of Public Health, 21*, 16.

Wolfe, B. E., Button, J. E., Santarelli, M., & Dutton, R. J. (2014). Cheese rind communities provide tractable systems for in situ and in vitro studies of microbial diversity. *Cell, 158*, 422–433. https://doi.org/10.1016/j.cell.2014.05.041

Wyness, L. A., Butriss, J. L., & Stanner, S. A. (2012). Reducing the population's sodium intake: The UK Food Standards Agency's salt reduction programme. *Public Health Nutrition, 15*, 254–261.

Chapter 14
Vegan Cheeses

Ridhi Pandey, Sonam Kumari, and Yogesh Khetra

Abbreviations

ES	Emulsifying salt
PCs	Protein Concentrates
PIs	Protein isolates
SFC	Solid fat content
VC	Vegan cheese

14.1 Introduction

Cheese is a complex dairy based food which is nutritional and also popular for its application in different types of food. Around the world, cheese is frequently chosen as an ingredient in different types of food. Vegan cheese (VC) are cheese analogue in which all the ingredients used are of non-dairy origin. Analogous cheeses are categorised into dairy, partially dairy, or non-dairy based on the source of the materials used in their composition. Several factors are responsible for the rise in demand for vegan cheese. This increase is due to shift in consumption towards plant based food which is mainly driven by awareness about environment sustainability as it is impractical for feeding the rising population with animal products, health concerns like lactose intolerance, milk protein allergy etc. Plant based substitutes have additional health benefits as replacement of saturated fat by unsaturated fats and

R. Pandey (✉) · Y. Khetra
Dairy Technology Division, ICAR-National Dairy Research Institute, Karnal, Haryana, India

S. Kumari
Sanjay Gandhi Institute of Dairy Technology, BASU, Patna, Bihar, India

S. Mansha Rafiq (ed.), *Technological Advances and Trends in Cheese Making*,
https://doi.org/10.1007/978-3-031-94476-5_14

incorporation of different types of dietary fibre. Furthermore, a high cost of animal milk due to its numerous other applications has led to a rise in the cost of cheese, making it inaccessible to people with low income. VC have sever benefit such as flexibility of using different ingredients, tailored made functionality as per its end application. It can be manufactured using diverse sources of vegetables. Also, the wide availability of plant sources, reduces the cost of production and makes it affordable to large number of people. The processing of VC typically takes less time when compared to conventional cheese as instead of going through the ripening stage, it involves mixing and cooking the ingredients to create a smooth mixture. This mixture is then moulded and chilled in the refrigerator until it forms a firm gel, resembling cheese (Fox et al., 2017; Guinee et al., 2004). Reduced processing time and cost is attractive for food companies that produces cheese based foods and ultimately contributing to increase in demand of VC. Production of VC is associated with different goals of the United Nations Sustainable Development Goals such as zero hunger, good health and well-being, and climate.

The perception of ingredients used in VC has changed overtime from ingredients of inferior quality to ingredients that can give tailor made functionality and nutrition. In addition, fortification of VC with minerals, vitamins and other additives can improve its nutritional quality. The concept driving the production of VC is to replicate the physical and sensory characteristics of traditional cheeses like cheddar or mozzarella. This is accomplished by employing various ingredients and techniques to achieve the desired properties. It is desirable that the processing of VC utilises the same equipment and facilities as the traditional cheese such that transition from conventional cheese to VC is easy and does not require additional cost. In practice, this can be quite challenging since plant-derived ingredients have distinct behaviours compared to those found in animal milk. Regular cheese production differs from plant-based cheese production in terms of ingredients and processing methods. Unlike plant-based cheese, regular cheese is made primarily from milk sourced from various animals, and the different types of cheese are created through different processing techniques. On the other hand, the creation of plant-based cheese analogues involves a variety of ingredients and therefore requires careful selection of both ingredients and processes. These versatile products have a wide range of uses, both directly for consumption and primarily as ingredients in various foods.

14.2 Types and Role of Ingredients

This section will focus on the main ingredients used to prepare VC. The basic ingredients needed to create VC include a mixture of water, protein source, starch, and fats/oils. Emulsifying salts (ES) and food additives are among other components that can be included. The type and physico-chemical properties of ingredients, as well as the interrelations among components, play a crucial role in shaping the rheology and functionalities of the final product.

14.2.1 Water

Water is the primary component of VC, usually comprising up around 48–52% of its composition (Fox et al., 2017). Solubilisation of different ingredients, dissolving of ES and hydration of protein occur in water. During the last step of cooling, water molecules entrapped enhances its ability to become more flexible and prone to breaking when compressed (Fox et al., 2017). Typically, water has the effect of reducing hardness and enhancing the meltability of VC. Excessive water can lead to protein over-hydration, which may result in the undesirable effect of over-emulsification. This can negatively impact the performance of VC, leading to issues such as poor meltability and reduced flow when heated (Masotti et al., 2018).

14.2.2 Proteins

Proteins are a key component of VC, with levels typically ranging from 18% to 24% in the formulation (Guinee & McSweeney, 2007). There is a wide range of protein sources that can be used in VC production, including pea, soy, lupin, potato, nuts and corn. Proteins play a crucial role in the hydration process during VC manufacture. The solubilized proteins enhance the emulsification of the free oil and elevate the viscosity. A fat-entrapping network is formed as the droplets of dispersed oil are stabilised. The characteristics of proteins have a significant impact on the physical, chemical, and cooking properties of VC. This includes factors such as type, amount, structure, mineral content, and solubility (Sołowiej et al., 2014). Various physicochemical and functional requirements, including emulsification, gelation, water retention, and flavour development in the product is achieved with proteins. Modification techniques may be applied to enhance protein properties. The approaches used for modification greatly impact the physicochemical properties of proteins in food matrices. Various modifications can be made to enhance the water solubility, emulsifying, foaming, and gelling properties of proteins. For example, various approaches can be advantageous in mitigating or eliminating the negative impacts of ant nutrients (Avilés-Gaxiola et al., 2018). There are significant difference in the composition of plant-based milk and dairy cheese, with protein casein being a key differentiating factor. When dairy cheese is heated, the cheese network begins to soften as the milk fat melts at around 40 °C. And as the heating continues, the casein network contracts because of increased hydrophobic interaction. The network contraction results in reduced contact between casein molecules, resulting in a weakening of the overall network. This ultimately leads to the cheese continuing to soften or melt (Lamichhane et al., 2018; Lucey et al., 2003). When plant-based proteins are heated, they undergo denaturation and form irreversible gels, resulting in the creation of more rigid structures. Plant proteins have distinct structural and functional characteristics compared to casein, mostly because of their greater molecular size and complex quaternary structure (Bachmann, 2001).

A brief description of different type of protein sources used in vegan cheese is discussed below:

Legume Protein Legume protein can be potential source of protein for production of VC as their protein content is high compared to whole grain cereal and their cost is low to that of nuts. Legumes are part of the Fabaceae family and consist of various types such as the common bean, fava bean, soybean, pea, cowpea, black-eyed pea, pigeon pea, chickpea, lupin, lentil, and peanut (Gepts et al., 2005). Legumes are abundant in proteins with a high biological value, carbohydrates, minerals such as calcium and iron, vitamins like thiamine and niacin, and bioactive substances. Additionally, they have a low fat content. Their high dietary fibre, oligosaccharides, slowly digesting starch, and resistant starch content make them a low glycemic meal. Legumes include a protein content of 14.9–52.0 g/100 g. This protein is made up of different types, including storage proteins that can be extracted with salt, globulins (which can be further separated into 11S and 7S subunits), albumin, prolamin, glutelin, and residual proteins (Boye et al., 2010). Legume protein isolates (PIs) and concentrates (PCs) are devoid of colour, flavours, odours, and antinutritional factors, making them a favourable choice for incorporation VC. The PDCAAS values for unprocessed legume products typically fall between 0.40 and 0.70 (Mefleh et al., 2022). While heat treatment does deactivate the primary antinutritional factors to some extent, it has minimal impact on the digestibility of certain legumes. Utilising a combination of several legume sources may be employed to provide enhanced technological and nutritional characteristics. Pea proteins are widely used in production of VC because of favourable qualities of pea protein in terms of functionality such as fat- and water-binding capabilities, emulsification properties and gelation, texture and nutritional values (Sandberg, 2011).

Zein Zein is a hydrophobic prolamin derived from the endosperm of corn, which is easily obtained as a by-product of corn starch and corn syrup production for both food and bioethanol purposes. Corn contains a natural protein content of 6–12%, with zein making up around 45–50% of that. The extraction process involves using various solvents to dissolve different fractions of milled corn, including α-zein, β-zein, γ-zein, and δ-zein. Among these fractions, α-zein is found in the highest concentration (Anderson & Lamsal, 2011). α-zein has hydrophobic core in its triple helix structure, which can be utilised to solubilize and stabilize hydrophobic bioactive components (Momany et al., 2006). It is also utilised to stablise oil-in-water emulsion through Pickering effect and also to encapsulate several hydrophobic components in emulsion (Fathi et al., 2018). In the study by Mattice and Marangoni (2020), where zein was used as source of protein in producing VC, it showed rheological behaviour similar to that of cheddar cheese with softening and increased viscosity with increasing temperature. This creates potential for using zein in different types of VC according to desired functionality.

Wheat The protein content of wheat varies between 8% and 15% depending on variety. Gluten protein are the main storage protein and it has distinctive functional properties. It creates strong, elastic network the main storage proteins in wheat grains are the gluten proteins. Gluten has a unique physical functional property that none of other plant proteins have. Gluten forms a cohesive, viscoelastic protein network that provides the unique ability of wheat to produce leavened products (Wrigley, 1996). Gluten is frequently added to VC for two main reasons: to enhance the protein level in the end product and to provide the stretchiness or fibrous texture found in cheeses like Italian Mozzarella and Stracciatella (Batista et al., 2005).

14.2.3 Starch

Starches are widely utilised polysaccharide-based components in VC. They are utilised in vegan cheeses because to their capacity to create gel when heated and cooled, which captures liquids and other components within the 3D-polysaccharide structure created (Kasprzak et al., 2018). Tapioca, potato, rice, maize and maize starches are the most often utilised in VC formulations. Starches exist in two forms: (a) native starch derived from agricultural raw materials and (b) modified starch acquired through physical, enzymatic, or chemical processing. The advantage of using modified starch instead of the natural form is its altered functional qualities, such as reduced viscosity, prevention of granulation, enhanced binding properties, and extended shelf-life. However, the choice of starch for VC formulation is determined by its qualities such as foam stabilisation, gelling, moisture retention, thickening, etc. During processing of VC, it is important that for the starches to undergo gelatinization, resulting in a malleable and elastic mass. Upon cooling, the starches should form a viscoelastic gel through partial retrogradation. Typically, a mixture of various forms of starch is used to achieve the desired level of retrogradation, softening temperature, final gel strengths, and viscoelastic qualities. These features are ultimately determined by the ratio of amylose to amylopectin, as well as any alterations made to the starch (Breuninger et al., 2009). For the application of starches in VC formulation, it is typical to choose mixes of high amylopectin starches. Amylopectin enhances the ability of the system to expand, increases its viscoelasticity, and contributes to the reversibility of the gel, leading to the system exhibiting traits of melting or softening. Furthermore, the inclusion of starches that contain amylose or modified starches is utilised to enhance the structural integrity of the network. Nevertheless, the apparent melting or softening found in existing plant-based cheese replacements is accompanied by a compromise in their nutritional characteristics. Commercial plant-based cheese products have negligible amounts of protein (Fresán & Rippin, 2021; Grasso et al., 2021). Resistant starch has been studied as a dietary fibre source to enhance the perception of VC as a health-promoting and nutritious food. Noronha et al. (2007) formulated a VC where 50% of the fat, composed of a mixture of hydrogenated palm oil and rapeseed oil in a 2:1 ratio by weight, was substituted with native resistant starch. In order to achieve an acceptable amount of hardness for the VC, moisture content was increased to 60%. By substituting starch for fat, it was feasible to achieve a cheese with only 2% fat content while maintaining suitable hardness and cohesiveness for slicing, shredding, and moulding. This was accomplished by keeping the moisture content at 60%. The AC with a low-fat content, enhanced by the addition of starch, was found to have a smoother and firmer texture compared to samples where up to 85% of the fat content was substituted.

14.2.4 Fats

Vegetable fat as a substitute of milk fat in VC is crucial for enhancing the texture, particularly the melting properties and mouth feel, of the final product, which resembles that of dairy cheese. Usually, the fat content ranges from 22% to 28% and plays a role in both the taste and consistency of VC (Mulsow et al., 2007). Many different types of plant-based fats have been utilised in the creation of VC, which includes avocado, canola, cocoa, coconut, corn, palm, vegetable, safflower, sesame, soybean, and sunflower oils (Nicolás Saraco & Blaxland, 2020; Sha & Xiong, 2020). Unflavoured coconut oil is widely utilised in the cheese-like industry due to its high content of saturated fatty acids and resulting high melting point, followed by palm and sunflower oils. When it comes to cheese, triglycerides are predominantly found in milk fat. Milk contains a variety of fatty acids, including saturated, mono-unsaturated, and polyunsaturated. These different types of fatty acids make up the composition of milk, with saturated fats being the most common, followed by mono-unsaturated fats and polyunsaturated fats. However, the composition of this may vary depending on factors such as the season, region, and breed. The milk fat contains a significant amount of long-chain saturated fatty acids, making it partially crystalline at temperatures below room temperature. This property plays a crucial role in the creation of solid-like structures in cheese. The fat crystals have a tendency to interact with one another, resulting in the formation of a 3D-network that plays a significant role in determining the mechanical properties of cheese. The relationship between temperature and the solid fat content (SFC) of milk fat is a significant functional characteristic that is desirable to replicate using plant-based fats. Plant-based fats and oils have varying fatty acid compositions based on their botanical origin and isolation procedures, resulting in varied melting properties. Typically, the melting point of fats rises as their chain lengths grow and the number of double bonds decreases. Most plant-based oils have high levels of unsaturated fatty acids, making them liquid at room temperature and unable to form solid fat crystal networks. In order to increase the melting point of a vegetable oil to match that of milk fat, it is necessary to partially hydrogenate or modify it through interesterification. Nevertheless, this procedure is not much desirable due to its tendency to transform beneficial unsaturated fats into detrimental trans- and saturated fats if not precisely monitored, which could contribute to a greater possibility of developing cardiovascular disease (Hu et al., 2001). Alternatively, the SFC versus temperature profile and textural characteristics of the VC can be enhanced for specific products by combining natural plant-based solid fats (such as coconut oils) with liquid oils. Additionally, it is important to consider that the nutritional characteristics and chemical stability of plant-based lipids are influenced by their fatty acid composition. Adding a vegetable oil that is high in omega 3 fatty acids, such as flaxseed, rapeseed, or soybean oil, can be advantageous for fortifying the VC with EPA and DHA. These long-chain polyunsaturated fatty acids are responsible for numerous physiological benefits (Dal Bello et al., 2017). Polyunsaturated fats may have advantageous impacts on health (Saini & Keum, 2018). However, they have a tendency to be in liquid form at room temperature and are very prone to lipid oxidation, which can decrease the desirability of a product as it produces unpleasant rancid smells (Arab-Tehrany et al., 2012). A study conducted by Mattice and Marangoni (2020), involved blending coconut oil (75%) with high oleic sunflower oil (25%) to replicate the saturated to unsaturated fat ratio found in cow milk. Maltodextrin, a type of fat replacer, can also be used to create low-fat VC. Inulin is a fat replacer that has rheological qualities similar to table fats and can be used as a texture modifier (Karimi et al., 2015).

14.3 Emulsifying Salts

Emulsifying salts facilitate the hydration and solubilisation of proteins by inducing various physico-chemical alterations in the cheese structure. Typically, the content in the formulation can vary from 0.5% to 3% (Guinee, 2011). Several different ES are utilised, such as citrates, lactates, phosphates, and tartrates, with ammonium, potassium, or sodium serving as a counter-ion.

14.4 Hydrocolloids

Hydrocolloids derived from polysaccharides are commonly used in the production of VC. These hydrocolloids can be classified into two categories: neutral gums, such as agar combination, basil seed, guar, konjac glucomannan, and locust bean gum (LBG); and anionic gums, which include alginates, pectins, κ-carrageenans, ι-carrageenans, λ-carrageenans, and xanthan gum (XG). Usually, these gums help with moisture control, which promotes a desirable texture and stability of the mix. Single or a combination of two hydrocolloids or vegetable microfibers can also be used. Carrageenan is most commonly used in combination with guar gum that increases the viscosity but does not form gel. While oat fibre is widely utilised as a plant fibre in the production of hard and extra-hard VC (Carpentier et al., 2021).

14.5 Food Additives

Flavour is crucial sensory characteristic that significantly influence the acceptance of VC. Flavour compounds are added at levels of 0.5–3%. It is mostly added at the end of processing to minimize the loss of volatile compounds during processing (Guinee, 2011). Flavouring which are artificial flavouring additives, are added to replicate the taste of various cheese varieties such as mozzarella, gouda, cheddar and others. For enchaning natural taste, vegetable-based ingredients such as carrot puree or onion powder are also used. Flavour enhancers such as glutamates and yeast auto lysates may also be used in VC formulation (Chandan & Kapoor, 2011). The colour of cheese plays a crucial role in attracting customers and influencing their preferences and purchasing. The majority of cheeses have a natural, buttery yellow colour that is derived from the presence of beta-carotene in cow's milk. Consumers have certain expectations when it comes to the colour of VC. Due to the absence of certain yellow pigments found in plant-based milk or proteins, it is likely that VC may have a white colour. Incorporation of colour in the formulation of VC are optional and may be added at a concentration about 0.04% (Guinee & McSweeney, 2007). Food colouring is incorporated to enhance its visual appeal for potential buyers especially in case where use of other plant based materials like legumes may give the final product, an undesirable colour. Mineral and vitamins can also be added to match the nutritional quality of conventional cheese at 0–0.5%. These may include magnesium oxide, zinc oxide, iron, vitamin A, palmitate, riboflavin, thiamine and folic acid (Chandan & Kapoor, 2011; Guinee, 2011). Preservatives are added to increase the shelf life by preventing growth of microbes. These may include potassium sorbet, nisin, propionate etc. at the rate of 0.1%. Acidifying agents is also added to increase the stability of VC. They decrease the pH of VC to value in the range of 6–6.5 such that its stability during storage is increased (Guinee & McSweeney, 2007). However, many customers consider additives concerning and view them as "unnatural" in products (Tangyu et al., 2019).

14.6 Production of VC

The production process of VC involves thoroughly mixing of dry ingredients with water under heat or mechanical treatment with addition oils or fats to produce an oil-in-water emulsion that forms a uniform cheese-lie mass. The ingredients for VC are selected on the basis of functionality desired in the end-product. An appropriate formulation of ingredients are required to produce VC similar in properties with cheese. Different ratios of different plant based ingredients along with different processing parameters can be used to produce wide array of functional properties which include melting, stretching, slicing and shredding (Fox et al., 2017). Many current studies involve determining the perfect combination and composition of elements and substances for a tailored application. The processing of vegan cheese is similar to processing of processed cheese and it does not involve ripening stage, resulting in shorter production time. Processing parameters involved in manufacture of VC such as time, temperature, agitation, type of cooker etc. have a significant impact on the functionalities of the final product. The process begins by melting the fat and gradually increasing the temperature to 70–85 °C. For the creation of a stable emulsion, a stabiliser system consisting of proteins and starches is added to it. The oil and water are mixed together and vigorously stirred to ensure the emulsion remains stable. The blending process results in their interaction, forming a uniform mixture to which salt, flavour compounds, and acid are subsequently added. The texture of the VC is determined by the pH of the final mass (Bachmann, 2001; Guinee et al., 2004). The molten mass is carefully poured into a mould and cooled at 4 °C to create a solid gel resembling cheese. Grossmann and McClements (2021) briefly described an overview of process involved in VC production based on phase transitions. The differentiation between the fractionation and tissue disruption paths is determined by the primary factor that triggers the sol-gel transition. In fractionation path, it is induced by different ingredients from different source whereas in tissue disruption path, transition is induced by the ingredients present in raw material. However, the tissue disruption path is more efficient in terms of energy usage as it utilises two phase transitions, unlike the fractionation path.

14.7 Physicochemical Properties of Cheese

The functional properties of vegan cheese plays an important role in its application as an ingredients. Cheese possess many functional and physical properties as an ingredients which are primarily utilized depending on the application for modifying flavour, alter their texture, and enhance viscosity, addition of mouth feel and for colour (Lucey, 2008). Hard cheese, for instance, can be grated and shredded to make it easier to as sprinkle on food products whereas soft cheese can only be spread and cannot be shredded and grated. These functional qualities are exhibited throughout the meal preparation and consumption process, influencing both taste and appearance of food prepared (Smith et al., 2018). Vegan cheese can be customised using various ingredients, formulas, and manufacturing conditions (Kapoor et al., 2007). The functional qualities include ability to melt, shed, slice and stretch and desirable texture. These specific properties must be imitated to produce VC which can have application in food products.

14.7.1 Texture

Texture plays a crucial role in distinguishing different varieties of cheese. It is also regarded as one of the factors that determine the overall quality of the cheese (Antoniou et al., 2000). Cheese texture is significantly influenced by its chemical composition, which is primarily determined by the milk's pH, protein, fat, and initial moisture content. The texture of cheese undergoes significant changes during ripening which is caused by proteolysis, moisture loss, salt absorption, pH fluctuations, and the gradual dissolution of residual calcium linked to casein particles (Lucey et al., 2003). Similarly, the texture of vegan cheese is determined by its ingredients and composition, along with processing parameters, which influences their functionality (Lobato-Calleros et al., 1997). Hardness refers to the cheese's ability to resist deformation when an external force is applied, particularly at ambient or low temperatures and it is determined by its moisture and fat content (Guinee et al., 2002). When creating a block cheese, a lower moisture content is required, whereas spreadable soft cheese has a higher moisture content. Brittleness refers to the property of cheese to fracture into tendency of a cheese to fracture into pieces upon subjecting an external force (Guinee et al., 2002). This property is important when cheese has application where it is to be shredded, grated (Kapoor et al., 2007). Another property is adhesiveness which is the resistance of cheese to separate from the material it comes into contact (Guinee et al., 2002). Block cheese with low adhesiveness is preferred because it is not sticky, making it easier to cut, slice, or shred. On the other hand, soft and spreadable cheeses require a high level of adhesiveness to ensure they stick together. In a study conducted by Mattice and Marangoni (2020) examined the melting and stretching capabilities of plant-based cheese prototypes that incorporated zein. Various types of Zein cheeses were created with different levels of zein content, ranging from 0% to 30% and were examined to assess their texture. Hardness, springiness, chewiness, and gumminess of zein cheeses were compared with cheddar cheese which was taken as control. Cheese with 30% zein exhibited a striking resemblance to cheddar cheese in terms of their textural characteristics. The recorded values and their changes with temperature increase were highly comparable. The texture of the sample exhibited notable changes as the temperature increased from 5 °C to 50 °C, with a decrease in hardness, chewiness, and gumminess. The decrease in each of these parameters is a result of the melting of solid fat at higher temperature, as well as the softening of the zein or casein networks. Both the 30% zein and cheddar cheese samples maintained their springiness even as the temperature increased. It appears that the protein content plays a significant role in determining the springiness of cheeses. Interestingly, the study found that the weakening of non-covalent interactions within the protein networks does not affect the overall springiness of the product.

14.7.2 Shreddability & Sliceability

Shreddability is the ability of cheese to cut into long thin strips and it is indicated by ease of shredding and shape and integrity of the shreds, the tendency of shreds to mat or remain free-flowing, and the tendency to form fines during shredding (Childs et al., 2007). The ability of cheese to be shredded is affected by a variety of material properties and is primarily determined by a delicate balance of its viscoelastic properties. It is crucial for cheese products like pizza cheese, as it facilitates their utilization and accelerates melting. Cheese hardness and adhesiveness significantly influence its shreddability. Cheeses with a hard texture and low adhesiveness are suitable for slicing and shredding, making them easier to consume and apply to other food products. Hard cheeses, with low moisture content and high storage moduli, may be too gummy and dry, causing shattering during shredding. Soft cheeses, with high moisture content, low storage moduli, and high surface energies, exhibit strong blade adhesion, decreasing their Shreddability (Childs et al., 2007). However, some soft cheeses, like Feta or Blue, are hard to shred due to their short, crumbly texture.

14.7.3 Stretchability

Strechability is the capacity of melted cheese to create elongated strands under force without breaking. This property varies with the type of cheese, while some of them have good strechability while other have no stretch at all. Like mozzarella cheese is known for its excellent stretchability when heated, making it a popular choice for baking pizzas. In a study by Mattice and Marangoni (2020), VC based on zein protein was prepared with a range of 0–30% and was compared for its stretchability with cheddar cheese. Also samples made of gluten and pea protein was compared. It was evaluated at temperature of 65 °C, by determination of the force used to elongate the sample, together with the distance at which the cheese sample fractures. Zein based cheese containing 30% zein showed stretchability similar to cheddar cheese. Samples made of gluten and pea protein tore suddenly during stretching. Peak of maximum force increased steadily and decreased when zein sample was tore but sharp decrease was observed in case of gluten and pea protein based cheese samples. Cheese containing 30% zein showed strechability which adds zein as potential ingredient for achieving cheese like properties in VC. Zein network softened similar to casein network with increase in temperature, which is likely due to weakening od non-covalent interactions (Gabriele et al., 2001). The strechability of both casein and zein is because of weakening of bonds rather than completely lost.

14.7.4 Meltabilty

Meltability of cheese refers to the extent to which it can melt and spread when subjected to heat. The requirement of this property differs upon their application. Foods such as cheese slices, pizza toppings requires high meltabilty whereas for application where food go through high temperature such as bakery, requires low melting resistance. A gel-sol transition is involved during melting of cheese where a solid cheese is converted into material which is viscous (Schenkel et al., 2013). Fats play major role in meltability. Gel structure of VC depends on the fraction of fat in the continuous phase, which is of two types. The first type is emulsion-filled gels, where the fat particles are evenly distributed within a continuous network, causing the structure to break (Sala et al., 2007). The second type is protein-stabilized emulsion gels, where the fat particles create a continuous network (Dickinson, 2012). Emulsion filled gels can melt due to presence of oil droplets which act as lubricants, enhancing flow of the melted continuous phase (Vogt et al., 2015). On the other hand, if the oil droplets come together to create a network, they have the potential to enhance the gel by encouraging the development of another network which can be used to adjust the melting properties (Dickinson, 2012). In conventional cheese, it is attribute to non-covalent bonds between casein and upon heating these are weaken. Nevertheless, the interactions are just diminished and not entirely eliminated, so preventing the substance from attaining full viscosity. This property is also highly affected by fat component of the cheese. Fat globules being larger in size than caseins, can disrupt their interaction. With increase amount of fat in between casein matrix leads to decreased strength of matrix and consequently high meltbailty of chees. Developing plant-based proteins that can replicate the functionality of casein in cheese has posed significant challenges (Bachmann, 2001). Most of the VC have very low protein content as there is inadequate information regarding plant protein functionality regarding meltability. Instead non-protein components are used for the structural requirement. Due to which many of VC contain substantial amount of palm oil, coconut oil and other fats from plant sources which will melt as the temperature rises, and can replicate the melting behaviour of cheese. Zein proteins is a source highly used in VC and it has a low content of cysteine which required for covalent linking and so the gel network formed by it is believed to occur through non -covalent bond and so it has ability to show comparable melting characteristic in VC. In a recent study it has been indicated that zein when used together with water, oil and starch, has the ability to imitate melting behaviour of traditional cheese (Mattice & Marangoni, 2020). Pea proteins forms heat induced gel but covalent bonds are involved which is not broken by heating. Therefore, pea proteins is not feasible for food applications that needs melting property but it can be used for development of semi-hard VCs. Cheese products that are made from starch and oil have superior melting characteristics, making them highly suitable for cuisines like pizza, nachos, and grilled cheese as they closely align with the required sensory attributes (Grasso et al., 2021).

14.8 Conclusion

Cheese are in the diet of humans since hundreds of years but with the rise in veganism and so there is a demand for a similar product made of plant ingredients. Though in achievement of technological properties similar to dairy cheese, the nutritional properties of VC should not be compromised. Nevertheless, creating plant-based analogues poses a significant challenge as it is difficult to replicate the intricate structure and composition of traditional cheese using plant-based ingredients.

Extensive efforts have been made over the past few years to develop a vegan cheese that meets the demands of consumers. Many challenges have been encountered when attempting to replicate the properties of natural cheese which include block-forming property, shreddability, stringiness, meltability, colour, taste, and flavour. Some cheeses that are easier to imitate have a smooth and creamy texture, like feta, ricotta, or cottage cheese. Additionally, cheeses with bold flavours, such as spicy or smoky varieties, can help mask the taste of the plant source. Consequently, studies are required to mitigate the problem related to different functionality, along with studies regarding enrichment of VC with nutrients.. In order to meet customer demands, it is crucial to tackle these problems and provide VC that is both clean-labelled and of good quality.

References

Anderson, T. J., & Lamsal, B. P. (2011). REVIEW: Zein extraction from corn, corn products, and coproducts and modifications for various applications: A review. *Cereal Chemistry, 88*(2), 159–173. https://doi.org/10.1094/CCHEM-06-10-0091

Antoniou, K. D., Petridis, D., Raphaelides, S., Omar, Z. B., & Kesteloot, R. (2000). Texture assessment of French cheeses. *Journal of Food Science, 65*(1), 168–172. https://doi.org/10.1111/j.1365-2621.2000.tb15974.x

Arab-Tehrany, E., Jacquot, M., Gaiani, C., Imran, M., Desobry, S., & Linder, M. (2012). Beneficial effects and oxidative stability of omega-3 long-chain polyunsaturated fatty acids. *Trends in Food Science & Technology, 25*(1), 24–33. https://doi.org/10.1016/j.tifs.2011.12.002

Avilés-Gaxiola, S., Chuck-Hernández, C., Rocha-Pizaña, M. D. R., García-Lara, S., López-Castillo, L. M., & Serna-Saldívar, S. O. (2018). Effect of thermal processing and reducing agents on trypsin inhibitor activity and functional properties of soybean and chickpea protein concentrates. *LWT, 98*, 629–634. https://doi.org/10.1016/j.lwt.2018.09.023

Bachmann, H.-P. (2001). Cheese analogues: A review. *International Dairy Journal, 11*(4–7), 505–515. https://doi.org/10.1016/S0958-6946(01)00073-5

Batista, A. P., Portugal, C. A. M., Sousa, I., Crespo, J. G., & Raymundo, A. (2005). Accessing gelling ability of vegetable proteins using rheological and fluorescence techniques. *International Journal of Biological Macromolecules, 36*(3), 135–143. https://doi.org/10.1016/j.ijbiomac.2005.04.003

Boye, J., Zare, F., & Pletch, A. (2010). Pulse proteins: Processing, characterization, functional properties and applications in food and feed. *Food Research International, 43*(2), 414–431. https://doi.org/10.1016/j.foodres.2009.09.003

Breuninger, W. F., Piyachomkwan, K., & Sriroth, K. (2009). Tapioca/cassava starch. In *Starch* (pp. 541–568). Elsevier. https://doi.org/10.1016/B978-0-12-746275-2.00012-4

Carpentier, J., Conforto, E., Chaigneau, C., Vendeville, J.-E., & Maugard, T. (2021). Complex coacervation of pea protein isolate and tragacanth gum: Comparative study with commercial polysaccharides. *Innovative Food Science & Emerging Technologies, 69*, 102641. https://doi.org/10.1016/j.ifset.2021.102641

Chandan, R. C., & Kapoor, R. (2011). Principles of cheese technology. In R. C. Chandan & A. Kilara (Eds.), *Dairy ingredients for food processing* (1st ed., pp. 225–265). Wiley. https://doi.org/10.1002/9780470959169.ch10

Childs, J. L., Daubert, C. R., Stefanski, L., & Foegeding, E. A. (2007). Factors regulating cheese shreddability. *Journal of Dairy Science, 90*(5), 2163–2174. https://doi.org/10.3168/jds.2006-618

Dal Bello, B., Torri, L., Piochi, M., Bertolino, M., & Zeppa, G. (2017). Fresh cheese as a vehicle for polyunsaturated fatty acids integration: Effect on physico-chemical, microbiological and sensory characteristics. *International Journal of Food Sciences and Nutrition, 68*(7), 800–810. https://doi.org/10.1080/09637486.2017.1301891

Dickinson, E. (2012). Emulsion gels: The structuring of soft solids with protein-stabilized oil droplets. *Food Hydrocolloids, 28*(1), 224–241. https://doi.org/10.1016/j.foodhyd.2011.12.017

Fathi, M., Donsi, F., & McClements, D. J. (2018). Protein-based delivery systems for the nanoencapsulation of food ingredients. *Comprehensive Reviews in Food Science and Food Safety, 17*(4), 920–936. https://doi.org/10.1111/1541-4337.12360

Fox, P. F., Guinee, T. P., Cogan, T. M., & McSweeney, P. L. H. (2017). *Fundamentals of cheese science* (2nd ed.). Springer.

Fresán, U., & Rippin, H. (2021). Nutritional quality of plant-based cheese available in Spanish supermarkets: How do they compare to dairy cheese? *Nutrients, 13*(9), 3291. https://doi.org/10.3390/nu13093291

Gabriele, D., De Cindio, B., & D'Antona, P. (2001). A weak gel model for foods. *Rheologica Acta, 40*(2), 120–127. https://doi.org/10.1007/s003970000139

Gepts, P., Beavis, W. D., Brummer, E. C., Shoemaker, R. C., Stalker, H. T., Weeden, N. F., & Young, N. D. (2005). Legumes as a model plant family. Genomics for food and feed report of the cross-legume advances through genomics conference. *Plant Physiology, 137*(4), 1228–1235. https://doi.org/10.1104/pp.105.060871

Grasso, N., Roos, Y. H., Crowley, S. V., Arendt, E. K., & O'Mahony, J. A. (2021). Composition and physicochemical properties of commercial plant-based block-style products as alternatives to cheese. *Future Foods, 4*, 100048. https://doi.org/10.1016/j.fufo.2021.100048

Grossmann, L., & McClements, D. J. (2021). The science of plant-based foods: Approaches to create nutritious and sustainable plant-based cheese analogs. *Trends in Food Science & Technology, 118*, 207–229. https://doi.org/10.1016/j.tifs.2021.10.004

Guinee, T. P. (2011). Cheese | Cheese analogues. In *Encyclopedia of dairy sciences* (pp. 814–821). Elsevier. https://doi.org/10.1016/B978-0-12-374407-4.00096-0

Guinee, T. P., & McSweeney, P. L. H. (2007). Cheese-like products. In *Cheese problems solved* (pp. 384–388). Elsevier. https://doi.org/10.1533/9781845693534.384

Guinee, T. P., Feeney, E. P., Auty, M. A. E., & Fox, P. F. (2002). Effect of pH and calcium concentration on some textural and functional properties of mozzarella cheese. *Journal of Dairy Science, 85*(7), 1655–1669. https://doi.org/10.3168/jds.S0022-0302(02)74238-0

Guinee, T. P., Carić, M., & Kaláb, M. (2004). Pasteurized processed cheese and substitute/imitation cheese products. In *Cheese: Chemistry, physics and microbiology* (Vol. 2, pp. 349–394). Elsevier. https://doi.org/10.1016/S1874-558X(04)80052-6

Hu, F. B., Manson, J. E., & Willett, W. C. (2001). Types of dietary fat and risk of coronary heart disease: A critical review. *Journal of the American College of Nutrition, 20*(1), 5–19. https://doi.org/10.1080/07315724.2001.10719008

Kapoor, R., Metzger, L. E., Biswas, A. C., & Muthukummarappan, K. (2007). Effect of natural cheese characteristics on process cheese properties. *Journal of Dairy Science, 90*(4), 1625–1634. https://doi.org/10.3168/jds.2006-746

Karimi, R., Azizi, M. H., Ghasemlou, M., & Vaziri, M. (2015). Application of inulin in cheese as prebiotic, fat replacer and texturizer: A review. *Carbohydrate Polymers, 119*, 85–100. https://doi.org/10.1016/j.carbpol.2014.11.029

Kasprzak, M. M., Macnaughtan, W., Harding, S., Wilde, P., & Wolf, B. (2018). Stabilisation of oil-in-water emulsions with non-chemical modified gelatinised starch. *Food Hydrocolloids, 81*, 409–418. https://doi.org/10.1016/j.foodhyd.2018.03.002

Lamichhane, P., Kelly, A. L., & Sheehan, J. J. (2018). Symposium review: Structure-function relationships in cheese. *Journal of Dairy Science, 101*(3), 2692–2709. https://doi.org/10.3168/jds.2017-13386

Lobato-Calleros, C., Vernon-Carter, E. J., Guerrero-Legarreta, I., Soriano-Santos, J., & Escalona-Beundia, H. (1997). Use of fat blends in cheese analogs: Influence on sensory and instrumental textural characteristics. *Journal of Texture Studies, 28*(6), 619–632. https://doi.org/10.1111/j.1745-4603.1997.tb00142.x

Lucey, J. A. (2008). Some perspectives on the use of cheese as a food ingredient. *Dairy Science and Technology, 88*(4–5), 573–594. https://doi.org/10.1051/dst:2008010

Lucey, J. A., Johnson, M. E., & Horne, D. S. (2003). Invited review: Perspectives on the basis of the rheology and texture properties of cheese. *Journal of Dairy Science, 86*(9), 2725–2743. https://doi.org/10.3168/jds.S0022-0302(03)73869-7

Masotti, F., Cattaneo, S., Stuknytė, M., & De Noni, I. (2018). Status and developments in analogue cheese formulations and functionalities. *Trends in Food Science & Technology, 74*, 158–169. https://doi.org/10.1016/j.tifs.2018.02.016

Mattice, K. D., & Marangoni, A. G. (2020). Physical properties of plant-based cheese products produced with zein. *Food Hydrocolloids, 105*, 105746. https://doi.org/10.1016/j.foodhyd.2020.105746

Mefleh, M., Pasqualone, A., Caponio, F., & Faccia, M. (2022). Legumes as basic ingredients in the production of dairy-free cheese alternatives: A review. *Journal of the Science of Food and Agriculture, 102*(1), 8–18. https://doi.org/10.1002/jsfa.11502

Momany, F. A., Sessa, D. J., Lawton, J. W., Selling, G. W., Hamaker, S. A. H., & Willett, J. L. (2006). Structural characterization of α-Zein. *Journal of Agricultural and Food Chemistry, 54*(2), 543–547. https://doi.org/10.1021/jf058135h

Mulsow, B. B., Jaros, D., & Rohm, H. (2007). Processed cheese and cheese analogues. In A. Tamime (Ed.), *Structure of dairy products* (1st ed., pp. 210–235). Wiley. https://doi.org/10.1002/9780470995921.ch8

Nicolás Saraco, M., & Blaxland, J. (2020). Dairy-free imitation cheese: Is further development required? *British Food Journal, 122*(12), 3727–3740. https://doi.org/10.1108/BFJ-11-2019-0825

Noronha, N., O'Riordan, E. D., & O'Sullivan, M. (2007). Replacement of fat with functional fibre in imitation cheese. *International Dairy Journal, 17*(9), 1073–1082. https://doi.org/10.1016/j.idairyj.2007.01.011

Saini, R. K., & Keum, Y.-S. (2018). Omega-3 and omega-6 polyunsaturated fatty acids: Dietary sources, metabolism, and significance—A review. *Life Sciences, 203*, 255–267. https://doi.org/10.1016/j.lfs.2018.04.049

Sala, G., Van Aken, G. A., Stuart, M. A. C., & Van De Velde, F. (2007). Effect of droplet–matrix interactions on large deformation properties of emulsion-filled gels. *Journal of Texture Studies, 38*(4), 511–535. https://doi.org/10.1111/j.1745-4603.2007.00110.x

Sandberg, A.-S. (2011). Developing functional ingredients: A case study of pea protein. In *Functional foods* (pp. 358–382). Elsevier. https://doi.org/10.1533/9780857092557.3.358

Schenkel, P., Samudrala, R., & Hinrichs, J. (2013). Thermo-physical properties of semi-hard cheese made with different fat fractions: Influence of melting point and fat globule size. *International Dairy Journal, 30*(2), 79–87. https://doi.org/10.1016/j.idairyj.2012.11.014

Sha, L., & Xiong, Y. L. (2020). Plant protein-based alternatives of reconstructed meat: Science, technology, and challenges. *Trends in Food Science & Technology, 102*, 51–61. https://doi.org/10.1016/j.tifs.2020.05.022

Smith, J. R., Carr, A. J., Golding, M., & Reid, D. (2018). Mozzarella cheese—A review of the structural development during processing. *Food Biophysics, 13*(1), 1–10. https://doi.org/10.1007/s11483-017-9511-6

Sołowiej, B., Cheung, I. W. Y., & Li-Chan, E. C. Y. (2014). Texture, rheology and meltability of processed cheese analogues prepared using rennet or acid casein with or without added whey proteins. *International Dairy Journal, 37*(2), 87–94. https://doi.org/10.1016/j.idairyj.2014.03.003

Tangyu, M., Muller, J., Bolten, C. J., & Wittmann, C. (2019). Fermentation of plant-based milk alternatives for improved flavour and nutritional value. *Applied Microbiology and Biotechnology, 103*(23–24), 9263–9275. https://doi.org/10.1007/s00253-019-10175-9

Vogt, S. J., Smith, J. R., Seymour, J. D., Carr, A. J., Golding, M. D., & Codd, S. L. (2015). Assessment of the changes in the structure and component mobility of Mozzarella and Cheddar cheese during heating. *Journal of Food Engineering, 150*, 35–43. https://doi.org/10.1016/j.jfoodeng.2014.10.026

Wrigley, C. W. (1996). Giant proteins with flour power. *Nature, 381*(6585), 738–739. https://doi.org/10.1038/381738a0

Chapter 15
Novel Cheese Making Technologies

Deepak Joshi and Syed Insha Rafiq

15.1 Introduction

Cheesemaking is an art although the primary objective of cheese making was to extend the shelf life of milk. During the course of time the cheese making has undergone many developments from the use of refrigeration and pasteurized milk to the use of coagulants other than enzyme rennin. Implementation of new technologies like microencapsulation, membrane concentration and separation processes came into existence and greatly enhanced plant capacity. Membrane filtration techniques, which include reverse osmosis, nano-filtration, ultrafiltration, and microfiltration has largely been used for the cheese making process. More recent development is use of precision fermentation for cheese making.

15.2 Microencapsulation

15.2.1 Encapsulation in Cheese Production

Encapsulation creates particles of varying sizes by encapsulating active substances (Vlahoviček-Kahlina et al., 2021; Lemic et al., 2021). The two main types of encapsulation systems are core-shell (capsules) and matrix (spheres) (Vinceković et al., 2017). The fill, core, internal, payload, or active phase is the encapsulated material.

D. Joshi (✉)
Department of Food Science and Technology, National Institute of Food Technology Entrepreneurship and Management, Sonipat, Haryana, India

S. I. Rafiq
Department of Food Engineering, Bhaigurdas College of Engineering and Technology, Sangrur, Punjab, India

S. Mansha Rafiq (ed.), *Technological Advances and Trends in Cheese Making*, https://doi.org/10.1007/978-3-031-94476-5_15

The coating membrane, capsule, shell, matrix, external phase, or carrier material encloses the substance (Kavitake et al., 2018). Cheesemakers need encapsulation technology to protect ingredients from oxygen, light, free radicals, and other environmental factors. In addition, it allows for precise component administration (Jurić et al., 2020; Fathi et al., 2021). Encapsulation in biodegradable polymers is expanding due to the popularity of natural materials. Biopolymers can be taken from nature, made with microorganisms, or synthesized precisely to have specific properties (Chakravarthi & Robinson, 2008). Using ionic gelation to create biopolymeric matrices for encapsulation is common due to its gentle conditions (Belščak-Cvitanović et al., 2017; Jurić et al., 2020). Encapsulation in iopolymeric microparticles is expanding to improve physicochemical, functional, and release properties. To achieve this, cost-effectiveness and sustainable materials are considered (Lozano-Vazquez et al., 2015). The physicochemical properties and composition of biopolymers affect the encapsulation process. To encapsulate active ingredients, biopolymer structure must be carefully selected and understood (Jurić et al., 2020). The effectiveness of encapsulation depends on the payload, matrix material, and method. By carefully choosing encapsulation conditions, core-enclosing systems can achieve their desired functionality and properties (John et al., 2011).

Address potential issues before encapsulating. Setting preparation conditions starts the process of making multi-active ingredient microparticles. Each encapsulation component must be assessed. Understanding the molecular interactions between microorganisms and their environment is crucial before encapsulation (Jurić et al., 2020). Biopolymeric matrices can enclose components like enzymes, lactic acid bacteria, and chemical agents (e.g., Ca2+ for cheese production) (Mrkonjic Fuka et al., 2021; Jurić, et al., 2021a). Encapsulating multiple ingredients is difficult. Both the inner core and outer shell contain bacterial cultures and enzymes to synthesize these particles. The outer shell is more desirable than the core material, so the active ingredient releases faster. There is little research on developing and integrating these microparticles into cheese manufacturing. A product that can be easily integrated into cheese production can be created by optimizing particles with multiple ingredients, simplifying the technological process. During cheese production, ripening is unpredictable. It takes time and enzymes to break down curd's carbohydrate, lipid, and protein (Azarnia et al., 2006). Flavorzyme encapsulation speeds cheese ripening (Jahadi et al., 2016). This is crucial because using the "free" enzyme usually degrades taste and texture (Jahadi et al., 2015). In addition, the conveniently enclosed enzyme can be physically separated from the curd mixture during cheesemaking and released into the matrix as the cheese ripens (Kailasapathy & Lam, 2005). This unique example shows cheese manufacturing using encapsulation technology. Encapsulation technology speeds up cheese ripening and improves quality by adding bioactive compounds or extending shelf life (Mohammadi et al., 2015). Table 15.1 shows recent examples of encapsulation improving cheese production by extending shelf life and increasing value. Probiotics can survive harsh

Table 15.1 Most recent examples of the encapsulation of various active agents (functional components) to aid in cheese production and increase its bioactive value and shelf-life

Particle encapsulated	Method of fabrication	Size of particle	Result	References
Bifidobacterium-bifidum	Method of internal gelation method (sodium alginate and k-carrageenan)	ND	The probiotic bacteria were encapsulated, leading to enhanced viability and increased safeguarding and stability within cheddar cheese	Afzaal et al. (2020)
Nisin derived from the *Lactococcus lactis*)	Method of electrospinning (pullulan and isolate of Amaranth protein)	120.0 nm	Nanofibers are used to protect nisin, reducing its contact with food particles & preserving its anti-microbial properties in freshly prepared cheese	Soto et al. (2019)
Lactobacillus rhamnosus	Spray aerosol method (CaCl2 and sodium alginate)	ND	The probiotic, when encapsulated, had a significant impact on decreased firmness of fat cream cheese Phytosterol and β-glucan emulsion helps preserve viable cells during storage Encapsulation successfully preserved the viable cells with minimal reduction over a storage period of 35 days	Ningtyas et al. (2019)
Peel extract of mandarin	Chitosan/lecithin/maltodextrin Multilayered nanoliposomes	250–450 nm	The nanoliposome enriched processed cheese containing extracts, maintained its quality unchanged throughout a 3-month period of refrigerated storage	El-Messery et al. (2019)
SLactobacillu paracasei LAFTT L26	Rennet/milk powder (skim) Enzymatic gelation Sodium caseinate/transglutaminase & skim milk powder		The addition of microparticles affected the texture of the cheese, specifically making it harder and more stringy. Additionally, it significantly increased the antioxidant capacity of Iranian UF Feta cheese	Moghaddas Kia et al. (2018)
Oil (essential) from oregano (Origanum vulgare L.)	Phase inversion temperature method for preparing nano-emulsions prepared Sunflower oil and surfactants (sorbitan monooleate, polyoxyethylene 4-lauryl ether and PEG-40 hydroxylated castor oil)	–	The nanoemulsions exhibited antifungal activity on Minas Padrão cheese slices against Penicillium sp. Fusarium sp., & Cladosporium sp., (inoculated with the same fungal genera)	Bedoya-Serna et al. (2018)

(continued)

Table 15.1 (continued)

Particle encapsulated	Method of fabrication	Size of particle	Result	References
Lactobacillus acidophilus Chr. Hansen A/S (Denmark)	Sodium alginate and $CaCl_2$ Emulsification, ionic gelation method	60.0–500.0 µm	Encapsulated L. acidophilus inside edible films successfully hinder mesophilic aerobic bacteria growth in refrigerated cheese (Manaba) throughout storage	Santacruz and Castro (2018)
Aromatic compounds extracted from Carnobacterium maltaromaticum LMA 28	CaCl2 and sodium alginate Ionic gelation method Paraffin coated cheeses without/with beads	0.15–1.69 mm	For imparting olfactory characteristics of 3-methylbutanal to cheese bacterial encapsulation is a more effective method	González Ariceaga et al. (2018)
Lactobacillus rhamnosus 6134	Maillard reaction products of isomaltooligosaccharides and whey proteins Cold gelation inside emulsification system	–	Bacteria found in white brined cheese is protected by encapsulation, without causing any noticeable changes to its sensory characteristics	Liu et al. (2017)
Flavourzyme	Heating methods to produce liposome Soy cholesterol/lecithin	180 nm	The brined cheese experienced a rise in the overall perception and a decline in the texture rating	Jahadi et al. (2016)
Epigallocatechin gallate and catechin	Liposomes via mechanical agitation Soy lecithin	133 nm	Improved durability of low-fat hard cheese	Rashidinejad et al. (2014)

gastrointestinal conditions, adhere to cells, and benefit cells when consumed in sufficient amounts (Hill et al., 2014). Food matrix composition—fat, protein, and moisture—and storage conditions affect probiotic microorganism survival. Probiotic bacteria must be stable during food processing and storage. Lactic acid bacteria and other microorganisms remain in food products and during production for a limited time. Encapsulation improves object preservation, as Santacruz et al. demonstrated in 2018. According to Afzaal et al. (2020), adding probiotic bacteria to cheese production increased its survival rate, stability, and protection. Adding probiotics to cream cheese with -glucan and phytosterol emulsion, Ningtyas et al. (2019) improved it. The cream cheese was stored with enough viable probiotic cells. Probiotics can improve cheesemaking. To preserve their existence, organic supplements can be added. This may boost bacteria survival and abilities. Given this, choosing a natural encapsulation material that can be economically and gently produced and compatible co-encapsulants is crucial. Adding biopolymeric microparticles for encapsulation has many advantages when integrating "unconventional" cheese production technologies. Specific sensory issues may still occur. Consumer perception can be affected by microparticle sizes, which range from 1 to 1000 m (Lengyel et al., 2019). When considering microparticles in cheese, there are many benefits. Microparticles improve food textures, properties, protection, and controlled release of encapsulated components. According to multiple reports, consumers preferred 250–500 micrometer products. Desirable attributes like crispiness and granularity were linked to this preference, unlike smaller size categories (Torri et al., 2015; Jurić et al., 2019, 2021b, 2021). Additionally, Narsaiah et al. (2019) found that food product sizes between 620 and 980 _m did not affect sensory acceptability. These factors and the many benefits of microencapsulating cheese production ingredients make it possible to improve cheese product sensory attributes.

Integrating materials that make particles more than transporters drives encapsulation science. Encapsulation science advanced complex systems. However, calcium alginate particles are best for microorganism encapsulation. Multiple "active" ingredients encapsulated together increase efficacy and may produce a synergistic effect. Microencapsulated formulations with multiple ingredients are expected to be easily integrated into normal cheese-making and other food production processes in the future (Jurić et al., 2020; Mrkonjic Fuka et al., 2021).

15.2.2 Application of Microencapsulated Bacterial Strains in Cheese Production

Food companies rarely use microencapsulated products, but pharmaceutical companies do. Multiple factors limit the effectiveness of probiotic strains in food products, as is well known. To protect microbes from harsh conditions, various methods were developed. Microencapsulation effectively protects against bacteria (Özer et al.,

2008). According to Gomes da Cruz et al. (2009) it can mitigate the viability loss from prolonged storage or salting. A protective coating is applied to small gaseous, liquid, and solid particles during microencapsulation (Calvo et al., 2011). Encapsulation benefits materials greatly. This product's-controlled release mechanism allows gradual and targeted release of active ingredients over time (Jeyakumari, 2016). This stabilizes adjunct cultures/starters, reducing ripening time. Too much attention is paid to food microencapsulation methods and bioactive ingredients. Gel-particle and spray-coating encapsulate probiotics (Champagne & Fustier, 2007). Ortakci et al. (2012) found that alginate confines probiotic bacteria best. Goderska et al. (2003) found that Lactobacillus rhamnosus in an alginate matrix survived for 48 h. In contrast, identical bacteria's free cells were completely inactive under identical conditions. In 2012, Abd-Elhamid found that microencapsulation improved Bifidobacterium adolescentis survival in Kariesh cheese stored at low temperatures. However, Godward and Kailasapathy (2003) claimed that microencapsulating probiotic cells in Feta cheese reduced microorganisms more. Suppressing cellular metabolite removal inside the capsule may have caused this. Microencapsulation raises microorganism viability and alters the physical and chemical properties of food products with bacteria. Özer et al. (2008) found that encapsulated probiotic bacteria increased cheese proteolysis. Although these facts are true, this technology has not yet been widely used to produce sustainable and functional food. Only thorough research can do this.

15.3 Membrane Filtration Techniques

Membrane processing has significantly upgraded the dairy industry, leading to remarkable progress in product development. These changes has advanced over a period of almost five decades and has encompassed all membrane processing systems that are commercially viable, namely microfiltration (MF), ultrafiltration (UF), nanofiltration (NF) and reverse osmosis (RO). The utilization of membranes in the production of cheese originated from the creation of the MMV technique by Maubois, Mocquot, and Vassal. This technique has facilitated progress in cheesemaking by improving factory efficiency, increasing cheese output, establishing a streamlined production method, and enabling the development of innovative cheese varieties. Consequently, many plants worldwide, particularly in Europe, currently employ this method to produce a diverse array of cheeses (Tamime et al., 2006). The MMV process has resulted in a worldwide proliferation of commercial membrane applications in the cheese industry, along with intensified research endeavors to create novel applications and improve existing ones. Consequently, there have been progressions in cheese production, encompassing the development of novel cheese varieties, improvements in existing cheese-making techniques, and the adoption of more sophisticated membrane systems and materials to enhance the effectiveness of the cheese-making process. In addition, membrane filtration techniques have been utilized in other cheese-making processes, including whey processing. Since 1969, the utilization of membranes in cheese manufacturing has proliferated across diverse domains. This involves the production of various types of cheeses using the milk of goats, cows water buffaloes or ewes. Membranes are utilized in the production of milk powders that possess desirable characteristics for cheesemaking. They have the ability to reinstate the coagulation properties of milk that has been subjected to UHT treatment. Membranes are utilized in the process of on-farm milk concentration, where they are used to remove bacteria and somatic cells from cheese milk through MF. Additionally, membranes are employed in the enrichment of cheese milk with casein using MF. The mentioned advancements were prompted by enhancing membrane constituents, particularly the production of ceramic & mineral membranes. Furthermore, the study of the physicochemical equilibria of ultrafiltration retentates, analysis of the rheological properties of milk enriched with proteins, examination of the growth and activity of starter bacteria in liquid precheeses and the resulting cheeses, and the implementation of innovative cheesemaking techniques in cheese plants and laboratories around the globe were essential in these advancements. The increase in the desire for cheeses, specifically low-fat cheeses of superior quality, has also propelled progress in the industry.

15.3.1 Applications

Ultra filtration is the primary membrane process used in cheese production and is extensively advanced. Industrial settings currently employ devised methods to eliminate bacteria, isolate milk fat globules, and enhance the concentration of micellar casein. To successfully manufacture cheese using MF or UF, it is essential to possess a comprehensive comprehension of the distinct attributes of the protein-enriched products. The properties mentioned have a substantial influence on both the quality of the cheese and the economic efficiency of using membrane technology. Cheesemaking using UF can be divided into three main categories:

- Protein-standardized milk,
- Medium or intermediate concentrated retentates, and
- Liquid precheeses

15.3.1.1 Protein-Standardized Milk

Seasonal factors like lactation stage, weather, feeding patterns, and cow breed affect dairy milk protein content. Milk composition changes require cheesemakers to adjust processing parameters. Rennet coagulation is weak when protein content is low, resulting in a large loss of caseins as fines in the whey. UF effectively addresses these challenges by slightly increasing protein content. Cheese plants with advanced machinery maintain a consistent protein content of 3.70–4.5% in cheese milk year-round (Govindasamy-Lucey et al., 2011). Camembert cheese production in Europe uses protein-standardized milk and an Alpma coagulator. It's also used to make hard, cooked cheese. Several pilot plant and industrial studies have used the LCR concept for cheddar and other hard cheeses by increasing concentration or adding ingredients. The best concentration ratio for making these cheeses is 1.70:1.0 to 1.80:1.0, according to research. Cheese production uses traditional equipment for protein standardization or LCR. Ultrafiltration in a cheese plant is easy to do. Recently, ultrafiltration (UF) equipment manufacturers have proposed using systems with protein and fat sensors. This will help assess the fat and protein content of the milk and adjust cheese milk protein and fat levels. Because UF increases manufacturing efficiency, cheese production per day increases, rennet needs decrease, cheese quality improves, and yield increases by 1% for most cheese varieties, the cost of using UF for this purpose is maintained. The increase in productivity is due to reduced casein and fat waste in whey and improved whey protein preservation in cheese liquid. Whey protein retention with LCR is moderately improved over MMV. This can be estimated using Vandeweghe (2000)'s formula for cheese production. Due to its higher protein and total solids content, the whey may appreciate in value. This is important, but these advantages are negligible in industrial settings. Interestingly, many ultrafiltration plants in Europe aim to standardize protein. By using permeate to lower fluid UHT milk protein concentration to legally required levels, LCR/protein standardization provides an indirect but significant benefit. Most EC countries have a minimum protein content of 2.8% w/w, but France has 3.15%. UHT milk processors profited from this practice, but the EU banned it in 1997. ROI was less than 6 months for the UF investment. Milk dilution with water can be detected, but permeate cannot. Different cheeses have been made using LCR. Cottage is one of the types listed. St. Paulin, Mozzarella, and others. Industrial and pilot-plant experiments have shown that a concentration ratio of 1.2 to 1.7 optimizes Cottage cheese yield, flavor, and body. Cheese solidifies and tastes bland when levels exceed these thresholds. Thermizing milk at 74 °C for 10 s before ultrafiltration (UF) yields the most cottage cheese. Low-concentration retentates at 1.75:1.0 can produce high-quality mozzarella cheese with low moisture content, remarkable elasticity, and melting properties. Higher concentration cheese was firmer and lost more fat in the brine. LCR makes directly- and starter-acidified Mozzarella cheese. The LCR Brick cheese had higher fat losses and lower whey pH than the controls. The cheese was denser and coarser than the control cheese and rated less preferred. Colby used ultrafiltration to avoid curd-washing. We had similar sensory acceptability to the control group. Reports showed reduced rennet use and cooking time. Edam cheese was made using 2:1, 4:1, and 6:1 retentates. The cheese made with 2:1 LCR had the fewest flaws and was of the highest quality. The degradation rate of αs1-casein in this cheese was comparable to the control, while β-casein degradation was slower. After adding 2:1 UF Danbo cheese made from diafiltered UF milk, there is a 50% reduction in rennet requirements, a slight yield increase, and a 40% increase in cheese vat capacity as per various studies. The LCR concept, which increases yield without extra equipment, is widely used in commercial cheese production like Cheddar. Ultrafiltering milk to 4.5% protein by weight was suggested to increase Cheddar cheese production and moisture. Oommen et al. (2000) found that homogenizing cream improves ultrafiltration (UF) in cheesemaking. To make cheddar cheese, the 35% fat cream was homogenized and mixed with UF milk and skim milk to achieve 6% protein. Cheese fat recovery was 96.8%, higher than the control cheese's 94.7%. Cheese meltability and texture could be improved. Low-fat cheese quality is improved by LCR. Low-fat cheddar, mozzarella, and processed cheese are made with UF sweet buttermilk.

15.3.1.2 Medium or Intermediate Concentrated Retentates

Cheeses of various types, ranging from hard to soft, is being developed using retentates with a moderate concentration. This method involves the production of cheese using specialized equipment that can cut and handle the solid gel formed from the coagulation of concentrated retentates. These retentates are then diluted with water, which can be pure, salted, or acidified. The concentration of retentates utilized varies between a ratio of 2:1 and 5:1. The main implementation, currently in commercial level is the manufacturing of structured Feta cheese.

APV-Siro Curd Process

The development of this process was the result of a collaborative endeavor between an Australian dairy research team and the APV company. Commercial plants were established in the USA and Australia; however, they have been shut down due to technical and economic reasons. Mistry and Maubois (2004) have conducted a comprehensive analysis of the complexities involved in the process. Under conditions of continuous and automated operation an increase of 6–8% in cheese production is possible by this process.

Structured Feta-Like Cheese

This method for producing Feta-like cheese was developed in Denmark in response to the demand from Mediterranean countries. These countries sought cheese that closely resembled traditional products in terms of their texture and appearance, particularly the presence of mechanical holes. UF Feta-like cheese developed from pre-cheeses lacked this particular characteristic. The milk, after undergoing pasteurization and standardization of its fat content, is commonly homogenized at a pressure of 18 MPa and a temperature of 60 °C. Subsequently, it undergoes ultrafiltration at 50 °C temperature. The final composition consists of 28.50% of total solids (TS) by weight, which is equivalent to a concentration factor of 3:1. Lipase, glucono-δ-lactone and starter culture are added to the UF retentate, which has been homogenized under 5 MPa pressure and 65 °C temperature. The mixture is heated at 80 °C for 60 s, then cooled down 34 °C. After storing in a tank for short period retentate is moved to Pasilac equipment or specially designed Alfa-Laval Alcurd using a pump. Rennet is incorporated into the production process. Rennet is thoroughly blended in both types of equipment, resulting in the coagulation of the ultrafiltration retentate within tubes. The formed coagulum is extracted from the tubes, cut into cubes, molded, and allowed to drain for 16–24 h at 10–14 °C until 4.8 pH level. The result of this procedure is nearly similar to the traditional product. The traditional process claims a 14% increase in yield on a solids basis, which is significantly less than the 30% increase achieved with the liquid precheese process. This discrepancy clarifies why only a restricted number of plants have the ability to produce structured cheese that resembles Feta.

General Considerations for Using Intermediate UF Retentates

The benefits obtained by utilizing intermediate UF retentates in manufacturing cheese must be enough to justify the replacement of conventional cheese-making techniques. Moreover, the product's sensory attributes must align with the consumer's expectations. Investments include both UF equipment and other equipment like curd makers. The correlation between the higher amount of cheese produced and the disparity in composition between the final cheese and UF retentate is extremely significant. Skim milk quantity conserved grows exponentially as the disparity diminishes, whereas the expenses associated with ultrafiltration also rise exponentially in relation to the rate of flow. When conducting an economic study to evaluate investments, cheese plants should also consider the expected value of by-products they obtain: drained whey, which has higher levels of protein and fat compared to regular whey, and permeate. Each of these secondary products is of interest to a specific industrial network located further along the supply chain. Additionally, there are minor benefits associated with reducing the amount of rennet used and requiring less floor space and volume.

15.3.1.3 Liquid Precheeses (LPC Concept)

This method involves the condensation of cheese milk using ultrafiltration (UF) until it achieves the desired composition of the drained curd. This step is completed prior to the addition of rennet. The amount of whey drainage is minimized, and there is no use of cheese vats (Maubois, 2002). This principle was first exhibited in the manufacturing of Camembert cheese (Maubois, 2002). Nevertheless, it has subsequently been effectively utilized in the production of diverse cheese varieties, such as "fromages frais" (Quarg), as well as semi-hard cheeses (Saint Paulin).

15.3.1.4 Fresh Unripened Cheeses

Early attempts to use UF to make this type of cheese condensed the milk before adding rennet and starter. Cheese tasted metallic, acidic, and bitter. Due to curd's high mineral content and buffering capacity, these defects occur. Brulé et al. (1975)

found that preacidification to pH 6.0 followed by ultrafiltration or excessive concentration followed by water dilution reduced milk acidity. Novel membranes like mineral membranes and engineered membrane supports have made UF of high-viscosity substances easier. This advancement allows process to fully resolve the organoleptic defect. After fermenting the milk with a mesophilic culture to pH 4.6, rennet is added. Ultrafiltration removes lactose and mineral salts while preserving whey proteins. Ultrafiltration of pH 4.6 milk showed a high initial flux rate, which decreased with concentration. Maubois', (2002) explanation of the milk's highly porous polarization layer explains these observations. UF was used to make Quarg, a 12%-protein German cheese. It was necessary to develop ultrafiltration equipment that reduced mechanical shear stress on the retentate to apply the same procedure to other French cheeses with lower protein content. Curds acidified to 4.6 pH lose viscosity when centrifuged or concentrated. UF technology produced fresh unripened cheeses from buttermilk. The high phospholipid content of this new starting material, which has poor renneting properties, makes it appealing. Cream's smooth, rich texture is enhanced by extensive high-temperature treatment before churning. Multiple types of unripened cheese are made using the LPC or MMV method. Ricotta, cream cheese, and mascarpone are examples. The complex precipitation process and precise flavor and texture requirements make Ricotta production difficult. They described an ultrafiltration (UF) method that acidified whole milk to 5.9 pH using food-grade acid, acid whey powder, or lactic starter. Acidified milk is ultrafiltered at 55–60 °C until the protein content reaches 12% by weight. A scraped-surface heat exchanger heats acidified liquid precheese to 80 °C. After that, packaging containers are filled. After ultrafiltration at 50 °C, the mixture has 30% solids by weight. The retentate is at 90.0 °C and 1.0–1.50 bar. Changing the pressure to atmospheric promotes curd formation. After cooling to 70 °C, the product is packaged and chilled to 10 °C. There's no whey drainage. Ultrafiltration (UF) creates high-moisture cheeses like "Faisselles" and "country cheese." This cheese is usually made from raw milk contaminated with rennet and mesophilic starter at 18–22 °C. To release whey slowly, small amounts of coagulated material are hand-poured into molds after overnight cooling to 12–16 °C. The expelled whey is manually removed from the molds and carefully placed in retail containers. Labor cost increases reduced production. Since UF eliminates draining, labor needs have decreased significantly. This cheese is produced at its highest level ever (Maubois, 2002).

15.3.1.5 Soft Cheese

French cheese Camembert, known for its surface mold, was the first to use the MMV principle (Maubois, 2002). Maubois (2002) proposed many recipes to optimize ultrafiltration in commercial cheese making, taking into account its limitations. These recipes aim for a precise calcium salt balance in the curd to achieve the desired texture and flavor of Camembert cheese. After 24 h of UF retentate production, a 16-h molding period followed. Alfa-Laval's "Camatic," continuous demolding and molding apparatus works as follows: Ultrafiltration at 50 °C gives pasteurized milk (HTST) a 5:1 precheese concentration ratio and 35% by weight total solids. The precheese is then cooled to 30 °C. In this phase, 2% mesophilic lactic starter and 0.75% NaCl are added. The mixture is acidified to 5.5 pH. The liquid is then automatically infused with rennet and poured into molds. Curd wheels mature faster and are delicately stirred in the Camatic apparatus for 45 min. Each cheese receives a continuous electric current after one inversion. The cheese is in a stainless-steel cup and in contact with a carbon electrode on its air-exposed side. Whey has little electrolysis, and an air injector makes cheese wheel removal from trays easy. Freshly made Camembert cheeses are then brined for 30 min. They are then removed and sprayed with Penicillium candidum spores. The cheeses are stored at 11–12 °C for 12 days, allowing the white mold to grow and cover the cheeses. Yield rises 12–15%. Germany has been a major Camatic equipment market. France's UF Camembert faces "psycho-commercial" challenges. UF Camembert tasted like regular cheese. UF cheese paste has a higher density due to the lack of mechanical apertures. French consumers, used to buying Camembert cheese individually but not by weight, may perceive UF Camembert as a lower value for their money (Maubois, 2002). Due to the failure of UF Camembert, French cheesemakers have created new varieties, increasing production of many of them. The cheese "Pavé d'Affinois," introduced in 1982, reached 5000 tons by 2000. This cheese is fermented from 4 parts ultrafiltered whole milk retentate and 1 part thermophilic lactic starter. Rennet is mixed and placed in a 5-cm-high rectangular tray. This tray is incubated at 43 °C for 6 h to accelerate acidification and coagulation. After reaching room temperature, an automated dividing knife cuts the cheese slabs into 96 pieces. Each piece is a compact rectangular paving stone measuring 7 × 5 × 5 cm. The cheese ripens for 10 days, similar to Camembert cheese. Blue cheese and other soft cheeses ripened by bacteria and mold on the surface are also developed using the same principles and are successful (Maubois, 2002). The MMV process has been successful in producing Feta-like cheese globally. Before EU market reforms in 1997, 35% of Denmark's cheese production was Feta, and UF made up over 90% of that. Most of this cheese was made for Iran. Ziabary and Hoffmann found in 2001 that Iran has been making UF Feta cheese locally for years. Cheesemakers can now directly package Feta cheese at retail thanks to LPC (Low Pressure Cooking). The reported 30% yield increases exceeded the expected 22% whey protein retention. The complete elimination of curd particles and prevention of fat losses during curd cutting and coagulation in retail boxes may explain the difference. The curd also retains all k-caseinomacropeptide. Egyptian Domiati cheese was made using the same method: The 5:1 ultrafiltration (UF) whole milk retentate was mixed with 5% NaCl, 2% lactic starter, and lipase/rennet. After blending, the mixture was placed in Tetra Pak packages Tamime et al. (2006) introduced Tetra Tebel Casofill Bacteriologically Acidified Feta-type, ideal for brined cheeses like Cream cheese, Feta, and Queso Fresco.

15.3.1.6 Semi-hard Cheeses

Saint Paulin is a French cheese that undergoes surface ripening through bacterial action. The moisture content of the substance is approximately 47%, while its salt content is 2.5%. In order to make this cheese using UF, it is crucial to attain a protein content of at least 21% by weight (equivalent to 45% solids by weight) (Maubois, 2002). Mineral membranes, as demonstrated by Goudédranche et al. (2000), are easier to obtain compared to polymeric membranes. Various methods have been developed to preserve cheese, including the use of brine salting and dry salting techniques. Goudédranche et al. (2000) reported that it is feasible to achieve a maximum increase in cheese yield of 19%, while simultaneously reducing rennet usage by 85%. In order to regulate the acidic taste and delay the maturation of UF Saint Paulin cheese, it is imperative to reduce the ash and lactose levels in the retentate to values lower than 1.9% each. In order to improve the flavor of UF Saint Paulin cheese, it is advantageous to add lysozyme at a concentration of 0.5–1.0 g/L. This addition enhances the proteolytic activity and reduces the population of mesophilic bacteria in the cheese. Another alternative is to incorporate damaged lactococci cells (Saboya et al., 2001). A French study (Maubois, 2002) examined a type of cheese that goes through fermentation by propionic bacteria. The process entails creating a 7.5:1 retentate through a two-step preparation. The initial stage entails ongoing diafiltration at a concentration ratio of 3.0:1, which is simulated by a refractometric sensor in the permeate line. Subsequently, a heat treatment is conducted at a concentration ratio of 4.0:1. The second step entails the use of continuous ultrafiltration to attain a concentration ratio of 7.5:1. Specialized equipment, such as positive displacement recirculation pumps and short cartridges, is utilized for the management of substances with high viscosity. Initial agitation devices and the addition of rennet were utilized. The molding equipment is furnished with a vacuum stage to eradicate solubilized gases & a specialized injection nozzle for dispensing LPC (renneted) to molds that have either a spherical or cylindrical configuration.

15.3.1.7 Cheese Quality

15.3.1.7.1 Texture

UF cheeses provides a significant productivity merits and have achieved widespread consumer approval. Nevertheless, they exhibit unique attributes in relation to their composition, ripening attributes, and texture excellence. Furthermore, there has been a suggestion to not simply copy existing types of cheese, but rather to develop a fresh selection of cheese varieties. Soodam et al. (2014) found that the concentration of milk protein has no effect on the texture and microstructure of cheese, as long as it is allowed enough time to mature. The high concentration of calcium salts in UF retentate, which has a pH of 6.7, is the main cause of texture issues in UF cheeses. These issues include a grainy texture, lack of firmness, and tendency to crumble. The mineralization process of the drained curd, which encompasses the overall calcium content and the allocation of calcium between the soluble phase and casein matrix, plays a vital role in determining the rheology of the particular cheese type. The mineral composition can be altered as previously described in the section labeled "Characteristics of UF Retentates." One noteworthy feature of UF cheese is the integration of whey protein into the cheese. The quantity of whey proteins that are preserved is determined by the specific type and degree of ultrafiltration concentration. Preserving all the whey proteins derived from milk will contribute to around 20% of the overall protein of cheese. Utilizing the LCR method results in the preservation of smaller quantities. Whey proteins have the potential to substitute for a portion of the casein in cheese, acting as a filler and potentially resulting in a softer texture of the cheese. On the other hand, whey proteins have a higher capacity to attach to water in comparison to casein, leading to UF cheeses being less susceptible to dehydration during the retail process in contrast to traditional cheeses. UF cheeses exhibit elevated concentrations of calcium phosphate and k-GMP, both of which possess captivating nutraceutical and nutritional characteristics (Maubois et al., 2001).

15.3.1.7.2 Proteolysis and Ripening Characteristics

Prior research has consistently shown that UF cheese experiences a slower maturation process in comparison to conventional cheese. The addition of more whey proteins typically slows down the process of flavor development. The quality and taste of UF cheese was found to vary significantly, which has been attributed to variations in the concentrations of proteose-peptones and immunoglobulin in the whey proteins. The influence of whey proteins on the formation of flavor is less apparent in LCR cheeses as a result of the lower quantities of whey proteins that are present. However, it gains greater importance in cheeses produced from retentate with higher concentrations and those that undergo a longer ripening process. The postponed development could be ascribed to multiple factors. The presence of a substantial quantity of β-lactoglobulin in cheeses with UF can impede the overall proteolytic activity) & plasmin to some extent. The undenatured whey proteins found in UF cheeses demonstrate resilience against proteolysis induced by both proteases and enzymes derived from starters. The main factor contributing to the high buffering capacity of UF cheeses produced from ultrafiltered milk with a pH of 6.7 is most likely this factor. Indeed, it hinders or even entirely ceize the rate at which mesophilic lactic starter undergoes self-digestion (Saboya et al., 2001), thereby impeding the breakdown of the casein network. Incorporating flavor-producing enzymes or cell extracts in UF cheese can accelerate the rate of flavor development. Unlike in conventional cheese making, these enzymes and extracts are completely retained in the retentate. Lactic microorganisms or nonstarter bacteria or that acidify slowly can be used in UF cheesemaking to take advantage of their ability to break down proteins and produce flavors. This is possible because UF cheesemaking allows for independent control of acidification, drainage, and the microorganisms involved in ripening.

15.3.1.8 Functionality

For using cheeses as ingredients, various characteristics are taken into account (like the melting characteristics, ability to be shredded, viscosity, and ability to stretch). Ultrafiltration (UF) of milk before cheese making changes the physico-chemical attributes of cheese, leading to distinct impacts on particular functional properties. In their study, Acharya and Mistry (2002) discovered that Processed cheese made from Cheddar cheese base derived from UF milk, which had a protein content of up to 6%, exhibited lower meltability compared to regular Cheddar cheese. Moreover, these cheeses demonstrated the highest calcium concentration. The aforementioned observations also applied to the original Cheddar cheese. This cheese (milk having 6% protein) had a melting value of 62 mm, whereas the control had a melting value of 77 mm. At a temperature of 80 °C, molten processed cheese has comparatively higher viscosity for the UF cheese (1043.0 cP) than the control (557.0 cP). By implementing techniques such as pre-acidification, it is possible to effectively reduce the calcium levels in cheese, thus improving its functionality. Madsen and Qvist (2001) ascribed the decelerated liquefaction of UF Mozzarella cheese to the existence of whey proteins and proposed the utilization of proteolytic enzymes to expedite the breakdown of casein, thereby enhancing its melting characteristics. The primary determinant of the functional alteration is the discrepancy in the calcium equilibrium and its correlation with the casein composition. Nevertheless, the previously mentioned texture and proteolysis characteristics also have an impact. Specifically, the reduction in the rate of protein degradation and the presence of whey proteins directly impact the melting characteristics of cheese. The assessment of cheese quality and the potential for introducing new cheese varieties or implementing advanced technologies, such as membrane separations, is highly dependent on consumer approval and economic viability.

15.4 Reverse Osmosis (RO) in Cheese Making

Richardson (1929) suggested adding evaporated milk to cheesemaking. After 30 years, revision of this idea using a novel method that combined Berridge's (1951) findings on cold-temperature rennet activity's early and late stages. Dehydrated milk or reverse osmosis can increase milk solids. Reverse osmosis (RO) is used to increase whole milk solids to 20–25% for cheddar cheesemaking. RO can increase cheese production by 2–3% by reducing milk volume by 20%. Whey solids increase fat reduction. Because whey solids processing partially homogenizes fat. An abrupt pressure drop during RO can hydrolyze cheese and milk fats. According to Richard et al. (2000), this impairs taste, smell, and texture. Cheese production may not use reverse osmosis (RO), but whey processing does. Many US cheese milk preconcentration plants use thermal evaporators. Rinsing curd removes lactose and minerals. The entire process's benefits must be assessed.

15.5 Applications of Microfiltration in Cheesemaking

Microfiltration (MF) is a filtration method that works under low pressure and employs pores with a size range of 0.2–5.0 μm to separate dissolved components from non-dissolved ones. The cheese industry has increasingly adopted the use of microbial purification of raw milk, selective fractionation of milk fat, casein enrichment, modification of the α and β-casein ratio, concentration of milk proteins, and brine decontamination due to the wide range of applications and benefits they offer (Mistry & Maubois, 2004).

15.5.1 MF for Microbial Purification of Raw Milk

Heat treatment is used to sterilize raw milk, but it can change its composition and quality. Effective metal forming (MF) reduces heat treatment. The multichannel geometry and highly permeable structure of MF ceramic membranes allow microfiltrate recirculation. The Bactocatch process, which uses 1.4 μm membranes, can reduce bacteria in raw skim milk by up to 99.6%. Double-layer microfiltration (MF) membranes remove 99.98% of specific bacteria (Saboya & Maubois, 2000). France has provisionally approved MF milk in AOP raw milk cheeses. Making cheese with "ultraclean" milk is complicated. Lactic starters and cheese-specific bacteria need more research (Maubois, 2002). Microbial fermentation (MF) reduced Domiati cheese rennet needs, according to Awad et al. (2010). Raw milk cheese had fewer bacteria, less moisture, higher pH, slower protein and fat breakdown, and better texture than pasteurized cheese. Milk and brine impurities are removed by microfiltration (MF), affecting cheese quality. Heat, kieselguhr filtration, UV, and preservatives are common brine treatments. However, microfiltration (MF) may replace these methods. Cheesemakers call MF milk "dead milk" because of its extensive sanitization, which produces bland cheese. Thus, more research is needed to understand lactic and non-starter lactic acid bacteria in microfiltered milk, especially cheese. MF for enriching Cheese milk with Casein.

Microfiltration (MF) can be used to specifically separate micellar casein from skim milk by using 0.1 μm pore size membrane. This process increases the quantity of casein, resulting in improved coagulation of rennet and a 2–4% rise in cheese production (Daviau, 2000). Additional processing can be applied to the microfiltrate to obtain highly purified whey protein isolates that possess valuable functional properties (Bacher & Konigsfeldt, 2000). Mozzarella and Cheddar cheese have been made using casein that has been enriched with MF. Mozzarella cheese was created by attaining a casein concentration of 17.9% through the utilization of 0.2 μm membranes. Neocleous et al. (2002a) described the process of microfiltering milk to enhance protein recovery specifically for production of Cheddar cheese. Amelia et al. (2013) have also utilized this technique in the creation of low-fat cheese, resulting in a more tender consistency but compromising on taste when compared to full-fat alternatives.

15.5.2 Selective Fractionation of Globular Milk Fat

Ceramic microfiltration (MF) membranes are capable of efficiently partitioning milk fat into both large and small globules, while maintaining the structural integrity of the original fat globule membrane (FGM). According to Goudédranche et al. (2000), cheese produced from milk that has small fat globules has a uniform texture and higher yield due to the interactions between the cheese casein matrix and fat globule membrane (FGM).

15.5.3 Modifying α and β-Casein Ratio by Micro Filtration

Microfiltration (MF) conducted at temperatures lower than 5 °C can efficiently separate soluble β-casein from milk. This process results in the creation of retentates that have a greater proportion of αs/β-casein and microfiltrates that are nearly entirely composed of β-casein. This procedure allows for the potential creation of new types of cheese with varying ratios of αs/β-casein.

The most recent advancements in ceramic membrane technology, such as the Membralox GP and Isoflux systems (offer cost-effective and energy-efficient solutions by eliminating the need for permeate recirculation loops).

15.5.4 Milk Protein Concentrates

Milk protein concentrates (MPC) with milk protein content of 35–85% are made using ultrafiltration (UF), microfiltration (MF), or both. MPCs have revolutionized cheesemaking by improving production and offering new composition options. Enriching milk with MPC increases cheese production and whey protein conservation (Mistry & Maubois, 2004). Microfiltration (MF) produces concentrated micellar casein powders. These powders can improve cheese production regardless of heat treatment. Cheesemakers need standardized production. Microfiltration (MF) separates caseins and whey proteins, allowing cheese milk casein-to-fat ratio control. Studies by Maubois et al. (2001) revealed that a 0.1 μm microfiltration membrane can isolate native calcium phosphocaseinate. Neocleous et al. (2002b) found that MF milk increased Cheddar cheese protein extraction without changing fat content. Brandsma and Rizvi (2001) made mozzarella cheese from milk using microfiltration (MF). Heino et al. (2010) made Edam cheese from microfiltered (MF) milk with 42 g/kg protein. The researchers recovered more solid substances and protein, but not casein. Researchers and companies are studying how microfiltration (MF) affects cheese and working to commercialize it. Modern cheesemaking relies on microfiltration to remove microorganisms, increase protein, and create new cheeses.

15.6 Applications

Nanofiltration (NF), a flexible membrane filtration method, is popular in cheese production. Nanofiltration membranes can concentrate and purify valuable compounds from dairy streams while preserving desirable properties. Nanofiltration (NF) separates and concentrates milk components in cheese making (Cassano et al., 2019; Salehi, 2014). Nanofiltration (NF) concentrates milk protein and fat, saving time and energy in cheesemaking (Cassano et al., 2019). Cheese quality and production can improve with concentration. Nanofiltration (NF) can also recover and recycle dairy byproducts like whey permeate for cheese production (Salehi, 2014; Cassano et al., 2019; Muthukumarappan, 2013). By reducing waste, recycling improves cheese production efficiency and the environment. NF can also remove microorganisms, minerals, and organic compounds from dairy streams. Purification can improve cheese safety, keeping quality, and taste. Nanofiltration (NF) technology in cheese production has shown promising results in terms of effectiveness, environmental friendliness, and product quality.

15.6.1 Concentration and Demineralization of Whey Permeate

Whey in casein and cheese production has BOD 5 30–50 g/L and low solids 5–6%. Waste disposal is expensive and difficult (Prazeres et al., 2013). After drying, whey became fertilizer and animal feed. Whey has been used more to make proteins in recent decades. Whey proteins are biologically, functionally, and nutritionally beneficial (Ganju & Gogate, 2017; Daufin et al., 2001). Selling recovered or processed whey reduces pollution and profits the dairy industry. Traditional whey demineralization uses evaporation or RO. Ion-exchange and electrodialysis (ED) demineralize concentrated whey, according to Chen et al. (2017). These processes are used in industries, but they require a lot of wastewater to be treated in an expensive wastewater treatment plant. Whey is concentrated (20–22%) and partially demineralized (25–60%, or 90% with DF) in one step by NF. Some lactose, organic acids, and monovalent ions pass through membranes (Daufin et al., 2001). Nanofiltration (NF) has many advantages over conventional methods. Cost reduction, energy savings, and taste, properties, and viscosity improvement. Low effluent compares to other demineralization methods. With its many benefits, NF replaces whey concentration and desalination. Its potential has been extensively studied here. A plate-and-frame setup Magueijo et al. (2005) tested Alfa-Laval NFT50 and DDS HR-95-PP NF membrane. A membrane was tested to extract valuable components from Serpa cheese and curd byproduct second cheese whey. Cheese whey was nanofiltered into lactose-rich retentate and permeate. The NFT750 membrane recovered 80% water at 30 bar. The wastewater organic content decreased and whey nutrient concentration increased fivefold. Nguyen et al. (2003) nanofiltered cottage cheese whey solids. The whey was processed using a nanofiltration (NF) unit with plate-and-frame membranes (XP-45, APV Company) after decanting, heating to 65C, and filtering through a 5 μm cheese bag Surface area of membranes was 0.29 m^2. A large pilot plant unit with two spiral-wound membrane modules (Osmonics Desal 5) was tested. For concentrating cottage cheese whey from 7 to 26 Brix, large unit operating pressure was increased from 20 to 34 bar in 3 steps. Three-quarters of sodium, potassium, and acid were removed. Retentate, with more fat, proteins, and lactose, is used to make ice cream and yogurt. Pan et al. (2011) used a Koch membrane Systems TFC 2540 SR2 spiral-wound nanofiltration (NF) membrane to concentrate and remove minerals from acidified whey. pH of feed solution affected whey desalination. Whey desalination worked best at 4.60 pH, its isoelectric point. Demineralization reached 72% with DF.

15.7 Precision Fermentation in Cheese Making

15.7.1 Precision Fermentation

Precision fermentation optimises fermentation processes using engineered microbial hosts as "cell factories". This method produces high-value functional food ingredients like enzymes, lipids, carbohydrates, vitamins, flavorings, colorants, antioxidants, and preservatives. High product yields and purity are desired (Mazumdar & Pichler, 2020). Traditional industrial methods like agriculture, husbandry, foraging, bulk extraction, and organic synthesis produce these ingredients in an unsustainable way (Teng et al., 2021). It's not new to ferment food products like enzymes and glutamate with natural microorganisms. According to its common definition, 'precision fermentation' is essentially metabolic engineering, which involves genetically modifying microbial organisms to produce substances they cannot naturally produce. Bioinformatics, next-generation sequencing, high-throughput and combinatorial library screening, synthetic biology, enzyme engineering, molecular cloning, comparative multiomics analysis, kinetic modeling, and machine learning are metabolic engineering methods. These references (Sun et al., 2021; Shi et al., 2022; Ko et al., 2020) discuss methods to improve microbial strain efficiency, metabolic pathways, product yields, and bioprocess scaling for large-scale production of valuable ingredients. The precision fermentation process commences by carefully choosing a suitable host microorganism for gene expression (Vieira Gomes et al., 2018). Subsequently, a strain engineering cycle is commonly employed to augment protein production (Sun et al., 2023). The subsequent stages entail the expansion of the manufacturing process and the purification and retrieval of the recombinant protein, which exhibits identical properties to those generated by the initial animal (Knychala et al., 2024).

15.7.2 Applications in Cheese Making

Precision fermentation, within the realm of cheese production, provides a sustainable and efficient approach to generating essential dairy constituents without the need for conventional animal sources. Precision fermentation is utilized in cheese making in the following manner:

Main Components Produced Through Precision Fermentation

Casein Proteins: Casein refers to a collection of proteins that play a crucial role in determining the structure and texture of cheese. Precision fermentation involves the incorporation of the casein production gene into microorganisms such as yeast or bacteria, enabling the production of casein proteins. Subsequently, these microorganisms generate casein through fermentation, which can be collected and employed in the production of cheese.

Whey Proteins: Whey proteins are a vital constituent of milk that is utilized in diverse dairy products, such as cheese. Like casein, whey proteins can be generated using precision fermentation, guaranteeing a reliable and cruelty-free source.

Rennet (Chymosin): Rennet is an enzyme that facilitates the coagulation of milk, causing it to separate into curds and whey. Traditional rennet is obtained from the gastric mucosa of young cattle, while precision fermentation can generate microbial rennet, particularly chymosin, by introducing the corresponding gene into bacteria or yeast. This microbial rennet is extensively utilized in contemporary commercial cheese production.

Flavor Compounds: Precision fermentation can also generate specific flavor compounds that contribute to the distinctive taste of cheese. Microorganisms can be genetically modified to synthesize these compounds, thereby augmenting the flavor characteristics of the cheese.

15.7.3 Advantages of Precision Fermentation in Cheese Production

Sustainability: Mitigates the environmental impact by minimizing the necessity for dairy farming, which requires significant resources and contributes to the emission of greenhouse gases. Demands reduced land, water, and energy resources in comparison to conventional dairy farming.

Animal Welfare: Removes the necessity for ingredients derived from animals, in accordance with ethical and animal welfare concerns.

Consistency and Quality: Guarantees a steady provision of top-notch ingredients, resulting in standardized cheese manufacturing. Mitigates the variability commonly observed in conventional dairy farming, such as fluctuations in the composition of milk.

Scalability and Efficiency: Can be expanded to accommodate growing demand without the constraints of traditional dairy farming. Provides a more regulated and streamlined manufacturing process (Knychala et al., 2024).

Precision fermentation-derived ingredients must undergo thorough regulatory approval to guarantee their safety and effectiveness. It is essential to educate consumers about the advantages and safety of precision fermentation products in order to gain market acceptance. Although overall costs are decreasing, the initial production costs may be higher in comparison to conventional methods. Nevertheless, the implementation of economies of scale and technological advancements is anticipated to gradually decrease costs in the long run (Chai et al., 2022).

15.8 Why Choose Precision Fermentation for Cheese Production?

The estimated value of the global precision fermentation ingredients market is $2.8 billion in 2023 and is projected to reach $36.3 billion by 2030. Once the cost of producing cheese from microorganism-derived milk matches that of traditional dairy, animal-free cheese will capture a market share of 33%. During the forecast period, the precision fermentation market is expected to experience the most significant growth in the Asia Pacific region. This growth can be attributed to the rising demand for alternative proteins, increasing disposable income, and growing awareness of environmental and animal welfare concerns. Moreover, a significant 79% of the market share in precision fermentation would be contributed by consumers who traditionally consume cheese. The indisputable potential of sustainable cheese production has attracted the attention of investors. The investment in precision fermentation companies worldwide surged to nearly $600 million in 2020 (State of Industry Report, 2022).

15.9 Conclusion

This chapter has demonstrated the significance of microencapsulation and membrane filtration in contemporary cheesemaking. The utilization of microencapsulated bacterial strains in the production of cheese enhances both its quality and functionality. Membrane filtration techniques such as reverse osmosis, nanofiltration, ultrafiltration, and microfiltration address the difficulties encountered in cheese production through inventive means. Analyzing the design and configurations of tubular and hollow fibers, plate and frame, spiral-wound, and vibrating membrane systems can lead to optimization. The versatility and practicality of cheesemaking membrane technologies are evident. Ultrafiltration enhances the efficiency and quality of cheese production in protein-standardized milk, liquid precheeses, and specialized processes such as the APV-Siro Curd Process. Reverse osmosis is used to increase the concentration of farm substances, while microfiltration is used to purify microorganisms, enhance casein, and selectively separate milk components. The utilization of nanofiltration, a process that concentrates and eliminates minerals from whey and UF whey permeate while also recovering lactic acid, demonstrates the significant effects of these advanced technologies. These advancements enhance the caliber of the product, optimize the efficiency of production, and promote sustainability within the cheese industry. These state-of-the-art technologies are introducing a new era of innovation and excellence in the cheese-making industry. The utilization of precision fermentation in cheese production provides distinct benefits in terms of sustainability, animal welfare, and product uniformity when compared to traditional dairy farming methods. Despite the obstacles, the fields of biotechnology and consumer awareness are propelling this inventive method forward.

References

2021 State of the Industry Report: Cultivated meat and seafood. Good Food Institute, 2022.

Acharya, M. R., & Mistry, V. V. (2002). Comparison of effect of vacuum condensed and ultrafiltered milk on pasteurized process cheese. *Journal of Dairy Science, 85*(Suppl. 1), 90.

Afzaal, M., Saeed, F., Ateeq, H., Ahmed, A., Ahmad, A., Tufail, T., Ismail, Z., & Anjum, F. M. (2020). Encapsulation of Bifidobacterium bifidum by internal gelation method to access the viability in cheddar cheese and under simulated gastrointestinal conditions. *Food Science & Nutrition, 8*, 2739–2747.

Amelia, I., Drake, M., Nelson, B., & Barbano, D. M. (2013). A new method for the production of low-fat Cheddar cheese. *Journal of Dairy Science, 96*, 4870–4884.

Awad, S., Ahmed, N., & El Soda, M. (2010). Influence of microfiltration and adjunct culture on the quality of Domiati cheese. *Journal of dairy science, 93*(5), 1807–1814.

Azarnia, S., Robert, N., & Lee, B. (2006). Biotechnological methods to accelerate Cheddar cheese ripening. *Critical Reviews in Biotechnology, 26*, 121–143.

Bacher, T., & Konigsfeldt, P. (2000). WPI by microfiltration of skim milk. *European Dairy Magazine, 5*, 14–16.

Bedoya-Serna, C. M., Dacanal, G. C., Fernandes, A. M., & Pinho, S. C. (2018). Antifungal activity of nanoemulsions encapsulating oregano (Origanum vulgare) essential oil: In vitro study and application in Minas Padrão cheese. *Brazilian Journal of Microbiology, 49*, 929–935.

Belščak-Cvitanović, A., Jurić, S., Đorđević, V., Barišić, L., Komes, D., Ježek, D., Bugarski, B., & Nedović, V. (2017). Chemometric evaluation of binary mixtures of alginate and polysaccharide biopolymers as carriers for microencapsulation of green tea polyphenols. *International Journal of Food Properties, 20*, 1971–1986.

Berridge, N. J. (1951). Enzymes and coagulation. I. Milk coagulation. In J. B. Sumner & K. Myrback (Eds.), *The enzymes* (Vol. I, pp. 1079–1105). Academic Press.

Brandsma, R. L., & Rizvi, S. S. H. (2001). Depletion of whey proteins and calcium by microfiltration of acidified skim milk prior to cheese making. *Journal of Dairy Science, 82*, 2063–2069.

Brulé, G., Maubois, J. L., Vandeweghe, J., Fauquant, J., & Goudedranche, H. (1975). Utilisation de l'ultrafiltration sur membrane pour la fabrication de fromages de type pâtes fraîches. *Revue Laitière Française, 328*, 117–122.

Calvo, P., Castaño, Á. L., Hernández, M. T., & González-Gómez, D. (2011). Effects of microcapsule constitution on the quality of microencapsulated walnut oil. *European Journal of Lipid Science and Technology, 113*, 1273–1280.

Chai, W. Y., Teo, K. T. K., Tan, M. K., & Tham, H. J. (2022). Fermentation process control and optimization. *Chemical Engineering & Technology, 45*(10), 1731–1747.

Chakravarthi, S. S., & Robinson, D. H. (2008). *Pharmaceutical manufacturing handbook*. Wiley.

Champagne, C. P., & Fustier, P. (2007). Microencapsulation for the improved delivery of bioactive compounds into foods. *Current Opinion in Biotechnology, 18*, 184–190.

Chen, G. C., Wang, Y., Tong, X., Szeto, I. M., Smit, G., Li, Z. N., & Qin, L. Q. (2017). Cheese consumption and risk of cardiovascular disease: a meta-analysis of prospective studies. *European journal of nutrition, 56*(8), 2565–2575.

da Cruz, A. G., Buriti, F. C. A., de Souza, C. H. B., Faria, J. A. F., & Saad, S. M. I. (2009). Probiotic cheese: health benefits, technological and stability aspects. Trends in Food Science & Technology, 20(8), 344–354.

Daufin, G., Gesan-Guiziou, G., Dresch, M., Manem, J., Chaufer, B., & Merin, U. (2001). Traitement et recyclage des effluents de l'industrie laitière. In G. Daufin, F. René, & P. Aimar (Eds.), *Les Séparations par Membrane dans les Procédés de l'Industrie Alimentaire* (pp. 343–359). Tech & Doc.

Daviau, C. (2000). *Effect des Caractéristiques Physico-chimiques du Lait sur l'Aptitude à la Coagulation Présure et à l'Egouttage dans une Technologie Pâte Molle*. PhD thesis. Ensar/Inra, p. 158.

El-Messery, T. M., El-Said, M. M., & Farahat, E. S. A. (2019). Production of functional processed cheese supplemented with nanoliposomes of mandarin peel extract. *Pakistan Journal of Biological Sciences, 22*, 247–256.

Fathi, M., Vinceković, M., Jurić, S., Viskić, M., Režek Jambrak, A., & Donsì, F. (2021). Food-grade colloidal systems for the delivery of essential oils. *Food Reviews International, 37*, 1–45.

Ganju, S., & Gogate, P. R. (2017). A review on approaches for efficient recovery of whey proteins from dairy industry effluents. *Journal of Food Engineering, 215*, 84–96.

Goderska, K., Zybała, M., & Czarnecki, Z. (2003). Characterisation of microencapsulated Lactobacillus rhamnosus LR7 strain. *Polish Journal Of Food And Nutrition Sciences, 53*, 21–24.

Godward, G., & Kailasapathy, K. (2003). Viability and survival of free, encapsulated and co-encapsulated probiotic bacteria in yoghurt. *Milchwissenschaft-Milk Science International, 58*, 161–164.

González Ariceaga, C. C., Afzal, M. I., Umer, M., Shah, S. M. U., Ahmad, H., Jacquot, M., & Cailliez-Grimal, C. (2018). Effect of alginate beads on olfactory sensory perception of paraffin coated cheese. *Czech Journal of Food Sciences, 36*, 255–260.

Goudédranche, H., Fauquant, J., & Maubois, J. L. (2000). Fractionation of globular milk fat by membrane microfiltration. *Le Lait, 80*, 93–98.

Govindasamy-Lucey, S., Jaeggi, J. J., Martinelli, C., Johnson, M. E., & Lucey, J. A. (2011). Standardization of milk using cold ultrafiltration retentates for the manufacture of Swiss cheese: Effect of altering coagulation conditions on yield and cheese quality. *Journal of dairy science, 94*(6), 2719–2730.

Heino, A., Uusi-Rauva, J., & Outinen, M. (2010). Pre-treatment methods of Edam cheese milk. Effect on cheese yield and quality. *LWT-Food Science and Technology, 43*(4), 640–646.

Hill, C., Guarner, F., Reid, G., Gibson, G. R., Merenstein, D. J., Pot, B., et al. (2014). The International Scientific Association for Probiotics and Prebiotics consensus statement on the scope and appropriate use of the term probiotic. *Nature Reviews Gastroenterology & Hepatology, 11*, 506–514.

Jahadi, M., Khosravi-Darani, K., Ehsani, M. R., Mozafari, M. R., Saboury, A. A., & Pourhosseini, P. S. (2015). The encapsulation of flavourzyme in nanoliposome by heating method. *Journal of Food Science and Technology, 52*, 2063–2072.

Jahadi, M., Khosravi-Darani, K., Ehsani, M. R., Mozafari, M. R., Saboury, A. A., & Zoghi, A. (2016). Modelling of proteolysis in Iranian brined cheese using proteinase-loaded nanoliposome. *International Journal of Dairy Technology, 69*, 57–62.

Jeyakumari, A. (2016). Microencapsulation of bioactive food ingredients and controlled release—A review. *MOJ Food Processing & Technology, 2*, 59.

John, R. P., Tyagi, R. D., Brar, S. K., Surampalli, R. Y., & Prévost, D. (2011). Bio-encapsulation of microbial cells for targeted agricultural delivery. *Critical Reviews in Biotechnology, 31*, 211–226.

Jurić, S., Ferrari, G., Velikov, K. P., & Donsì, F. (2019). High-pressure homogenization treatment to recover bioactive compounds from tomato peels. *Journal of Food Engineering, 262*, 170–180.

Jurić, S., Jurić, M., Siddique, M. A. B., & Fathi, M. (2020). Vegetable oils rich in polyunsaturated fatty acids: Nanoencapsulation methods and stability enhancement. *Food Reviews International, 32*, 32–69.

Jurić, S., Jurić, M., Ferrari, G., Režek Jambrak, A., & Donsì, F. (2021a). Lycopene-rich cream obtained via high-pressure homogenisation of tomato processing residues in a water–oil mixture. *International Journal of Food Science & Technology, 56*, 4907–4914.

Jurić, S., Tanuwidjaja, I., Mrkonjic Fuka, M., Vlahovićek Kahlina, K., Marijan, M., Boras, A., et al. (2021b). Encapsulation of two fermentation agents, Lactobacillus sakei and calcium ions in microspheres. *Colloids and Surfaces B: Biointerfaces, 197*, 1–11.

Kailasapathy, K., & Lam, S. H. (2005). Application of encapsulated enzymes to accelerate cheese ripening. *International Dairy Journal, 15*, 929–939.

Kavitake, D., Kandasamy, S., Devi, P. B., & Shetty, P. H. (2018). Recent developments on encapsulation of lactic acid bacteria as potential starter culture in fermented foods—A review. *Food Bioscience, 21*, 34–44.

Knychala, M. M., Boing, L. A., Ienczak, J. L., Trichez, D., & Stambuk, B. U. (2024). Precision fermentation as an alternative to animal protein, a review. *Fermentation, 10*(6), 315. https://doi.org/10.3390/fermentation10060315

Ko, Y. S., Kim, J. W., Lee, J. A., Han, T., Kim, G. B., Park, J. E., & Lee, S. Y. (2020). Tools and strategies of systems metabolic engineering for the development of microbial cell factories for chemical production. *Chemical Society Reviews, 49*(14), 4615–4636.

Lemic, D., Orešković, M., Mikac, K. M., Marijan, M., Jurić, S., Vlahovićek-Kahlina, K., & Vinceković, M. (2021). Sustainable pest management using biodegradable apitoxin-loaded calcium-alginate microspheres. *Sustainability, 13*, 6167.

Lengyel, M., Kállai-Szabó, N., Antal, V., Laki, A. J., & Antal, I. (2019). Microparticles, microspheres, and microcapsules for advanced drug delivery. *Scientia Pharmaceutica, 87*, 20.

Liu, L., Chen, P., Zhao, W., Li, X., Wang, H., & Qu, X. (2017). Effect of microencapsulation with the Maillard reaction products of whey proteins and isomaltooligosaccharide on the survival rate of Lactobacillus rhamnosus in white brined cheese. *Food Control, 79*, 44–49.

Lozano-Vazquez, G., Lobato-Calleros, C., Escalona-Buendia, H., Chavez, G., Alvarez-Ramirez, J., & Vernon-Carter, E. J. (2015). Effect of the weight ratio of alginate-modified tapioca starch on the physicochemical properties and release kinetics of chlorogenic acid containing beads. *Food Hydrocolloids, 48*, 301–311.

Madsen, J. S., & Qvist, K. B. (2001). The effect of added proteolytic enzymes on meltability of Mozzarella cheese manufactured by ultrafiltration. *Le Lait, 78*, 259–272.

Magueijo, V., Minhalma, M., Queiroz, D., Geraldes, V., Macedo, A., & Pinho, M. D. (2005). Reduction of wastewaters and valorisation of by-products from "Serpa" cheese manufacture using nanofiltration. *Water science and technology, 52*(10-11), 393–399.

Maubois, J. L. (2002). Membrane microfiltration: A tool for a new approach in dairy technology. *Australian Journal of Dairy Technology, 57*, 92–96.

Maubois, J. L., Fauquant, J., Famelart, M. H., & Caussin, F. (2001). Milk microfiltrate, a convenient starting material for fractionation of whey proteins and derivatives. In B. Behr's Verlag (Ed.), *The importance of whey and whey components in food and nutrition. Proceedings of the 3rd international whey conference* (pp. 59–72). Hamburg, Munich, Germany.

Mazumdar, S., & Pichler, D. (2020, January). 2020 state of the industry report. *Composites fabrication*.

Mistry, V. V., & Maubois, J. L. (2004). Application of membrane separation technology to cheese production. In P. F. Fox, P. L. H. McSweeney, T. M. Cogan, & T. P. Guinee (Eds.), *Cheese: Chemistry, physics and microbiology* (Vol. 1, 3rd ed., pp. 261–285). Elsevier.

Moghaddas Kia, E., Alizadeh, M., & Esmaiili, M. (2018). Development and characterization of probiotic UF Feta cheese containing Lactobacillus paracasei microencapsulated by enzyme based gelation method. *Journal of Food Science and Technology, 55*, 3657–3664.

Mohammadi, R., Mahmoudzadeh, M., Atefi, M., Khosravi-Darani, K., & Mozafari, M. R. (2015). Applications of nanoliposomes in cheese technology. *International Journal of Dairy Technology, 68*, 11–23.

Mrkonjic Fuka, M., Zgomba Maksimovic, A., Hulak, N., Kos, I., Marusic Radovcic, N., Juric, S., et al. (2021). The survival rate and efficiency of non-encapsulated and encapsulated native starter cultures to improve the quality of artisanal game meat sausages. *Journal of Food Science and Technology, 58*, 710–719.

Narsaiah, K., Sharma, M., Sridhar, K., & Dikkala, P. (2019). Garlic oil nanoemulsions hybridized in calcium alginate microcapsules for functional bread. *Agricultural Research, 8*, 356–363.

Nguyen, M., Reynolds, N., & Vigneswaran, S. (2003). By-product recovery from cottage cheese production by nanofiltration. *Journal of Cleaner production, 11*(7), 803–807.

Neocleous, M., Barbano, D. M., & Rudan, M. A. (2002a). Impact of low concentration factor microfiltration on the composition and aging of Cheddar cheese. *Journal of Dairy Science, 85*(10), 2425–2437.

Neocleous, M., Barbano, D. M., & Rudan, M. A. (2002b). Impact 9 of Low Concentration Factor Microfiltration on Milk Component Recovery and Cheddar Cheese Yield. *Journal of Dairy Science, 85*(10), 2415–2424.

Ningtyas, D. W., Bhandari, B., Bansal, N., & Prakash, S. (2019). The viability of probiotic lactobacillus rhamnosus (non-encapsulated and encapsulated) in functional reduced-fat cream cheese and its textural properties during storage. *Food Control, 100*, 8–16.

Oommen, B. S., Miśtry, V. V., & Nair, M. G. (2000). Effect of homogenization of cream on composition, yield, and functionality of Cheddar cheese made from milk supplemented with ultrafiltered milk. *Le Lait, 80*(1), 77–91.

Ortakci, F., Broadbent, J. R., McManus, W. R., & McMahon, D. J. (2012). Survival of microencapsulated probiotic lactobacillus paracasei LBC-1e during manufacture of Mozzarella cheese and simulated gastric digestion. *Journal of Dairy Science, 95*, 6274–6281.

Özer, B. U., Demir, A., & Ferrari, J. R. (2009). Exploring academic procrastination among Turkish students: Possible gender differences in prevalence and reasons. *The Journal of social psychology, 149*(2), 241–257.

Özer, B., Uzun, Y. S., & Kirmaci, H. A. (2008). Effect of microencapsulation on viability of Lactobacillus acidophilus LA-5 and Bifidobacterium bifidum BB-12 during Kasar cheese ripening. *International Journal of Dairy Technology, 61*, 237–244.

Pan, K., Song, Q., Wang, L., & Cao, B. (2011). A study of demineralization of whey by nanofiltration membrane. *Desalination, 267*(2–3), 217–221.

Prazeres, A. R., Carvalho, F., Rivas, J., Patanita, M., & Dôres, J. (2013). Pretreated cheese whey wastewater management by agricultural reuse: Chemical characterization and response of tomato plants Lycopersicon esculentum Mill. under salinity conditions. *Science of the total environment, 463*, 943–951.

Rashidinejad, A., Birch, E. J., Sun-Waterhouse, D., & Everett, D. W. (2014). Delivery of green tea catechin and epigallocatechin gallate in liposomes incorporated into low-fat hard cheese. *Food Chemistry, 156*, 176–183.

Richard, J., Desmazeaud, M., Lenoir, J., Remuef, F., Schneid, N., Hermier, J., et al. (2000). Cheesemaking milk. In A. Eck & J. C. Gillis (Eds.), *Cheesemaking: From science to quality assurance* (2nd ed., pp. 189–301). Lavoisier Publishing.

Richardson, W. D. (1929). *Method of making cheese*. U.S. Patent 1,711,032.

Saboya, L. V., & Maubois, J. L. (2000). Current developments of microfiltration technology in the dairy industry. *Le Lait, 80*, 541–553.

Saboya, L. V., Goudédranche, H., Maubois, J. L., Lerayer, A. L. S., & Lortal, S. (2001). Impact of broken cells of lactococci and propionibacteria on the ripening of saint-Paulin UF cheeses: Extent of proteolysis and GC-MS profiles. *Le Lait, 81*, 699–713.

Salehi, F. (2014). Current and future applications for nanofiltration technology in the food processing. *Food and Bioproducts Processing, 92*(2), 161–177.

Santacruz, S., & Castro, M. (2018). Viability of free and encapsulated lactobacillus acidophilus incorporated to cassava starch edible films and its application to Manaba fresh white cheese. *LWT, 93*, 570–572.

Shi, S., Wang, Z., Shen, L., & Xiao, H. (2022). Synthetic biology: A new frontier in food production. *Trends in Biotechnology, 40*(7), 781–803.

Soodam, K., Ong, L., Powell, I., Kentish, E., & Gras, S. (2014). The effect of milk protein concentration on the microstructure and textural properties of full-fat Cheddar cheese during ripening. *Food and Bioprocess Technology, 7*, 2912–2922.

Soto, E., Rus, F., & Ostroff, G. R. (2019). Yeast cell wall particle encapsulation of pro-terpene payloads. *TechConnect Briefs*, 435–8.

Sun, L., Xin, F., & Alper, H. S. (2021). Bio-synthesis of food additives and colorants-a growing trend in future food. *Biotechnology Advances, 47*, 107694.

Sun, M., Gao, A. X., Liu, X., Yang, Y., Ledesma-Amaro, R., & Bai, Z. (2023). High-throughput process development from gene cloning to protein production. *Microbial Cell Factories, 22*(1), 182.

Tamime, A. Y., Robinson, R. K., & Kiers, G. L. J. (2006). Industrial manufacture of feta-type cheese. In A. Y. Tamime (Ed.), *Brined cheeses* (pp. 77–116). Blackwell Publishing.

Teng, T. S., Chin, Y. L., Chai, K. F., & Chen, W. N. (2021). Fermentation for future food systems: Precision fermentation can complement the scope and applications of traditional fermentation. *EMBO Reports, 22*(5), e52680.

Torri, L., Piochi, M., Lavelli, V., & Monteleone, E. (2015). Descriptive sensory analysis and consumers' preference for dietary fibre- and polyphenol-enriched tomato purees obtained using winery by-products. *LWT Food Science and Technology, 62*, 294–300.

Vandeweghe, J. (2000). Cheese yield. Evaluation and measurement. In A. Eck & J. C. Gillis (Eds.), *Cheesemaking: From science to quality assurance* (2nd ed., pp. 738–747). Lavoisier Publishing.

Vlahoviček-Kahlina, K., Jurić, S., Marijan, M., Mutaliyeva, B., Khalus, S. V., Prosyanik, A. V., & Vinceković, M. (2021). Synthesis, characterization, and encapsulation of novel plant growth regulators (PGRs) in biopolymer matrices. *International journal of molecular sciences, 22*(4), 1847.

Vieira Gomes, A. M., Souza Carmo, T., Silva Carvalho, L., Mendonça Bahia, F., & Parachin, N. S. (2018). Comparison of yeasts as hosts for recombinant protein production. *Microorganisms, 6*(2), 38.

Vinceković, M., Viskić, M., Jurić, S., Giacometti, J., Kovačević, D. B., Putnik, P., ... & Jambrak, A. R. (2017). Innovative technologies for encapsulation of Mediterranean plants extracts. *Trends in Food Science & Technology, 69*, 1–12.

Chapter 16
Dairy Ingredients in Cheesemaking

Syed Mansha Rafiq, Shamim Hossain, and Syed Irshaan

16.1 Introduction

Cheese production is indeed a complex process that transforms milk into a concentrated protein gel, where fat and moisture are retained (Law & Tamime, 2011). Milk is the fundamental ingredient in cheesemaking, as it contains the proteins, fats, and sugars necessary for cheese production. The milk used can come from different animals, including cows, goats, sheep, and buffalo. Each type of milk has a distinct composition, which affects the cheese's texture and flavor. Sometimes cream is added for specialized cheese making. Besides this rennet, starter culture, salt, lactose are the other ingredients. Cheesemaking involves a balance of science, art, and timing, with each stage contributing to the final flavor, texture, and appearance of the cheese. More recently, whole milk powder (WMP), skim milk powder (SMP), milk protein concentrates (MPC), whey protein concentrates (WPC), casein, caseinates, and buttermilk are increasingly being used in cheese production due to their functional properties and ability to enhance cheese texture, flavor, and nutritional content.

S. Mansha Rafiq (✉)
Department of Food Science and Technology, National Institute of Food Technology Entrepreneurship and Management, Sonipat, Haryana, India

S. Hossain
Dairy Technology Division, AR-National Dairy Research Institute, Karnal, Haryana, India

S. Irshaan
Department of Food Technology, School of Agriculture and Food Technology, Vignan's Foundation for Science, Technology and Research, Guntur, India

S. Mansha Rafiq (ed.), *Technological Advances and Trends in Cheese Making*,
https://doi.org/10.1007/978-3-031-94476-5_16

16.2 Dairy Ingredients for Cheesemaking

16.2.1 *Whole Milk Powder*

It is known that approximately one million tons of milk powder is used in the world annually for cheese making. Some researchers suggest that fresh cheeses can be more easily made from recombined milk (milk that has been reconstituted from milk powder), but the production of hard and semi-hard cheeses from recombined milk poses additional challenges. The quality of the raw milk determines the quality of milk powder which in turn affects the cheese making properties. A low heating temperature is necessary to have good rennetability of the recombined milk. The use of whole milk powder has been reported for the manufacture of brine and Domiati cheese. However, whole milk powder has limited solubility which causes fat destabilization and sediment formation and to prevent that instant whole milk powders could be used. The soft brine cheese prepared from reconstituted instant whole milk powder showed acceptable quality characteristics except for salt content which was higher than cheese made from raw milk at the end of ripening. The structure of the cheese as revealed by electron microscopy showed a slightly aggregated state of the protein in the cheese made from reconstituted milk in comparison to the cheese made from raw milk which showed a homogeneous structure. Domiati cheese was made from dried milk and the ripening characteristics were compared to the fresh milk cheese. The elution pattern of the proteinaceous fractions and the pattern of free fatty acids were similar in both cheeses but the nitrogen concentrations were higher in fresh cheese than those made from dried milk. Domiati cheese from reconstituted milk was found to show less flavour intensity and lower levels of free fatty acids compared to cheeses made from fresh milk. However, cheese flavour can be improved by treatment with enzymes at a level of 0·02%, which enhances the formation of free fatty acids and improves flavour (Omar, 1987).

16.2.2 *Skim Milk Powder*

Skim milk powder requires careful reconstitution and recombination techniques to achieve the desired quality in cheese production, more so than whole milk powders. In areas without modern equipment or processing facilities, using whole milk powders might be a more practical option. Skim milk powder has been used successfully in the production of Karish cheese, Domiati cheese, and processed cheese. The addition of skim milk powder, along with milk proteins and stabilizers, improved the yield and quality of Karish cheese, making it more acceptable (Awad et al., 2015). Initial attempts to produce Domiati cheese from recombined milk (using skim milk powder) resulted in an inferior product, with a firm texture and weak flavor despite maturation. However, later improvements were made by directly adding skim milk powder to buffalo or cow milk (up to 35% total solids) and coagulating the concentrated milk with brine. This method produced a more acceptable taste and flavour after a month of storage. According to FDA recommendations, non-fat dry milk (such as skim milk powder) is commonly used in processed cheese foods and spreads. However, when processed cheese is made with a cheese base from reconstituted skim milk powder, the result is a firmer cheese compared to the control. Skim milk powder has also been used for fortifying non-traditional white soft cheese in Egypt, improving its nutritional profile and quality (El-Alfy et al., 2010). Overall, while skim milk powder presents certain challenges in reconstitution, it is widely used in various cheese types to improve yield, texture, and quality, especially when the right processing techniques are applied. The impact of heat treatments on the cheese-making properties of recombined milks, specifically those produced using milk powders is discussed below.

Heat treatments applied during the concentration and drying processes of milk powders (such as medium and high heat milk powders) affect their ability to form cheese. These treatments inhibit the rennet hydrolysis of κ-casein (a key protein involved in cheese curd formation) and cause an increase in the micelle electronegativity, leading to the formation of complexes between β-lactoglobulin and κ-casein at the micelle surface. This process results in the formation of insoluble calcium phosphate precipitates, which reduces the availability of native micellar calcium phosphate when the milk cools, ultimately affecting cheese-making efficiency. To improve the cheese-making properties of recombined milk, it is suggested to reduce the intensity of heat treatments or adding ionic calcium. Furthermore, a treatment involving partial removal of β-lactoglobulin from skim milk, followed by a low or medium heat treatment, can result in milk powder with enhanced cheese-making abilities. Heat stability is influenced by the type of non-fat dry milk (NDM) powder used (low or medium heat) and the materials used for standardization, such as permeate powder or edible lactose powder. The combination of low-heat NDM and permeate powder provides the best heat stability in reconstituted milk. The addition of calcium salts, like calcium citrate and calcium carbonate, further improves heat stability. Specifically, calcium carbonate in reconstituted skim milk powder (SMP) with a 1.75% protein level significantly enhances heat stability, likely due to its neutralizing effect. When reconstituted milk is subjected to ultrafiltration, its permeation rate is typically higher than that of pasteurized fresh milk with the same total solids, and this rate increases as the milk undergoes more intensive heat treatment during powder production. This indicates that heat treatments affect the filtration and texture properties of the recombined milk, impacting its suitability for various applications.

16.2.3 Milk Protein Concentrates (MPC)

Milk protein concentrates (MPC) are important products for cheese making. These are produced using various technologies like ultrafiltration (UF), diafiltration and spray drying (Mistry, 2002). This dried or liquid product is used to increase the protein content of cheese milk that results in enhanced rennet coagulability, reduction in gelation time and increased curd firmness which improves the consistency and efficiency of cheese-making (Havea, 2006; Guinee et al., 2006). Milk Protein Concentrates (MPC) has rennet coagulation properties similar to milk when the protein concentration and ionic strength are comparable. Rennet coagulation, a critical step in cheese making, depends on the casein micelles and calcium content in the serum. However, reconstituted MPC doesn't coagulate unless supplemented with calcium. This is due to the removal of minerals during ultrafiltration/diafiltration. Adding sufficient calcium restores the rennet coagulation properties of MPC, bringing it close to that of raw skim milk. Calcium bridges the casein micelles and reduces soluble caseins in the serum, which improves rennet coagulation. The gelation properties of MPC are positively influenced by soluble and insoluble calcium and negatively affected by soluble caseins (Sandra & Corredig, 2013). Studies have been reported for different cheese varieties. Table 16.1 discusses the effect of addition of milk protein concentrates on the yield and properties of various cheese types. MPC is beneficial when used to standardize whole milk for pizza cheese production. It increases yield, enhances the quality of the whey stream (by reducing lactose levels), and improves cheese meltability and browning resistance during baking. It also increases calcium content in cheeses made through culture acidification, but the effect is less pronounced with direct acidification (Shakeel-Ur-Rehman et al., 2003). The addition of MPC to milk for Mozzarella cheese preparation significantly increases the yield (from 13.8% to 16.7%) by improving the recovery of milk solids and proteins. A level of 5.38% MPC improves protein retention and yield (Harvey, 2006). Fortifying Gouda cheese with 1% MPC increases cheese yield by reducing casein losses and improving whey protein retention (Mistry & Pulgar, 1996). MPC is also used in manufacturing process cheese, where it is ultrafiltered and diafiltered, creating a pre-cheese base that does not require a cheese base. This method results in reduced-calcium pre-cheese, which is then used in traditional process cheese production (Mistry & Maubois, 2004). The use of MPC to standardize milk for reduced-fat Cheddar cheese production has been shown to double the cheese yield,

Table 16.1 Effect of addition of milk protein concentrates on the yield and properties of various cheese types

Cheese type	Cheese properties
Reduced-fat cheddar	Doubled cheese yield Decreased proteolysis
Pizza cheese	Increased cheese yield by 61% Improved meltability during storage
Gouda cheese	Increased cheese yield.
Cheddar	Increased yield.
Mozzarella and feta	Increased yield No sensory defects
Mozzarella	Good sensory attributes Improved fat recovery

Source: Masotti et al. (2017)

significantly increase total solids recovery, and have little impact on fat recovery. MPC cheeses also showed slower secondary proteolysis during ripening compared to control cheeses. In other studies, adding MPC at levels of 4.56% and 6.48% increased protein retention and yield, though higher protein levels negatively affected fat retention and cheese texture (Harvey, 2006). MPC enhances rennet coagulation properties, cheese yield, and quality across various cheese types, including Mozzarella, Gouda, Cheddar, and pizza cheese. Its impact is particularly notable in improving protein retention and reducing losses in casein and whey proteins. The addition of calcium is crucial for restoring rennet coagulation in reconstituted MPC, and MPC helps increase yield while affecting cheese texture and ripening processes.

16.2.3.1 MPC and Rennet Coagulation

The addition of MPC does not affect the gelation times of rennet-induced gels significantly. However, reconstituted MPC dispersions exhibit a lower gelation time compared to reconstituted skim milk with the same protein concentration. This suggests that MPC gelates faster than skim milk, despite similar protein content. Reconstituted MPC does not coagulate on its own unless supplemented with calcium chloride. This is due to the mineral removal during ultrafiltration/diafiltration, a process used to concentrate MPC. Adding calcium restores the rennet coagulation kinetics and gel strength of reconstituted MPC to levels similar to raw skim milk. This emphasizes that the ionic equilibrium, particularly calcium levels, is crucial for ensuring the desirable coagulation properties of MPC. High-protein MPCs (like MPC 80) often require added calcium chloride to maintain good coagulation behaviour (Kuo & Harper, 2003; Martin et al., 2011). MPC requires calcium supplementation to restore its rennet coagulation properties and gel strength, with the ionic balance being a key factor.

The renneting properties of MPC are influenced by storage conditions such as storage time and temperature. It was found that renneting properties decrease exponentially with longer storage time, with the effect being more pronounced at higher temperatures. This decrease is linked to the secondary stage of renneting (the aggregation and coagulation of rennet-treated casein micelles). However, the primary phase of renneting (cleavage of κ-casein) was not affected by storage time or temperature. Recombined milk made from MPC showed a slower increase in the storage modulus (G') of rennet-induced gels compared to raw milk. This means that the gel formed by MPC took longer to set and become firm. Additionally, gelation time decreased, and the yield force (the force needed to fracture the gel during heating) also decreased. This suggests that gels made from MPC are more fragile and less resistant to breaking when subjected to heat. The extent of whey protein denaturation (caused by heat treatment) was associated with the decrease in the G' value and yield force of the gels. When more whey protein was denatured due to heat, the gel structure became weaker, resulting in lower gel strength and higher susceptibility to fracture (Pomprasirt et al., 1998). Storage conditions (time and temperature) negatively affect the secondary stage of rennet coagulation (aggregation of casein micelles), while the primary stage (cleavage of κ-casein) remains unaffected. MPC-based gels have slower gelation, lower gel strength, and reduced yield force during heating, which can be influenced by whey protein denaturation caused by heat treatments.

16.2.4 Whey Protein Concentrates (WPC)

Whey protein incorporation into the cheese offers several benefits and challenges, particularly in terms of yield, texture, flavor, and functional properties. Table 16.2 depicts the effect of addition of whey protein concentrates on the properties of various cheese types. The addition of whey proteins generally increases cheese yields across various cheese types, including cream cheese, Cheddar, Gouda, processed cheese, and Domiati cheese. However, it may result in a slightly poorer flavor and texture (Hinrichs, 2001). In cheeses like Camembert and Iranian white cheese, the addition of whey proteins results in a softer texture. This softening can be beneficial for certain types of cheese, but it may affect the overall mouthfeel and structure of

Table 16.2 Effect of addition of whey protein concentrates on various cheese types

Cheese type	Cheese properties
Camembert	Increased cheese yield Resulted in soft, uneven ripened texture with bitter flavour
Cheddar	Increased the yield by 12–15% Normal flavour with minimal softer texture
Iranian white cheese	Increased the cheese yield Decreased cheese hardness and chewiness and resulted in softer texture WPC added >15% reduced overall quality
Low-moisture part-skim mozzarella	Reduced melting properties
Fresh cheese	Increased cheese yield Higher MWP reduced sensorial acceptability Increased stability during storage
Semi-hard cheese	Increased cheese yield by 12% Resulted in sour flavour Resulted in less holes
Gouda	Increased cheese yield Resulted in crumbly texture
Emmentaler	Increased yield by 14% No sensory defects
Reduced-fat gouda-type cheese	Increased cheese yield Changed meltability
Low-fat fresh Kashar cheese	Improved acceptability Sensory scores decreased during storage

Source: Masotti et al. (2017) and Hinrichs (2001)

the cheese. In a rennet casein process cheese model (with 17% casein, 24% fat, and 2% emulsifying salt), replacing up to 2% of casein with whey protein was found to slightly increase the firmness of the cheese while decreasing meltability. This indicates that while whey protein can enhance some aspects of cheese texture, it may compromise others, like meltability (Mleko & Foegeding, 2000). Whey protein concentrates are often used as fat replacers in processed cheese, leading to a reduction in cheese hardness. This makes the cheese less dense, which could be useful in certain cheese formulations. The addition of WPC-38 at 4.5% improved the spreadability of cheese spreads, enhancing their usability in various applications. In Domiati cheese, replacing some of the cheese milk with whey proteins and carboxymethyl cellulose (from whey) resulted in Increased cheese yield, Reduced weight loss during pickling, Enhanced body (the texture and consistency) of the cheese, Flavour intensity remained unaffected by the addition of whey proteins (Ashour et al., 1986), Replacing some rennet casein with whey protein concentrate in mozzarella cheese analogue resulted in greater firmness and meltability and fat leakage and moderate chewiness and lower cohesiveness. This suggests that whey protein addition can improve some desirable properties, such as meltability and firmness, but might slightly affect cohesiveness and fat retention. Whey proteins, particularly whey protein concentrates, enhance the yield of various cheeses and improve certain functional properties, such as spreadability and firmness. However, their addition may slightly alter texture (making cheeses softer) and meltability (in processed cheeses). In some cases, whey proteins improve the body of cheeses (like Domiati), but flavour intensity is usually not affected. The use of whey proteins also shows promise in fat replacement and in improving the overall texture of cheese products like cheese spreads and Mozzarella analogues.

Incorporating whey proteins into cheese with high total solids content can be enhanced through "partial" milk concentration, although retention is not fully complete. The rennet-induced gel network mechanically traps the denatured whey proteins in the cheese milk. While adding whey protein boosts yield, it can negatively impact the cheese's flavour and texture (McClements & Grossmann, 2022). The increased yield results from the addition of whey proteins and better serum retention within the cheese matrix. Specifically, well-hydrated and denatured whey proteins reduce syneresis, limiting water loss during cheesemaking. The process of "articulation" involves simultaneously heating and shearing whey proteins into particles within a whey concentrate, causing denaturation and aggregation. The efficiency of this process depends on the composition and specific parameters, such as process simplicity. The effect of the degree of denaturation on whey protein retention in soft and semi-hard cheeses was investigated using two whey protein concentrates (WPC 1 and WPC 2) and a whey protein isolate (WPI), which were subjected to heat and shear treatments to produce particulate whey proteins. Protein additions of WPC or WPI ranged from 0.3% to 1.25%. Increased denaturation led to a significant rise in retention, with the semi-hard cheese achieving a maximum retention of 70% at over 90% denaturation. More significant whey protein recovery can be completed in soft cheese, primarily when WPI is used. Whey protein particles (WPP), like fat globules, are passively integrated into the pores of the casein network (Deeth & Bansal, 2018).

The native and added undenatured whey proteins from the milk are lost into the whey. The network's pore size, around 10 μm, emphasizes the critical diameter for particle inclusion. Larger particles may disrupt the network's uniformity and decrease its firmness, while whey protein particles (WPP) are passively integrated into the matrix at sizes ranging from 1 to 10 μm. An advantage of WPP is that it enhances cheese quality, especially in low-fat varieties, due to its shape and size, which allows it to function similarly to integrated fat globules. However, when recycling whey proteins as particles, adjustments to the cheese-making process are necessary. These adjustments include reducing the water content and removing lactose and other milk components from the cheese, which is recycled using WPP. Key steps in producing soft, semi-hard, and hard cheeses can be optimized by increasing the fermentation and renneting temperatures by 2–3 °C, a technique frequently used for high-fat cheeses, to enhance the renneting process (Limbaugh, 2015).

To produce mild cheese, lowering lactose levels, increasing cheese curd processing, removing whey, and washing water are essential. Whey protein particles derived from WPC can also raise lactose levels, necessitating more thorough washing. The options for modifying the cheese-making process are limited, which restricts the extent of whey protein particle recycling (Limbaugh, 2015). The quantity of recycled whey protein particles is influenced by factors like moisture content and the texture of specific cheese varieties. This method enables the production of high-quality cheese with increased whey protein content. The dry matter content of the cheese affects the amount of added whey protein particles and their incorporation into the cheese matrix. As total solids content rises, the integration of whey protein particles diminishes. Modifications to the cheese-making or processing techniques are necessary to fully leverage whey protein incorporation while maintaining quality standards (Masotti et al., 2017).

16.2.5 Caseins

Caseins, particularly rennet casein, sodium caseinate, have been used as primary ingredients in dairy-based imitation and analogue cheese products. Rennet casein is the preferred protein source in imitation cheeses due to its functional properties and ability to contribute to desirable flavor. It has a high calcium-to-casein ratio, which allows for easy control of calcium sequestration and para-casein aggregation. This, in turn, facilitates the desired fat emulsification, flow, and firmness when the cheese is cooked. In imitation cheese, sodium caseinate is often used but can cause issues like over-creaming when the oil phase is over-emulsified. This can lead to poor meltability and firmness, making it unsuitable for some applications like pizza toppings (O'Riordan et al., 2011). Therefore, sodium caseinate is not ideal as the sole protein source in cheese analogues. Due to the high cost of rennet casein, there is a growing trend to find low-cost substitutes to replace it partially or totally in analogue cheeses. Whey protein concentrate (WPC) is one such substitute. However, excessive addition of WPC can lead to undesirable effects: Adding too much WPC makes the cheese harder and more cohesive (Sołoweij et al., 2010). In sodium caseinate-based imitation cheese, adding more than 1.5% WPC can negatively impact the color and meltability of the product (Hosseini et al., 2014).

Increasing the casein content in milk by using native micellar casein has become common to produce new varieties of cheese. This process improves technological properties, such as faster development of curd firmness, increased firmness and higher yield (Simov et al., 2005). Adding soluble caseins can inhibit the aggregation of casein micelles, which is a key factor in cheese making. Adding calcium chloride ($CaCl_2$) before renneting enhances rennet aggregation in samples containing additional sodium caseinate (Gaygadzhiev et al., 2012). The addition of casein at varying levels (3%, 4%, 5%, or 6%) to low-fat Cheddar cheese made from diafiltered microfiltered retentate (DMF), commercial calcium caseinate (CaCN) powder, and ultrafiltered retentate (UF) has been found to have a significant effect on cheese yield (St-Gelais et al., 1998). When preparing Mozzarella cheese, partial removal of whey proteins through a combination of microfiltration and ultrafiltration before heat treatment prevents the formation of a β-lactoglobulin-κ-casein complex during processing. This resulted in similar composition and properties as Mozzarella made from raw milk. 7.3% higher cheese-making yield, while maintaining comparable cheese-making properties to fresh milk (Garem et al., 2000). Rennet casein remains the gold standard for cheese analogues due to its functional properties and flavor, although it is costly. Sodium caseinate is a more affordable option but has limitations in terms of meltability and firmness. Enriching cheese milk with casein and techniques like microfiltration and ultrafiltration to remove whey proteins can improve cheese yield and functionality while maintaining similar properties to fresh milk.

16.2.6 Butter Milk Powder

Buttermilk (BM) has high content of phospholipids that makes it an important functional ingredient in an array of products (Rombaut et al., 2007). The main goal of fortification with buttermilk is to increase the yield without altering the cheese quality. This is achieved through enhanced solvation, mainly ascribable to inclusion of phospholipids and denatured WP (Vanderghem et al., 2010). Buttermilk addition into cheese has emerged as a latest technique and in the recent year's different cheese varieties like Cheddar, Pizza and processed have been fortified with powdered, condensed, ultra-filtered or traditional BM's. The addition of buttermilk powder (BMP) into Cheddar-type cheese is often a more favorable option compared to adding buttermilk (BM) itself, especially when the goal is to boost the phospholipid levels in the cheese. Phospholipids play a crucial role in the texture, flavor, and nutritional profile of cheese. Phospholipids are important for the texture, mouthfeel, and overall quality of cheese, and the addition of BMP helps in enhancing these levels. However, adding BM itself comes with several negative attributes that affect the cheese's overall quality (Hickey et al., 2018). The addition of BMP in low-fat Cheddar cheese in comparison to skim milk powder (SMP) produces a smoother, more homogenous protein network with smaller voids and a more uniformed distribution of fat globules within the cheese matrix (Romeih et al., 2012; Martinovic et al., 2013). Addition of concentrated BM to low-fat cheese milk produces a high moisture cheese with a soft texture (Turcot et al., 2002). Condensed sweet cream BM addition decreases the curd strength and flowability in pizza cheese however increases the yield (Govindasamy-Lucey et al., 2007). The addition of concentrated sweet buttermilk (BM) to processed cheese spreads can enhance certain organoleptic parameters, such as flavor, texture, and overall sensory appeal. However, while it can improve the sensory qualities, it also comes with some trade-offs, particularly in terms of the cheese's meltability over time (El Sayed et al., 2010).

16.3 Conclusion

Dry milks can be utilized successfully in cheese making depending on the renneting properties of recombined milks. While reduction of minerals affects the renneting properties of reconstituted milk protein concentrates, addition of calcium chloride solves the renneting problems and increases the structural properties and strength of gel. Utilizing skim milk powder will solve the problem of storage of excess skim milk powder but cheese produced from skim milk powder has higher lactose which is problematic in ripened cheese. Further research can solve the problems of high lactose cheese. Using milk protein concentrate which has high protein and reduced salt and lactose will be beneficial for ripened cheeses. Moreover, incorporating milk proteins in cheese improves nutritional quality and yield and can be beneficial in producing low fat and low calorie cheese varieties. Milk proteins possess excellent technological functionalities which make them suitable ingredients for cheese making besides value addition. Current research is focused on the incorporation of buttermilk powder in cheese and cheese products to enhance yield. This is economical and also helps in dairy by-product utilization.

References

Ashour, M. M., Baky, A. A., El Neshawy, A. A., & Salem, O. M. (1986). Improving the quality of Domiati cheese made from recombined milk. *Food chemistry, 20*(2), 85–96.

Awad, R. A., Salama, W. M., & Ragb, W. A. (2015). Enhancing yield and acceptability of Kareish cheese made of reformulated milk. *Annals of Agricultural Sciences, 60*(1), 87–93.

Deeth, H. C., & Bansal, N. (Eds.). (2018). Whey proteins: From milk to medicine. Academic Press.

El Sayed, M. M., Askar, A. A., Hamzawi, L. F., Fathy, F. A., Mohamed, A. G., El Sayed, S. M., & Hamed, I. M. (2010). Utilization of buttermilk concentrate in the manufacture of functional processed cheese spread. *Journal of American Science, 6*(9), 876–882.

El-Alfy, M. B., Shenana, M. E., Ismail, E. A., Gafour, W. A., & Roshdy, A. M. (2010). Improvement of non-traditional white soft cheese made from fresh milk fortified with adding skim milk powder and vegetable oils using different ratios of starter culture. *Egyptian Journal of Applied Science, 25*(10B), 495–518.

Garem, A., Schuck, P., & Maubois, J. L. (2000). Cheese making properties of a new dairy-based powder made by a combination of microfiltration and ultrafiltration. *Le Lait, 80*(1), 25–32.

Gaygadzhiev, Z., Massel, V., Alexander, M., & Corredig, M. (2012). Addition of sodium caseinate to skim milk inhibits rennet-induced aggregation of casein micelles. *Food Hydrocolloids, 26*(2), 405–411.

Govindasamy-Lucey, S., Jaeggi, J. J., Johnson, M. E., Wang, T., & Lucey, J. A. (2007). Use of cold microfiltration retentates produced with polymeric membranes for standardization of milks for manufacture of pizza cheese. *Journal of Dairy Science, 90*(10), 4552–4568.

Guinee, T. P., O'Kennedy, B. T., & Kelly, P. M. (2006). Effect of milk protein standardization using different methods on the composition and yields of Cheddar cheese. *Journal of Dairy Science, 89*(2), 468–482.

Harvey, J. (2006). Protein fortification of cheese milk using milk protein concentrate-yield improvement and product quality. *Australian Journal of Dairy Technology, 61*(2), 183.

Havea, P. (2006). Protein interactions in milk protein concentrate powders. *International Dairy Journal, 16*(5), 415–422.

Hickey, C. D., O'Sullivan, M. G., Davis, J., Scholz, D., Kilcawley, K. N., Wilkinson, M. G., & Sheehan, J. J. (2018). The effect of buttermilk or buttermilk powder addition on functionality, textural, sensory and volatile characteristics of Cheddar-style cheese. *Food Research International, 103*, 468–477.

Hinrichs, J. (2001). Incorporation of whey proteins in cheese. *International Dairy Journal, 11*(4–7), 495–503.

Hosseini, M., Habibi Najafi, M., & Mohebbi, M. (2014). Modification in the functional properties of sodium caseinate-based imitation cheese through use of whey protein and stabilizer. *Journal of Agricultural Science and Technology, 16*(6), 1313–1324.

Kuo, C. J., & Harper, W. J. (2003). Effect of hydration time of milk protein concentrate on cast Feta cheese texture.

Law, B. A., & Tamime, A. Y. (Eds.). (2011). *Technology of cheesemaking*. Wiley.

Limbaugh, M. D. (2015). Replacement of saturated fats in a cream cheese product (Master's thesis, The Ohio State University).

Martin, N. H., Murphy, S. C., Ralyea, R. D., Wiedmann, M., & Boor, K. J. (2011). When cheese gets the blues: Pseudomonas fluorescens as the causative agent of cheese spoilage. *Journal of Dairy Science, 94*(6), 3176–3183.

Martinovic, A., Moe, K. M., Romeih, E., Aideh, B., Vogensen, F. K., Østlie, H., & Skeie, S. (2013). Growth of adjunct Lactobacillus casei in Cheddar cheese differing in milk fat globule membrane components. *International Dairy Journal, 31*(2), 70–82.

Masotti, F., Cattaneo, S., Stuknytė, M., & De Noni, I. (2017). Technological tools to include whey proteins in cheese: Current status and perspectives. *Trends in Food Science & Technology, 64*, 102–114.

McClements, D. J., & Grossmann, L. (2022). Dairy alternatives–cheese, yogurt, butter, and ice cream. *Nextgeneration plant-based foods: Design, production, and properties*, 443–521.

Mistry, V. V. (2002). Manufacture and application of high milk protein powder. *Le Lait, 82*(4), 515–522.

Mistry, V. V., & Pulgar, J. B. (1996). Use of high milk protein powder in the manufacture of Gouda cheese. *International Dairy Journal, 6*(2), 205–215.

Mistry, V. V., & Maubois, J. L. (2004). Cheese: Chemistry, Physics and Microbiology, Vol. 1.

Mleko, S., Foegeding, E.A. (2000) *Milchwissenschaft, 55*(9), 513–515.

Omar, M. M. (1987). Microstructure and chemical changes in Domiati cheese made from ultrafiltered milk. *Food Chemistry, 25*(3), 183–196.

O'Riordan, E. D., Duggan, E., O'Sullivan, M., & Noronha, N. (2011). Production of analogue cheeses. *Processed cheese and analogues*, 219–244.

Pomprasirt, V., Singh, H., & Lucey, J. A. (1998). Effect of heat treatment on the rennet coagulation properties of recombined high total solids milk made from milk protein concentrate powder. *International Journal of Dairy Technology, 51*(3), 65–71.

Rehman, S. U., Farkye, N. Y., Considine, T., Schaffner, A., & Drake, M. A. (2003). Effects of standardization of whole milk with dry milk protein concentrate on the yield and ripening of reduced-fat Cheddar cheese. *Journal of Dairy Science, 86*(5), 1608–1615.

Rombaut, R., Dejonckheere, V., & Dewettinck, K. (2007). Filtration of milk fat globule membrane fragments from acid buttermilk cheese whey. *Journal of Dairy Science, 90*, 1662–1673.

Romeih, E. A., Moe, K. M., & Skeie, S. (2012). The influence of fat globule membrane material on the microstructure of low-fat Cheddar cheese. *International Dairy Journal, 26*(1), 66–72.

Sandra, S., & Corredig, M. (2013). Rennet induced gelation of reconstituted milk protein concentrates: The role of calcium and soluble proteins during reconstitution. *International Dairy Journal, 29*(2), 68–74.

Simov, J., Maubois, J. L., Garem, A., & Camier, B. (2005). Making of Kashkaval cheese from bovine micellar casein powder. *Le Lait, 85*(6), 527–533.

Solowiej, B., Mleko, S., Gustaw, W., & Udeh, K. O. (2010). Effect of whey protein concentrates on texture, meltability and microstructure of acid casein processed cheese analogs. *Milchwissenschaft, 65*(2), 169.

St-Gelais, D., Roy, D., & Audet, P. (1998). Manufacture and composition of low fat Cheddar cheese from milk enriched with different protein concentrate powders. *Food Research International, 31*(2), 137–145.

Turcot, S., St-Gelais, D., & Turgeon, S. L. (2002). Affinage de fromages allégés de type Cheddar fabriqués à partir de laits enrichis en phospholipides. *Le Lait, 82*(2), 209–223.

Vanderghem, C., Bodson, P., Danthine, S., Paquot, M., Deroanne, C., & Blecker, C. (2010). Milk fat globule membrane and buttermilks: from composition to valorization. Base.

Index

A
Antimicrobial packaging, 202
At-line testing, 170–173

C
Cheese, 1, 11, 33, 48, 63, 81, 107, 135, 151, 169, 191, 229, 266, 293
Cheese cultures, 82, 178
Cheese making, 1–8, 11–27, 33–44, 49, 53, 55, 63–66, 68, 69, 71–72, 81, 85, 107, 108, 125, 146, 151, 152, 159–161, 169, 178–180, 216, 235, 236, 265–286, 293–296, 300–302
Cheese ripening, 15, 24–27, 71, 81–98, 151, 154, 156, 176–178, 266
Chromatographic techniques, 89, 154
Coagulation, 1, 5, 11, 12, 15–18, 21–25, 27, 33–35, 40, 42, 43, 47, 63–75, 83, 84, 108, 110, 116, 121–123, 125, 129, 130, 158, 163, 164, 172, 174–179, 184, 271, 273, 274, 276, 281, 284, 296–298

D
Dairy ingredients, 125, 293–302
Dairy Science, 34, 41, 43
Dietary salt intake (DSI), 229
Dry cheese, 38, 130–132
Dry milk cheese, 129, 145, 163–164, 172, 173

E
Edible films, 192–194, 199, 268
Electronic nose (E-nose), 85, 86, 153, 154, 156, 157
Electronic tongue (e-tongues), 86–87
Enzymes, 2, 5, 15, 17, 18, 20, 25, 26, 36, 56, 58, 59, 64–71, 74, 75, 82–84, 92–96, 109, 110, 113, 123, 130, 144, 153, 169, 183, 184, 198, 209, 216, 223, 265, 266, 278, 279, 283, 284, 294

F
Fermentation, 12, 22, 25, 35–37, 43, 47, 51–59, 63, 68, 81, 85, 108, 110–114, 116, 119, 124, 126, 142, 163, 183, 210, 211, 215, 277, 280, 283, 284, 300
Fermented foods, 48, 52, 56–59
Food processing, 269
Fourier transform infrared (FTIR), 88, 89, 155, 156, 159, 161, 176
Functional properties, 95, 97, 119, 126, 135–148, 176–178, 217, 220, 252–253, 256, 257, 279, 281, 293, 298–301
Functionality, 20, 126, 137, 138, 140–145, 148, 201, 208, 216–218, 224, 250, 252, 256–258, 260, 261, 266, 279, 286, 301, 302

S. Mansha Rafiq (ed.), *Technological Advances and Trends in Cheese Making*,
https://doi.org/10.1007/978-3-031-94476-5

G

Gel, 16, 18, 23, 36, 63, 70, 71, 73–75, 89, 113, 117, 118, 178, 200, 221, 222, 250, 253, 255, 257, 259, 260, 273, 293, 297–299, 302

H

Hybrid cheese, 125–129, 145, 147, 163, 173

I

In-line testing, 170–173, 184, 185
Intelligent packaging, 192, 202, 203

L

Lactic acid bacteria (LAB), 2, 4, 5, 15, 20, 22, 25, 26, 36, 37, 47, 48, 50, 53, 55–57, 82–85, 92, 114, 126, 130, 136, 142, 214, 218, 240, 266, 269, 280
Lipolysis, 26, 27, 65, 82, 88, 91, 93, 94, 96, 97, 132, 137, 144, 236, 239
Low-fat cheese, 15, 94, 139, 207–224, 271, 273, 281, 302
Low-sodium cheeses (LSCs), 230, 231, 233, 239–242

M

Manufacturing process, 20, 27, 48, 107–132, 145, 171, 176, 177, 223, 284, 285, 296
Market potential, 6, 7, 17, 49, 69, 97, 107, 115, 120, 128–129, 151, 153, 197, 219, 222, 243, 285
Melting properties, 112, 138–139, 254, 260, 272, 298
Microfiltration (MF), 13–14, 42, 123, 216, 265, 270, 280–282, 286, 301
Milk, 1, 11, 33, 47, 63, 81, 108, 139, 153, 169, 191, 229, 265, 293
Milk protein concentrates (MPC), 209, 281–282, 293, 295–298, 302
Milk quality, 12–13, 33–37, 39–44, 153, 172
Modified atmosphere packaging (MAP), 192, 194, 197, 198

N

Nano filtration, 265
Nuclear magnetic resonance (NMR), 88, 174, 176–178
Nutritional benefits, 59, 119, 125, 128

P

Plant-based, 69, 107, 108, 115–121, 125–128, 147, 160–163, 183, 195, 249–251, 253–256, 258, 260
Precision fermentation, 128, 265, 283–286
Proteolysis, 8, 13, 15, 25, 54, 66, 67, 70, 71, 82, 83, 87–89, 91, 93–98, 132, 144, 159, 175, 211, 215, 217–219, 230, 234, 235, 239, 258, 270, 278–279, 296, 297

R

Rennet, 1, 2, 5, 12, 14, 15, 17, 18, 21–24, 27, 33, 34, 38, 63–75, 81–84, 93, 108, 109, 113, 122, 126, 130, 143, 144, 158, 163, 164, 169, 210, 213, 234, 235, 267, 271, 272, 274–277, 279–281, 284, 293, 295–301
Reverse osmosis (RO), 265, 270, 279, 280, 282, 286
Rheology, 72–73, 118, 135–148, 176, 217, 241–242, 250, 278
Ripening, 1–4, 11–16, 18, 20, 22, 24–27, 48, 50, 53, 54, 69, 71, 72, 81–90, 93, 95–98, 108, 123, 127, 140, 142–144, 153, 157, 176–178, 208, 212, 214, 215, 218, 223, 231, 236, 239–241, 250, 257, 258, 266, 270, 277–279, 294, 297

S

Sodium chloride (NaCl), 20, 83, 177, 230–232, 234–236, 238–243, 276
Spectroscopic method, 156
Starter cultures, 4–5, 11–13, 15, 16, 19, 22, 25, 27, 43, 47–59, 65, 83, 93, 95, 110, 113, 114, 126, 130, 142, 158, 159, 164, 197, 212, 224, 234, 236, 273, 293
Stretch cheese, 185
Stretched cheese, 111–115, 145, 146, 158–160, 173

U

Ultrafiltration, 15, 42, 63, 123, 124, 126, 130, 216, 265, 270–279, 281, 286, 295–297, 301
Utilization, 8, 25, 56, 63, 66, 84, 92, 114, 124, 171–173, 177, 201, 259, 270, 271, 279, 281, 286, 302

V
Varieties, 1, 3, 6–9, 11, 15–20, 22, 23, 26, 27, 48, 49, 51, 53, 55, 57, 59, 68–70, 81, 82, 85–87, 89, 96, 98, 107–109, 111, 112, 115, 126, 135, 136, 139–141, 144, 146, 148, 169, 172, 173, 182, 184, 191, 192, 194, 200, 207, 216, 221, 223, 224, 232, 233, 237, 238, 250, 252, 254, 256–258, 261, 270, 272, 274, 276, 277, 279, 296, 300–302
Vegan, 64, 107, 108, 111, 115, 116, 119, 145–148, 157, 160, 161, 170, 172, 173, 183, 195, 253
Vegan cheese (VC), 108, 115–121, 145, 146, 160, 161, 173, 183, 249, 252, 256–258, 261

W
Whey cheese, 121, 123, 124, 145, 146, 161, 162
Whey protein concentrates (WPC), 196, 214, 221, 293, 298–301

The manufacturer's authorised representative in the EU is Springer Nature Customer Service Centre GmbH, Europaplatz 3, 69115 Heidelberg, Germany. If you have any concerns regarding our products, please contact ProductSafety@springernature.com

Printed and bound by CPI Group (UK) Ltd, Croydon, CR0 4YY
05/10/2026
02228126-0008